BK 628.5 R451S
SOURCEBOOK ON THE ENVIRONMENT: THE SCIENTIF
PERSPECTIVE /REVELL
1974 6.25 FV
3000 348303 30014
St. Louis Community College
W9-COT-897

Sourcebook on the Environment

Sourcebook on the Environment

The Scientific Perspective

Charles ReVelle
Johns Hopkins University

Penelope ReVelle
Essex Community College

HOUGHTON MIFFLIN COMPANY

BOSTON Atlanta Dallas
Geneva, Ill. Hopewell, N.J.
Palo Alto London

To our parents

Printed in the U.S.A.

Library of Congress Catalog Card Number:
73-8074

ISBN: 0-395-17018-4

Contents

Foreword

Headlines about environmental problems are evidence of a growing awareness that humankind occupies a delicate position in the web of life which envelops the earth. We know the web is there, though we do not completely understand its pattern. An abrupt movement sends shudders through it and tests its strength. Too violent a motion can destroy part and weaken portions nearby. All plant and animal species are suspended in this web. From the human vantage point, we can see the species upon which we depend directly, but our vision becomes cloudy when we try to view more distant dependencies. Even man's impact upon himself is not well understood, although we are trying to understand.

The needs of humankind—mechanical and electrical energy, water and food, raw materials—seem at first glance to be separate and distinct and unconnected. In fact, they are tightly bound to one another, so tightly bound that a decision about one has frequent and profound repercussions on the others.

Dr. Abel Wolman, one of the nation's eminent environmental scientists whose involvement extends to the era when we were still making our water safe to drink, captured our dilemma in testimony before Congress:

> The impact of man upon his environment has existed since man himself walked the earth. Whatever man does changes the ecology of his surroundings for good or evil. As populations grow, as urbanization and industrialization move forward, and as science and technology burgeon, the potentiality for ecological disturbance and degradation increases. . . . Given the thesis that man creates and modifies his environment for good and evil, what are the means available to him for avoiding the bad and multiplying the good consequences of his existence? . . . The most valuable tool is in his better understanding of the environment, and how his actions affect it.

The need for understanding can, however, be turned against those who seek environmental improvement. Some people argue that scientific evidence of dangerous effects must be unassailable before action is taken. Rarely do we experience such a situation. Instead, there is a weight of scientific evidence, which must be judged. In addition, the economic costs of control and social consequences must be considered. The balancing of these costs and consequences against the benefits of a healthy environment is the responsibility not simply of the scientific community but of all society.

We have attempted to provide a better understanding, from a scientific perspective, of the principal environmental issues that civilization confronts. Such an understanding is the logical precedent of sound decisions.

Acknowledgments

The people who assisted us in the preparation of this book are numerous. Helpful suggestions and criticisms evolved from many conversations with friends and colleagues. We wish to thank everyone who encouraged us and bore with us as we discussed our progress.

We had the best of secretarial help. First, there was Mary Perks, who gave her time to type portions of the initial manuscript. Berni Sciuto cheerfully and flawlessly typed much of the first draft and the revised version. Ann Flynn lent a helping hand too. Then Helen Klager, in addition to typing much of the original manuscript, polished off the last half of the revised text. Even as the book was going to press, Thelma Cole typed our final entry.

We had help in the library as well. Miss Kai-Yun Chiu at the documents section of the Johns Hopkins Library and her staff, Jean Bell and Nancy Mees, gave us unparalleled assistance in culling environmental literature from the vast outpourings of the federal government. That aid kept us abreast of some of the very latest articles in a number of fields.

As the writing drew to a close, Peter and Linda Rottmann proofread major portions of the text with us. Although any remaining errors or organizational idiosyncracies are acknowledged as our own, the reader has been spared many obscure meanings by their efforts.

Our editors at Houghton Mifflin played important roles in the shaping of this book. Their discerning criticisms helped us communicate our aims to the reader and led us to new insights as well.

Professor G. Fred Lee of the the University of Wisconsin was kind enough to review and criticize Figure E.1, which describes the aging of lakes.

Finally, we would like to thank Professor John Bushnell of the University of Colorado for a number of extremely useful criticisms and constructive suggestions.

Sourcebook on the Environment

INTRODUCTION

Using This Book

One may legitimately inquire why this book is organized alphabetically, rather than in the more conventional chapter format. A number of elements favor an encyclopedic format; some of them make the book easier to use, and others serve a learning function. Alphabetic listing allows quick retrieval of information. Readers concerned about environmental issues have varied levels of background knowledge. An encyclopedic format allows entry into the material for any interested person, no matter how much or how little grounding he possesses in environmental problems. Thus, if one wishes to read a discussion on **Strip Mining,** an entry is found under that title. It is not necessary to know that strip mining is associated with the use of coal for electric power. Similarly, **Acid Mine Drainage** is set under its own title and defined directly, and the issues on which it bears are indicated.

Straightforward definitions or problem statements often precede the detailed presentation of material. Several subject headings exemplify this approach and provide entry into the material as a whole. **Water Pollution, Air Pollution, Electric Power,** and **Pesticides** begin with short definitions or summaries of major environmental concerns and then cover specific aspects of the topic in detail. Cross-references, which appear in **bold** type, direct the reader to related entries and sections.

The organization of this book can serve several valuable learning functions. Cross-referencing makes the reader aware of the interconnectedness of environmental problems and prevents the compartmentalizing of environmental hazards into "water pollution," "air pollution," "solid waste disposal." The preliminary definitions with references to related entries allow the reader to learn at several different levels.

Interdependency and Interconnectedness

Environmental issues are not isolated instances of technology gone astray. They are fundamental. Many are pieces of larger puzzles. Cross-referencing makes it possible to assemble the pieces into many different wholes, all of which have validity. It is hoped that readers will thus gain the ability to see the larger pictures.

Perhaps interdependency and interconnectedness are most vividly displayed in the issues which intertwine with **Electric Power.** You may arrive at **Electric Power** from numerous starting points. If you start with **Strip Mining,** you will realize that this low-cost but potentially devastating mode of coal extraction provides its product primarily to the electric utilities. If you begin with **Oil Pollution,** you will see that tanker spills are increasing as the utilization of oil rises. The growing demand for oil is due in large part to its growing use at electric generating stations as a substitute for coal. The substitution is a method of decreasing **Air Pollution.** If you scan **Air Pollution,** you will be directed to **Particulate Matter** and **Sulfur Oxides.** Under **Particulate Matter** you will find that in the United States coal-fired electric power plants are the largest source of atmospheric particles. Substituting oil for coal in electric power generation is one remedy for the problem and accounts in part for the growth of the **Oil Pollution** problem. In **Sulfur Oxides** you will discover that coal-fired power plants are again the principal enemy and that oil is being substituted to reduce the emission of sulfur oxides as well as to control the dispersal of particles.

Read **The Automobile and Pollution** and notice the role of nitrogen oxides in **Photochemical Air Pollution.** Jump to the latter entry and discover that the burning of coal for **Electric Power** is an important source of nitrogen oxides. Consult **Thermal Pollution** and learn that this problem is tied to all electric power generation except hydropower. The listing could continue, but the notion should be clear. Looking at neat headings without cross-references to related topics will not increase environmental understanding to the desired degree.

Consider the examination of the colors of the spectrum through a prism. If you stare at the middle of the blue band, you gain an appreciation for blue. Only when your eye wanders to the periphery do you see blue shade into green. The area of shading and transition is crucial to understanding environmental problems. When we realize that environmental problems do shade into each other, we begin to suspect that they have a common root. **Electric Power** is such a root. Attacking problems individually, one by one, may not accomplish the desired ends. It may be necessary to intervene at the main source of many problems in order to make meaningful changes. But before intervention must come identification. We need to train ourselves to look for commonalities. Our hope is that the organization of the book will foster this skill.

A false feeling of security accompanies the sorting of environmental issues into compartments. **Sulfur Oxides** fit in the air pollution slot, aluminum cans are part of the problem of disposing of **Solid Wastes,** and **Mercury** is a water pollutant. But even a brief survey of the subjects reveals that such divisions are artificial.

It would be more nearly correct to say that hazards cycle in the environment, that there is a flow from one compartment to the next.

Consider, for example, the hazards presented by mercury and its compounds. Mercury is released to the air by the burning of the fossil fuels, coal and oil. Atmospheric mercury so derived is washed down from the air by rain and becomes a form of water pollution. The casings of mercury batteries disposed of in sanitary landfills can rust and release the mercury from their cells. Soil and water bacteria can convert this mercury to organic mercury compounds which find their way by groundwater to streams and rivers. Other sources, such as industrial sewage, add to the water's burden of mercury. Minute aquatic creatures ingest and absorb mercury; these organisms then serve as food, either directly or indirectly, for other water dwellers. Thus, oysters, swordfish, and tuna may ingest the mercury.

What sort of environmental problem is mercury? It is a water pollutant; it is an air pollutant; and it is a problem of solid waste disposal. The compartments into which one can put environmental problems are not isolated from one another. They meet at common boundaries, which pollutants are free to cross.

The problem of pesticides, perhaps even more than mercury, impinges on all the environmental compartments. Pesticides are most often applied to growing plants. Whether the plants are sprayed or whether chemicals are plowed into the ground, the soil becomes a major repository for pesticidal chemicals. Some of them break down quickly, but others, like arsenic and lead, can accumulate to such levels that crops will no longer grow in the contaminated soil. Because crops are commonly sprayed from the air, pesticides become an air pollutant as they drift from the intended target. Dust storms can also transport pesticides long distances; the pesticides are carried with the minute particles of earth on which they have been adsorbed. A severe dust storm in Texas in 1965 caused elevated pesticide levels at air monitoring stations in cities as far east as Cincinnati. Thus, pesticides contaminate air and soil.

When dust laden with pesticides is washed down by rain, surface waters are contaminated. Runoff from agricultural lands is another source of pesticides in water. Additionally, drift from crop spraying and the discharge of industrial wastes pollute water. Finally, pesticides may be added directly to water to control aquatic pests. Thus, pesticides contaminate water as well as soil and air. Frequently, pesticidal pollution of one compartment is a direct result of the presence of the substance in another compartment. For instance, water is polluted when airborne pesticides are washed down by rain and when pesticides in the soil reach the water in runoff.

Attempts to categorize pollutants within a rigid framework are not merely difficult; a strict assignment of environmental problems may cause errors of judgment. Action to correct a problem in one area may lead to new problems in another. For example, a utility company building a new power plant in the northeast was concerned that river water used to condense the steam from the turbines would become too hot to return to the river. In order to avoid this potential **Thermal Pollution** problem, a cooling tower was built to remove heat by vaporization. After the plant began operations, it was found that the water vapors from the cooling tower mixed with sulfur dioxide in the stack gases from an adjacent power plant. The mixture formed a mist of sulfuric acid. The utility company had compartmentalized, seen a water pollution problem, corrected it, and created a serious air pollution problem.

Another example of the hazards of compartmentalization is the suggestion to diminish

Carbon Monoxide levels in cities by speeding up traffic. Frequent changes of gear characterize stop-and-go driving and vehicle emissions. To speed the flow of cars, traffic lights could be timed and streets made one-way. Superficially, the suggestion is reasonable, but a knowledge of the generation of noise should temper enthusiasm for it. Urban areas already tolerate high levels of background noise. A doubling of speed and traffic volume could be expected to lead to a perceived noise intensity nearly double the original level.

The organization of this book, alphabetically by subject rather than by the more conventional large divisions, is meant to discourage easy sorting. Rather, we hope to encourage the reader, with the help of the cross-references, to explore all the ramifications of actions affecting the environment.

The organization serves another function. The summary entries, which precede or direct one to more comprehensive discussions, allow the reader to learn at the level he chooses. Two entries in particular exemplify this—**Air Pollution** and **Water Pollution.** The former begins with brief descriptions of the occurrence, effects, and control of each of the major air pollutants. Upon completing this entry, one may stop reading about the subject. Alternatively, since each description refers to a more thorough discussion elsewhere, the reader may go on to consider **Sulfur Oxides, Particulate Matter, Carbon Monoxide, Photochemical Air Pollution,** and **Lead.**

The same system is utilized in **Water Pollution;** brief entries point the way to comprehensive treatments of the subjects. Thus, the reader may go on to become more knowledgeable about **Bacterial and Viral Pollution, Organic Water Pollution, Eutrophication,** and **Thermal Pollution.**

The material in this book may be tailored to fit the character and intent of the reader's interest. It may be used to derive a superficial acquaintance with broad areas such as air and water pollution. Or it may be used to obtain a thorough understanding of a relatively narrow problem. The material may be consulted sequentially, so that not only does a single area become well known but the interdependency and interconnectedness of environmental issues are also recognized.

Acid Mine Drainage

The environmental impact of coal cannot be taken lightly. When it is burned to boil water in steam power plants for electricity, it produces air pollutants such as **Sulfur Oxides,** nitrogen oxides, and **Particulate Matter.** Coal miners risk their lives and live in fear of explosions and cave-ins. They are subject also to incurable **Black Lung Disease.** Where coal mines have been abandoned or where **Strip Mining** has left vast depressions in the earth, water seeps in and acid mine drainage flows away. Over 5,000 miles of streams in ten states are polluted by acid mine drainage. Pennsylvania's streams account for about half of this total.

One of the components of coal is iron pyrite, which chemically is iron sulfide ("fool's gold"). Much coal is left in an abandoned mine and much is left in the stripping trench. Water comes in contact with this coal in several ways. First, in strip-mined areas natural drainage and seepage form pools. Seepage also allows water to enter the deep mines. Additionally, when a portion of an abandoned mine collapses, the surface may subside, leaving a sinkhole which becomes a source for accelerated seepage. When the earth above an abandoned mine is strip-mined, still further sources for water entry are created.

Oxygen in the water oxidizes the ferrous iron of iron sulfide to ferric iron and the sulfur to sulfate. The water dissolves these substances. Then ferric iron precipitates as ferric hydroxide, which gives the characteristic yellow-brown color of acid mine drainage. The precipitation leaves the water highly acidic and corrosive. Bacteria seem to play a role in the oxidation.

Acid wastes draining from a coal mine near Rico, Colorado.

The acid waters are detrimental to many forms of life; the oxidation may leave the water devoid of dissolved oxygen and threaten the natural species of the stream (see **Organic Water Pollution** for more detail on the effects of low dissolved oxygen). The ferric hydroxide precipitate may coat the bottom of a stream, depriving bottom feeders of food. The drainage water may also contain manganese and aluminum and beryllium (see Table D.1, "Trace Metals in U.S. Waters"). Use of the waters for recreational purposes such as swimming, fishing, and boating may be curtailed. Water supplies of domestic and industrial users may be jeopardized by the acid and minerals.

There are numerous approaches to control. Since strip mining trenches and sinkholes aggravate acid mine drainage problems by providing entry points for water, one method of control is to reclaim the strip-mined lands and to fill the sinkholes. A combination of fly ash, limestone, and slag has been used to backfill the sinkholes. Water which percolates through these materials will become alkaline and will to an extent neutralize the acidity of the mine drainage. Another technique is to route streams around sinkholes by providing a system of ditches. These methods, however, cannot prevent the natural seepage of groundwater into the mines.

During periods of high acid concentrations, it would be advantageous to be able to dilute the acid stream with relatively neutral waters. Consider two streams which join—one carrying acid mine drainage, the other relatively neutral. A reservoir above the point of juncture placed on the neutral stream may be used to provide dilution water to mix with the acid stream during periods of high acid flow. The methodology is known as "low flow augmentation" and is often humorously referred to as "the dilution solution to pollution." The sealing of mines is another approach, but its effect is controversial. Greater benefit is probably derived from this approach when the mine opening is above the level of the mine than when the opening is below the level of the mine.

Where mines are still in production and water seeps in, the resulting acid waters need to be removed to allow production to continue. Pumping the water out promptly will reduce the amount of acidity and iron it may accumulate, but treatment of the water may be required prior to its release into a stream. The addition in a detention basin of crushed limestone and dolomite will neutralize the acid waters but at the cost of increasing their hardness. (The neutralizing effect of limestone explains its use in the backfilling of sinkholes.)

Some observations indicate that the acid concentrations in drainage from a given source will diminish over the years. This assumption provides only slight cheer. The problem appears to be relatively intractable and may get worse if strip mining grows, unchecked and unregulated.

REFERENCES

1. *Surface Mining and Our Environment*, U.S. Dept. of the Interior, 1967.
2. H. L. Barnes and S. B. Romberger, "Chemical Aspects of Acid Mine Drainage," *Journal of the Water Pollution Control Federation* 40, no. 3 (March 1968): 371.

Air Pollution

See also Air Quality Standards; Carbon Monoxide; Lead; Noise Pollution; Particulate Matter; Photochemical Air Pollution; Sulfur Oxides.

No single insult to the environment has so captured the nation's attention in many years as air pollution. True, **Water Pollution** concerns us, for we want clean waters in which to bathe and on which to sail and fish, but the water we *drink* is pure. True, **Mercury** in swordfish worries us, but swordfish can be removed from the marketplace. Air pollution has achieved a special status of concern because we *must* breathe the air. And the air is polluted; it damages materials, vegetation, and man's health.

The Clean Air Act of 1970 charged the Environmental Protection Agency with the responsibility of formulating **Air Quality Standards.** These are listed under that title below.

The first of the following sections sketches the events that brought the air pollution problem to public recognition. There is an unrecorded history also of years of itchy eyes, raw throats, coughing, poor visibility, and millions of dollars' worth of economic damage. Attention is next focused on the specific types of contaminants which afflict our cities; there is a discussion of their sources, effects, and methods of control. The last section details the meteorological factors affecting pollution levels and emphasizes the role of inversions.

EPISODES

By *episode* we mean an event of striking character—striking in diminished visibility, increased illnesses and deaths, or in measured concentrations of air pollutants. Episodes represent the peak occurrences of air pollution, those of which we were most vividly aware; they do not represent the day-to-day effects which taken together could prove to be more detrimental than all the episodes combined.

If one traces the improvement of our water supplies through time, he will see that epidemics[1] provided the impetus for **Water Treatment** and for the protection of water supplies. In the same way, the episodes of air pollution give impetus to the movement to improve the quality of our air.

Although air pollution has existed since industrialization began, the first episode to be well studied took place in December 1930. A dense fog blanketed the Meuse Valley in Belgium for five days. The incidence of respiratory symptoms rose, and 63 individuals died. It is estimated that sulfur as oxides and as sulfuric acid may have risen to 25,000 micrograms per cubic meter.

In the United States, only a score of years have passed since the incident at Donora, Pennsylvania. In October 1948, an inversion trapped a layer of air over the community, and the atmosphere filled with pollutants. Twenty deaths were attributed to air pollution.

Minor episodes served to point out the existence of an air pollution problem. The London fog of December 1952 dramatized the effects of air pollution in a most tragic way. Over 4,000 excess deaths occurred as a result of the five-day fog. The deaths, which were distributed over an 18-day period, seem to have been largely among people with histories of respiratory or cardiac disease, and the illnesses fell heavily among the elderly. The sulfur dioxide level reached about 4,000 micrograms per cubic meter. In the next decade, air pollution incidents occurred year after year in London, although none was so massive as the fog of 1952. New York City had air pollution incidents in 1953, 1962, 1963, and 1966, but, again, none with such extreme effects as the 1952 London fog.

Abundant data beyond these episodes relate air pollution levels to health in a statistical way. Many studies attempt to find the association of health with specific pollutants. For this reason the studies are referred to under **Photochemical Air Pollution, Sulfur Oxides, Particulate Matter,** and **Carbon Monoxide.**

[1] See **Bacterial and Viral Pollution of Water.**

CONTAMINANTS

SULFUR OXIDES

From the combustion of sulfur-bearing coal and petroleum products come the oxides of sulfur. Power generation from coal-fired power plants stands as the most voluminous contributor of these corrosive gases and particles. In the late 1960's about 40 percent of the annual emissions of sulfur oxides derived from this source. The auto is relatively innocent in the sulfur oxides picture.

One of the oxides of sulfur (the trioxide) reacts with water vapor to produce sulfuric acid mist, which attacks building materials such as limestone, marble, and mortar. The same oxide of sulfur may combine with metal oxides to give sulfates, which, falling on surfaces, accelerate corrosion. Not only are green plants and

Air pollution hanging over Manhattan.

trees damaged by these gases, but it is becoming clear that human health may be seriously affected by them. A number of air pollution episodes were characterized by elevated sulfur oxide levels as well as by high concentrations of **Particulate Matter.** Numerous studies have associated respiratory symptoms with the atmospheric concentration of sulfur oxides. One recent investigation found excess mortality in New York City during periods not characterized by episodes or even by noticeably unusual pollution levels. The excess mortality was linked to the concentration of air pollutants including sulfur oxides.

There are many approaches to controlling sulfur oxides, most of them centering on the power plant source. One involves substituting low-sulfur coal or low-sulfur oil (both more costly) for coal currently in use. Natural gas is likewise a candidate for substitution because of its very low sulfur content, but shortages and higher prices are in prospect. Different energy sources such as **Nuclear Power Plants** and hydroelectric power plants may also be utilized, and plants may be set at greater distances from population concentrations. Several processes to remove sulfur oxides from the stack gases of power plants are being developed. These are discussed under **Sulfur Oxides.**

PARTICULATE MATTER

Solid or liquid particles dispersed in air (sometimes called aerosols) arise from many sources. Those that plague our cities originate primarily from combustion processes. Peak concentrations are frequent in winter when combustion for power and heating are elevated, but even at other times, especially in the fall, inversions may create conditions in which particle concentrations build up to high levels. **Electric Power** generation, in which coal and fuel oil are combusted, accounts for about 25 percent of the annual particle emissions in the United States. Of the two fuels coal and oil, coal is by far the worse particle source. Incineration and heating each contribute about 8 percent of the total annual particles. Industrial sources may produce about 50 percent of the total, but combustion is included with this source. Motor vehicle operation produces particles, principally **Lead** salts in areas without photochemical smog, but also liquid organic aerosols in areas subject to **Photochemical Air Pollution.**

Particles which are acidic salts (such as sulfates) may damage vegetation. Nonmetallic surfaces may be soiled by particles, but metallic surfaces may experience accelerated corrosion due to particle deposition. Costs of cleaning, painting, resurfacing, or replacing surfaces exposed to particles may be large.

Weather may be inadvertently modified by airborne particles; the particles may serve as nuclei upon which water droplets or ice particles form. Solar radiation reaching the earth's surface may be markedly reduced in winter in urban areas as the particles disperse the radiation. Visibility too is decreased by particle scattering. The heat balance of the world may be undergoing a change as particles reflect away solar radiation.

When an individual inhales particles in the air, they may lodge in the lungs. Alternatively, they can be swept up in the mucus coating the respiratory tract and then swallowed. Thus not only can substances be absorbed into the bloodstream from the lungs; they can also be absorbed in the stomach or small intestine. In fact, the incidence of deaths from stomach cancer has been correlated with particle levels in the air. Attempts to correlate deaths from lung cancer and bronchitis with particles have given inconclusive results. However, a class of compounds has been found in the benzene-soluble fraction of particulate matter which causes cancer in experimental animals and which has been linked statistically with lung cancer. These compounds are the polynuclear aromatic hydrocarbons, of which benzo(a)pyrene is the prime suspect.

Particles may be controlled by switching fuels from coal to oil or by switching power sources from **Fossil Fuel Power Plants** to hydropower and **Nuclear Power Plants.** Also a wide range of devices are available for removing particulate matter from a gas stream. These are indicated under **Particulate Matter.**

PHOTOCHEMICAL POLLUTION

The Clean Air Act of 1970 specifies reduction in hydrocarbon and nitrogen oxide emissions from automobiles. While it is true that these compounds are present in significant concentrations in urban air, there is only limited evidence that either substance alone has harmful effects on man's health. Yet the action of the Congress is well taken and timely, for the chemical reactions which nitrogen oxides and hydrocarbons undergo in the atmosphere produce some of the more noxious air pollutants.

In the last years of the 1960's, motor vehicles accounted for about 50 percent of the total nationwide emissions of hydrocarbons, industry for about 14 percent, and solvent uses for 10 percent. Other sources make smaller contributions. Thus motor vehicles are an important target for control. Hydrocarbon emissions from automobiles are already being controlled by positive crankcase ventilation, which captures gases that blow by the piston during compression and returns them to the combustion chamber. Better fuel metering, which produces better mileage, also diminishes hydrocarbons.

With the 1971 models came a recycling of evaporation losses from the carburetor and fuel tank. The automobile industry is aiming to install a catalytic converter to further oxidize hydrocarbons in time for the 1975 or 1976 models; this step is in response to the specifications of the Clean Air Act.

When fuels are oxidized at high temperatures, nitrogen from the air (mainly) and from the fuel combines with oxygen. While motor vehicles accounted for about 35 percent of the total annual emissions of nitrogen oxides in the latter portion of the 1960's, the combustion of fossil fuels at power plants furnished another 20 percent of the total emissions. Peak concentrations of nitrogen oxides coincide with rush hours, although the morning peak tends to be higher because of less atmospheric circulation at that time of day. There are currently no control devices for nitrogen oxides on cars, but the catalytic converter is expected to convert nitrogen oxides to nitrogen as well as combust hydrocarbons.

The photochemical reactions which these compounds undergo are not well understood. They are described under **Photochemical Air Pollution.** Ozone, peroxyacyl nitrate (PAN) compounds, and aldehydes are among the secondary contaminants produced by the photochemical reactions. They are responsible for a wide range of effects. Ozone and PAN compounds are known to damage many plants including cash crops. Ozone and aldehydes are also known to irritate the nose and throat. Eye irritation in photochemical smog is caused by the aldehydes and PAN compounds. Although photochemical air pollution has been linked to the aggravation of asthma conditions, few firm connections to other aspects of health have yet been established because of the difficulty of setting up studies which control other factors. We do know that in a survey of Los Angeles residents, three-fourths of the study population said they were "bothered" by air pollution, and 13 percent had considered moving because of it.

CARBON MONOXIDE

Carbon monoxide is so pervasive in our urban atmospheres that in the latter half of the last decade its annual tonnage equaled the estimated annual emissions of all other air pollutants combined. The villain in the piece is the motor vehicle. Incomplete combustion of hydrocarbon fuels such as gasoline gives rise to vast quantities of the gas. Daily the carbon monoxide levels rise and fall in the periods around the two traffic rush hours. In a number of our largest cities over 90 percent of the carbon monoxide is derived from transportation activities. Power generation plays only a little role in this pollutant's levels.

While the general public is being exposed to high levels on the streets, certain occupational groups face even more severe exposures. Especially exposed are individuals working in underground garages, at loading platforms, and in tunnels. Carbon monoxide reacts with the hemoglobin in blood to form carboxyhemoglobin. The capturing of hemoglobin deprives oxygen of its carrier to the tissues. A carbon monoxide level of 70 micrograms per cubic meter for 8 to 12 hours will lead to a level of carboxyhemoglobin equal to about 10 percent of all the blood's hemoglobin. Such atmospheric levels are not unheard of in tunnels and parking garages.

Many tests have attempted to link carbon monoxide levels with impairment of judgment, such as the inability to discriminate the lengths of "beeps" in tests. At blood levels of carboxyhemoglobin as low as 3 percent, the ability to distinguish light intensities has deteriorated. Such levels could occur as the result of exposures in the nation's larger cities. No link has yet been established between carbon monoxide levels and illnesses, but a tenuous link exists between its levels and automobile accidents.

Exhausts of carbon monoxide are being diminished under Congressional pressure. Technological changes being made to accomplish this reduction include increasing the air-to-fuel ratio to a precise quantity and mixing air with the exhaust stream for final combustion prior to exhaust. By 1975 or 1976 a catalytic converter on autos may further oxidize the carbon monoxide in the exhaust stream prior to ejection.

LEAD

There is no doubt that the lead in the air is from the lead tetraethyl and lead tetramethyl antiknock additives in gasoline. Lead exposures via air are supplemented by lead in the diet and in water. In fact, because the exposures via diet and air are so unregulated, the Public Health Service set its **Drinking Water Standards** for lead at a lower level than it would otherwise have done.

As a result of lead levels in the air, lead levels in urban dwellers are elevated, as is shown by blood measurements. However, urban children constitute a population especially subject to the risk of developing lead poisoning. Because of a child's high food consumption per unit of body weight and rapid respiration, both food and air supply lead more rapidly to children than to adults. Pica, the habit of ingesting nonfood items, places the urban child in real peril if he has access to peeling lead paint, as is commonly the case in deteriorating prewar dwellings. Investigators estimate that 2 percent of children in pre–World War II dwellings have lead poisoning *already.* Whether lead levels short of toxic concentrations have any harmful effects is not known.

The amount of lead in gasoline had been limited to a maximum of 4.23 grams of lead or 4 milliliters of lead tetraethyl per gallon. In 1973 the Environmental Protection Agency issued regulations requiring stations above a certain size to provide at least one grade of lead-free gasoline by July 1974. The rules would also force a gradual decrease in the allowable lead concentrations to 1.25 grams of lead per gallon; the latter rules have not been finalized as of this writing. In issuing the regulations, the agency cited both the health effects of lead and the fact that lead "poisons" the catalyst which is to be used in the exhaust system of 1976 model cars.

Lead poisoning from air pollution is not likely among adults, but urban children living in environments where there is peeling lead paint are already exposed to the threat of lead poisoning. Lead in the air may be viewed as a dangerous supplemental source for a population even now at great risk.

NOISE POLLUTION

Noise pollution is a type of air pollution that subjects people to unwanted sound. Transportation vehicles such as planes, cars, buses, and trucks are a principal source of noise. Some 9 million people live in areas exposed to levels of highway and aircraft noise felt to be unsuitable for residential living. For most others, transportation activities contribute the major portion of the steady background noise that pervades the cities. Each year approximately 30 million Americans are affected by noise from construction activities. Construction and transportation noises combine to deny from 5 to 10 million people who live or work in cities the use of outdoor areas for conversation.

Noise adversely affects people in both physiological and psychological ways. Hearing loss can consist in obvious deafness or in a subtle diminution of the ability to discriminate the higher-pitched sounds. Psychological effects include such responses as loss of sleep and interference with task performance. In economic terms, noise pollution entails medical costs and loss of manpower due to hearing damage and due to increased errors or accident rates. Noise may lower property values; it may also decrease airport capacity by preventing use of an airport during nighttime hours.

The Environmental Protection Agency estimates that noise from highway vehicles will double by the year 2000. Furthermore, the average level of background noise can be expected to rise from 1970 levels of 46 decibels (A) to almost 50 decibels (A) unless an aggressive program of federal legislation and consumer education is undertaken. Such a program could be expected to decrease the residual level to 42 decibels (A) and provide a higher-quality, quieter environment.

METEOROLOGICAL ASPECTS

The air pollution episodes discussed earlier represent periods in which high pollutant emissions and certain meteorological phenomena converged. Day-to-day pollution levels as well are frequently aggravated by weather conditions.

The dominant mechanism for dispersing pollution is the mixing of the lower, polluted layers of the atmosphere with higher, relatively unpolluted layers. If this mechanism is blocked, dilution of contaminants will be markedly diminished, and the stage will be set for the buildup of pollutants in the atmosphere. The condition which can block the upward mixing is an inversion.

To understand an inversion it helps to consider the usual pattern that temperature follows as elevation above the surface increases. Typically, the air closest to the earth is warmest; it gets gradually cooler as distance from the ground increases. Since the air at ground level is warmest, it has a tendency to rise because of its buoyancy. Its movement makes room for cooler air from above. The circulation of warm and cool air is called *convection* and produces winds known as *thermal convection currents,* or "thermals." Thermals may arise from paved areas, sandy areas, stony surfaces, or dry fields because of more intensive warming in those places. Under such conditions, pollutants emitted at or near the ground will be readily diluted with the upper and cleaner masses of air.

An inversion represents the opposite (or inverse) condition—temperature increasing, rather than decreasing, with elevation. The cooler air resides below the warmer air and, because it is more dense, has a tendency to remain in that position. Because the lower air is more dense, pollutants dispersed within it will tend to accumulate because of the lack of mixing with upper layers. Inversions may occur at all levels; that is, the region of temperature increase may occur at all levels. Whatever the level, the circulation of the air within the inversion layer will be impeded by its lack of buoyancy, and pollutants will be concentrated within it.

There are two important kinds of inversion. The *radiation inversion* usually forms at night because of the radiational cooling of the earth; the inversion layer is commonly confined to the first few hundred feet above ground. The *subsidence inversion* is the result of subsiding air masses from high pressure areas which become compressed (and are heated as a consequence). In the United States, such inversions occur principally on the West Coast; the inversion base may vary in elevation from the surface to several thousand feet in the air. Inversions confined to the first few hundred feet are often termed *surface inversions.* Those elevated above ground may be termed *inversions aloft.*

Low-level inversions, which are most characteristic of the fall season, are more common in the morning when air near the surface has cooled and cool air from higher ground has "drained" into the lowlands. Pollutants from industrial and domestic sources may collect at lower levels during the night because there is no sunshine to set thermal currents in motion. The result may be an early morning peak in air pollution. Because of the draining effect, smokestacks in valleys ought to be just below or at the level of surrounding hills to ensure that the smoke plume will escape the stable layer of air. Large-scale inversions are capable of moving many hundreds of miles without being broken up by atmospheric turbulence. For this reason, pollution from urban sources may be a threat in distant, even rural, areas.

REFERENCES

1. Richard Scorer, *Air Pollution*, Pergamon Press, Elmsford, N.Y., 1968.
2. W. Faith and A. Atkinson, Jr., *Air Pollution*, Wiley, New York, 1972.
3. A. C. Stern, ed., *Air Pollution: A Comprehensive Treatise*, vol. 1, *Air Pollution and Its Effects*, vol. 2, *Analysis, Monitoring and Surveying*, vol. 3, *Sources of Air Pollution and Their Control*, 2nd rev. ed., Academic Press, New York, 1968.

Other references are given at the ends of the companion entries listed at the beginning of this topic.

Air Quality Standards

Standards for air quality, to apply nationwide, were published for the first time in 1971. The standards were issued by the Environmental Protection Agency (EPA), which was charged with this responsibility by amendments to the Clean Air Act of 1970. There were two phases in the issuing of the standards. In the first phase, proposed standards were published in the *Federal Register* (January 30, 1971) and comments were invited. Then on April 30, 1971, with comments at hand, revisions were made and the standards appeared in the *Federal Register* as a portion of the Code of Federal Regulations (Part 410). In the list of pollutants and their maximum allowable concentrations it is noted where revisions were made from the first publication of the standards.

The agency set primary and secondary standards for six classes of pollutants: **Sulfur Oxides, Particulate Matter, Carbon Monoxide,** photochemical oxidants, nitrogen oxides, and hydrocarbons. (All the contaminants are discussed under the **Air Pollution** heading. More detailed discussions are available under the separate headings for the first three pollutants and under **Photochemical Air Pollution** for the last three contaminants.)

Primary standards are designed to protect the public health. Secondary standards are designed to protect the public welfare. Presumably the secondary standards are meant to limit the effects on vegetation, materials, weather, soil, water, visibility, etc., in order to prevent economic damage.

Each state was required by the Clean Air Act of 1970 to present a plan to EPA which is intended to implement, maintain, and enforce the standards. The primary standards are to be achieved within three years of plan approval by EPA. Secondary standards are to be achieved under the plan within a "reasonable time."

The administrator noted in promulgation of the standards on April 30, 1971, "Current scientific knowledge of the health and welfare effects of these air pollutants is imperfect. . . . [however], the need for increased knowledge of the health and welfare effects of air pollution cannot justify failure to take action on knowledge presently available."

The standards are:

Sulfur Oxides

1. Primary Standards

   80 micrograms per cubic meter, average annual concentration.

   365 micrograms per cubic meter, maximum 24-hour concentration, not to be exceeded more than once per year.

2. Secondary Standards

   60 micrograms per cubic meter, average annual concentration.

   260 micrograms per cubic meter, maximum 24-hour concentration, not to be exceeded more than once per year, "as a guide to be used in assessing implementation plans to achieve the annual standard."

   1300 micrograms per cubic meter, a maximum 3-hour concentration, not to be exceeded more than once per year.[2]

Particulate Matter

1. Primary Standards

   75 micrograms per cubic meter, annual geometric mean.[3]

   260 micrograms per cubic meter, maximum 24-hour concentration, not to be exceeded more than once per year.

2. Secondary Standards

   60 micrograms per cubic meter, annual geometric mean, "as a guide to be used in assessing implementation plans to achieve the 24-hour standard."

   150 micrograms per cubic meter, maximum 24-hour concentration, not to be exceeded more than once per year.

Carbon Monoxide: Primary and Secondary Standards

   10 milligrams per cubic meter, maximum 8-hour concentration, not to be exceeded more than once per year.

   40 milligrams per cubic meter,[4] maximum 1-hour concentration, not to be exceeded more than once per year.

Photochemical Oxidants: Primary and Secondary Standards

   160 milligrams per cubic meter,[5] maximum 1-hour concentration, not to be exceeded more than once per year.

Hydrocarbons: Primary and Secondary Standards

   160 milligrams per cubic meter,[6] maximum 3-hour concentration, not to be exceeded more than once per year, "for use as a guide in devising implementation plans to achieve oxidant standards."

Nitrogen Dioxide: Primary and Secondary Standards

   100 milligrams per cubic meter,[7] average annual concentration.

---

[2] Added in revised standards.

[3] The geometric mean is the antilog of x where x is sum of the logarithms of the data divided by the number of data points.

[4] This figure was revised upward from 15 milligrams per cubic meter in the second publication of standards with the comment that levels of carboxyhemoglobin in the blood would reach no higher than 2 percent with this revision. Two percent is the level which a concentration of 10 milligrams per cubic meter would produce after about an eight-hour exposure. Heavy smokers regularly tolerate a carboxyhemoglobin level of 2.5 percent and higher.

[5] This is revised upward from 125 milligrams per cubic meter in the first publication of the standards. The administrator noted that the new figure is still below the level at which asthmatic conditions are thought to be aggravated (200 MPCM). The upward revision was apparently in response to a questioning of data on impaired athletic performance.

[6] This is a revision upward from 125 milligrams per cubic meter. The administrator justified it on the basis of the elevated photochemical oxidant standard (note 5). Since hydrocarbon control is aimed at diminishing oxidant levels, a higher allowable oxidant concentration admits a higher allowable hydrocarbon level.

[7] Although the same figure was in first publication of the standards, the administrator eliminated the daily standard for nitrogen dioxide, which was "250 milligrams per cubic meter, 24-hour concentration, not to be exceeded more than once per year." The daily standard was removed because, according to the administrator, no health effects have been associated with short-term exposures. Although this may be the case, one should note the key involvement of nitrogen dioxide in **Photochemical Air Pollution**—where short-term exposures have yielded health effects. A three-hour standard for hydrocarbons was retained, the justification being their contribution to photochemical air pollution; direct health effects due to short-term exposures

TABLE A.1

EPA DEFINITIONS OF AIR POLLUTION EPISODES

POLLUTANT	ALERT[a, c]	WARNING[b, c]	EMERGENCY[b, c]	ENDANGERMENT[b]
Sulfur dioxide (micrograms per cubic meter, 24-hour average)	800	1,600	2,100	2,620
Particulate matter (micrograms per cubic meter, 24-hour average)	375	625	875	1,000
Product of sulfur dioxide and particulate matter	65,000	261,000	393,000	490,000
Carbon monoxide (milligrams per cubic meter, 8-hour average)	17	34	46	57.5 (8-hour average) 86.3 (4-hour average) 144 (1-hour average)
Photochemical oxidants (micrograms per cubic meter, 1-hour average)	200	800	1,200	1,400 (1-hour average) 1,200 (2-hour average) 800 (4-hour average)
Nitrogen dioxide (micrograms per cubic meter)				
1-hour average	1,130	2,260	3,000	3,750
24-hour average	282	565	750	938

[a] *Federal Register*, August 14, 1971, p. 15503.

[b] *Federal Register*, October 23, 1971, p. 20513.

[c] An alert, warning, or emergency is declared when any *one* of the pollutants exceeds the value listed in the table.

In example regulations the agency also indicated for state agencies preparing implementation plans the air pollution levels at which an "alert," "warning," "emergency," or "endangerment" would be announced. An alert is the least severe episode; a warning is more severe. An emergency is the most serious category, short of "endangerment," which is listed as posing imminent harm to health. At each of these levels, control actions designed to abate the pollution were noted (*Federal Register*, August 14, 1971, pp. 15504–05). Table A.1 summarizes air pollutant levels for each of the episode categories.

The control actions suggested become successively more severe as the stage of the air pollution episode worsens. The following is not an exhaustive list of actions but provides an idea of the control plans to be employed. For instance, open burning of tree waste, etc., is banned at the alert stage and at all other stages. At the alert stage incinerators are curtailed in hours of operation and at all worse stages are prohibited from operation. At the alert stage,

to hydrocarbons are not known. The removal of the daily standard for nitrogen dioxide appears to be inconsistent with the philosophy used in setting hydrocarbon standards.

fossil fuel electric plants are to utilize low ash and low sulfur fuels and are to import electric power from stations outside the alert area. Other steam-generating facilities are also required to switch to less polluting fuels and are to reduce steam demands as much as possible. Manufacturing industries are to curtail or postpone operations in order to make *substantial* reductions in emissions. At the warning stage, fossil fuel electric plants and other steam-generating plants are to achieve the *maximum* possible reduction in emissions by the means outlined above. Manufacturing industries are to achieve *maximum* reductions by deferring operations, accepting reasonable economic hardships in the process.

It is interesting to examine the successive controls on vehicle use. At the alert level the operation of motor vehicles is to be limited only to necessary operations. At the warning level, the use of car pools and public transportation is recommended. At the emergency level, the curtailment of vehicle use is achieved by closing all government offices (except those vital for public safety), most wholesale and retail trade establishments (pharmacies and food stores are excepted), banks, auto repair services, amusements, schools at all levels, and most other services. Moreover, only approved emergency use of motor vehicles is allowed.

Arsenic

Arsenic has been a lethal instrument in many dramas, fictional and real. In general, however, the use of chemicals like arsenic has been on a discrete basis, taking one victim at a time. We now find ourselves in the somewhat imprudent position of introducing quantities of arsenic into waters we may eventually have to drink.

SOURCES

Arsenic occurs naturally in a number of forms. Concentrations in rocks range from 1.5 to 10.0 milligrams per kilogram. The natural weathering of arsenic-containing rocks is the major source of trace amounts of arsenic commonly found in unpolluted waters.

Higher concentrations of arsenic in water can occur where arsenic is leached from mineral deposits or mining wastes, and serious problems have arisen when drinking water supplies were contaminated in this way. Two such occurrences were in Córdoba Province, Argentina, and in Reichenstein, Silesia. In the latter case the contamination was linked with a high incidence of skin and liver cancers in the population.

Major sources of arsenic pollution related to civilization include industrial wastes,[8] increased land erosion, mining processes, and the burning of coal. The use of arsenic-based insecticides and **Herbicides** in agriculture is a further source of environmental contamination. The annual world production of arsenic is close to 13,000 tons per year, most of this quantity being used in insecticides and herbicides.

Land erosion has almost tripled in response to increased agricultural and land development. Since soil contains arsenic, the amount of arsenic entering waters from this source has probably risen proportionately. In addition, mining processes expose wastes from which arsenic can leach into watercourses.

Arsenic is often a contaminant in phosphate compounds. As such, it is incorporated to varying extents in detergents, enzyme presoaks, and fertilizers. In terms of polluting natural waters, the amount of arsenic added by these

[8] Arsenic is used in the manufacture of pigments, pesticides, and medicines; it is also used to alloy certain metals and in chemical warfare agents, such as the nerve gas lewisite.

materials is probably not significant compared to that from other civilization-related sources.

Arsenic may be released to the atmosphere by the refining of certain sulfide-containing minerals and by the burning of coal. The arsenic content of coals has been determined to range from 3 to 45 parts per million. If one assumes an average value of 5 milligrams per kilogram and estimates that half of the arsenic is given off when the coal is burned, the combustion of the 3.3 billion tons of coal produced in 1970 would have given off 8,250 tons of arsenic. In the United States, most of the 600 million tons of coal burned in 1971 was used in the generation of **Electric Power.** The burning of petroleum, on the other hand, does not release significant quantities of arsenic to the atmosphere.

TOXICITY

Arsenic is a cumulative poison. Small doses can accumulate and eventually poison an organism even though the single individual doses may not be harmful. Drinking waters with 0.21 to 10.0 milligrams of arsenic per liter have been reported to cause chronic arsenic poisoning. A dose of about 100 milligrams will cause acute poisoning. Chronic poisoning is characterized by weakness, loss of appetite, stomach pains, weight loss, and bleeding gums. Kidney, liver, and nerve damage is common. Breathing arsenic can lead to laryngitis and bronchitis.

Two forms of arsenic appear to be especially poisonous to humans: the arsenites, such as sodium arsenite, and the derivatives of arsine, such as trimethylarsine. Arsenates are also poisonous but less so than arsenites. It is believed, however, that arsenates can be converted to arsenites in the body. Arsenites and arsenates are absorbed easily from the air through the lungs or from drinking water through the gut. Arsenite in the body apparently combines with thiol (sulfhydryl—SH) groups on enzymes and thus prevents the enzymes from taking part in vital chemical reactions.

Arsine derivatives such as trimethylarsine have proved poisonous to humans when breathed. In several cases arsenic poisoning occurred when microorganisms acted on arsenic-based pigments in wallpaper or plaster to produce trimethylarsine. Such pigments are no longer recommended for use on wallpaper. Trimethylarsine is also a common form of arsenic in seafood. Shrimp and lobsters, for example, absorb and concentrate relatively large amounts of trimethylarsine (up to 200 milligrams per kilogram). Although this form of arsenic is known to be toxic to humans when it is breathed, there is some doubt about whether it is poisonous in food.

The belief that arsenic can cause human cancers is supported by epidemiological data showing correlations between incidences of cancer and high levels of arsenic in water supplies, and by industrial studies. The carcinogenic potential of arsenic is still subject to doubt, however, since supporting experimental data on animals are lacking.

Aquatic organisms are sensitive to arsenite at concentrations of 1 to 45 milligrams per liter (measured as arsenic). Plants are also sensitive to arsenic; this is the basis for the substance's herbicidal effects. Some plants are even injured at concentrations as low as 1 milligram per liter. Since arsenic is persistent, it can accumulate in the soil. Thus, continuous use of arsenic-containing **Pesticides** in some orchards has resulted in soil accumulations which are toxic to the fruit trees.

OCCURRENCE IN WATER

Arsenic reaches water by a number of mechanisms related to man's earthmoving activities.

In addition, agricultural runoff may contain arsenic-based pesticides. Rainwater may wash down arsenic in the atmosphere (due, for instance, to the burning of coal). Herbicides like sodium arsenite are used to kill weeds in lakes and reservoirs. Background levels of arsenic in seawater average around 3 micrograms per liter. Arsenic levels in fresh waters vary. Some natural wells and springs contain up to 4 milligrams per liter, but fresh waters generally contain only 1 to 10 micrograms per liter.

Arsenic can be concentrated by aquatic organisms from the environment in which they live. Unlike **Mercury,** however, the concentration of arsenic does not increase as it moves up the food chain. Thus, concentrations of arsenic in the plankton, fish, and birds in a given area are generally of the same order of magnitude. The World Health Organization has set a permissible limit of 50 micrograms per liter for arsenic in public water supplies. The U.S. Public Health Service recommends a maximum of 10 micrograms per liter and states that values over 50 micrograms per liter would indicate water unfit to drink.

In a five-year study of trace metals in U.S. waters it was determined that arsenic was the second most frequent substance to violate Public Health Service **Drinking Water Standards.** Significant arsenic concentrations were found in almost 7 percent of the samples taken, but a study which examined the arsenic concentrations in 969 public drinking water supplies found that 2 percent of the water supplies had more than 50 micrograms per liter. Either sources of arsenic pollution are not at present contaminating drinking water supplies, or water treatment processes are removing the arsenic. Both cold-lime softening and charcoal filtration remove a large portion of the arsenic in water (85 percent and 70 percent respectively). These processes, however, are not used by all water treatment facilities, nor is the entire population served by treated water.

In summary, although widespread danger to the general population is not apparent at the present time, if current upward trends in arsenic levels continue, arsenic has the potential to become a major environmental problem.

REFERENCES

1. J. F. Ferguson and J. Gavis, "A Review of the Arsenic Cycle in Natural Waters," *Water Research* 6 (1972): 1259.
2. J. F. Kopp and R. C. Kroner, *Trace Metals in Waters of the United States*, U.S. Dept. of the Interior, Federal Water Pollution Control Administration, Division of Pollution Surveillance, Cincinnati, Ohio, 1970.
3. "Toxicological and Environmental Implications on the Use of NTA as a Detergent Builder," Staff Report, Committee on Public Works, U.S. Senate, December 1970.
4. *Public Health Service Drinking Water Standards*, U.S. Dept. of Health, Education and Welfare, Public Health Service, 1962.
5. W. M. Gafafer, ed., *Occupational Diseases: A Guide to Their Recognition*," U.S. Dept. of Health, Education and Welfare, Public Health Service, 1966.
6. Report of the Committee on Water Quality Criteria, U.S. Dept. of the Interior, Federal Water Pollution Control Administration, 1968.
7. W. C. Heuper, "Cancer Hazards from Natural and Artificial Water Pollutants," *Proceedings* of the Conference on Physiological Aspects of Water Quality, U.S. Public Health Service, 1960.
8. L. J. McCabe et al., "Survey of Community Water Supply Systems," *Journal of the American Water Works Association* 62 (1970): 670.

Asbestos

Among the more difficult health hazards to control are those whose effects do not become visible immediately, but rather require most of a man's lifetime to become apparent. Many carcinogenic substances fall into this category. That is to say, there is a long time lapse after initial exposure and before the cancer is diagnosed. The danger involved in breathing airborne asbestos particles illustrates the phenomenon well. Diseases which result from breathing asbestos dust are often not apparent until 20 to 40 years after the victims are first exposed.

A case in point might be that of a 55-year-old woman who was found to have developed pleural mesothelioma. This rare form of lung cancer seems to be almost exclusively associated with exposure to asbestos dust. The woman, however, had never worked at an asbestos-related occupation. She had been born in the area of the Cape asbestos fields in South Africa but had left when she was 5 years old. The woman had briefly attended school there, near an asbestos dump. The children often slid down the dump on their way home from school. Two of her classmates have subsequently died from the same rare cancer. Thus, a period of some 50 years passed before the cancer, apparently caused by a short but intensive exposure to asbestos, was diagnosed.

MINERAL FORMS

Asbestos is the general name for a group of fibrous minerals all of which are basically hydrated silicates. The common industrial asbestoses are chrysotile, crocidolite, amosite, tremolite, anthophyllite, and actinolite. They vary in metal content and heat resistance as well as many other properties. These variations determine the industrial uses of the different types and influence their biological effects.

DISEASES RELATED TO ASBESTOS EXPOSURE

Almost all the procedures necessary to mine asbestos, to mill it, and to manufacture usable articles from it cause some release of fibers and bundles of fibers into the air.

ASBESTOSIS

Before the hazardous nature of asbestos dust was generally recognized (around the 1930's) workers in asbestos mining and manufacturing operations were subjected to very high concentrations of asbestos in the air. Some time after they began such work, possibly as little as 7 to 9 years but more generally 20 to 40 years later, they were struck with a disease characterized by severe breathing difficulties. The name *asbestosis* was given to the disease. Some of the fibers these people had inhaled had become embedded in their lung tissues. The body's usual reaction when such fibers become embedded seems to be to coat them with a protein-iron complex, forming asbestos bodies. However, excessive inhalation of the fibers leads to a progressive formation of fibrous tissue, which eventually covers much of the lungs. This tissue interferes with absorption of oxygen from the air. Once it begins, the disease is progressive, even if there is no further exposure to asbestos dust. Death, from respiratory or cardiac failure, usually occurs 2 to 10 years after onset of the disease.

Several studies have shown that workers exposed to high concentrations of dust for 20 years or more are almost certain to develop asbestosis. From 65 to 80 percent of the workers exposed under severe conditions have evidence

Asbestos dust contaminating the air around an asbestos-processing plant in California.

of asbestosis. However, studies of pipe coverers in new ship construction and workers in textile plants, where asbestos levels were generally low, have indicated that even low levels of airborne asbestos can cause a significant incidence of asbestosis. Dust reduction procedures in many industries now ensure that asbestosis is no longer the major cause of death in occupations involving the handling of asbestos. Nevertheless, methods do not yet exist to protect all asbestos workers.

LUNG CANCER

The major occupationally related cause of death among asbestos workers at the present time is lung cancer.[9] A direct relationship between breathing asbestos dust and the subsequent development of lung cancer has not been easy to establish. The long period of time (an average of 39 years in one study) which intervenes between initial exposure to the dust and a clinical diagnosis of cancer obscures the processes involved. The possible roles of other carcinogenic agents (for instance, cigarette smoke) also complicate the picture. In any event, it is now generally accepted that a moderate exposure to asbestos dust will increase the risk of lung cancer to 10 times that of the general population. If a person smokes in addition to breathing asbestos dust, his risk of developing lung cancer may be 90 times that of the general population. It has not yet been established whether there is an ambient concentration of dust below which the risk of lung cancer is negligible. One industrial study did seem to indicate that workers exposed to low concentrations of dust suffered no more risk of cancer than the general population.

Mesothelioma, an unusual type of cancer,

[9] Specifically, bronchogenic cancer (originating in the bronchi of the lungs).

involves tumors of the pleura. The pleura is the membrane which covers the outside of the lungs and lines the chest cavity. Available evidence points to the breathing of asbestos dust as a cause of mesotheliomas. There is again a time lag between exposure to asbestos and clinical recognition of the cancer, the lag being commonly on the order of 30 years. Recent studies have been directed toward the diagnosis of mesotheliomas. The increase in the number of cases found is, of course, in part due to the more intensive search, but the absolute incidence of this previously rare cancer does seem to be rising. In addition, it has been noted that the level of asbestos dust capable of inducing mesotheliomas may be much lower than that which induces bronchogenic lung cancer or asbestosis. This conclusion is supported by the high proportion of mesothelioma victims having only light exposure to asbestos dust.

Although it is generally agreed that asbestos dust can cause cancers, there is little evidence indicating why it does so. Asbestos may, of itself, cause the formation of cancerous tissue, or it might instead be a cocarcinogen, a substance which acts with some other necessary ingredient to cause cancer. Asbestos fibers might function only as carriers for carcinogenic substances. The carcinogens chromium, nickel, and iron complexes are present as impurities in some types of asbestos. Benzo(a)pyrene, another carcinogen, is an asbestos impurity, notably in crocidolite and amosite.

RELATIVE HAZARDS OF DIFFERENT ASBESTOS TYPES

A report of the National Academy of Sciences states that there is insufficient evidence at the present time for designating any form of asbestos as more or less dangerous than any other. All forms are capable of causing asbestosis and all must be considered capable of inducing cancers until further studies are made.

SOURCES OF ASBESTOS DUST

The world production of asbestos increased from a few thousand tons in 1900 to a few million tons in 1968. Most of this is mined in the USSR, Canada, and Africa. Only 3 percent is mined in the United States[10] but twenty-five percent of the world's output is used in the United States.

OCCUPATIONAL EXPOSURE

It was estimated in 1969 that approximately 100,000 workers were directly exposed to asbestos dust in occupations such as the mining, milling, and trucking of asbestos as well as the fabrication of asbestos into various products. A much larger group of workers (about 3.5 million) was indirectly exposed to asbestos dust, mainly through the construction and demolition of buildings. Asbestos has been sprayed as insulation during construction in order to fireproof buildings. It is also a component of construction materials like wallboard and furnace duct linings, which must be cut during construction. These same materials release quantities of asbestos dust when buildings are demolished.

Controls are available for the protection of workers in some, although not all, asbestos-related occupations. Demolition operations and the spraying of asbestos fireproofing insulation are the least controllable procedures at present. Spraying has been prohibited in Boston and Chicago and restricted in several other cities. It is possible to apply the asbestos in sheet or solid form instead of spraying it.

New occupational safety and health standards for asbestos exposure were published in the *Federal Register* in June 1972. Involved is a two-step reduction in the standard for worker exposure to asbestos dust. As of July 1972, the

[10] California (66%), Vermont (31%), Arizona, North Carolina.

airborne asbestos concentration during an 8-hour workday may not exceed five fibers per cubic centimeter, longer than 5 microns.[11] In July 1976, the limit will be lowered to an 8-hour exposure not exceeding two fibers per cubic centimeter, longer than 5 microns in length.

In addition, a ceiling limit of ten fibers per cubic centimeter, longer than 5 microns, has been set, designed to prevent short exposures to very high levels of asbestos dust. A report of the Department of Health, Education and Welfare [8] notes that "the effect after several decades of a one-time acute dose of limited duration which overwhelms the [body's] clearing mechanism, and is retained in the lungs, may be as harmful as the cumulative effect of lower daily doses of exposure over many years of work." The same source also states that efforts should be made to decrease asbestos exposures to as low a level as possible. There are not enough data, the report points out, to determine what level is low enough to prevent asbestos-caused cancers, if indeed such a level does exist.

NONOCCUPATIONAL EXPOSURE

The general public can be divided into two groups on the basis of relative asbestos exposures. The group exposed most heavily includes people living in the vicinity of an asbestos mine, mill, or manufacturing plant. In addition, those living on roads where asbestos is trucked in open trucks and those living near construction or demolition sites are seriously exposed. Members of the families of workers in asbestos-related occupations also fall in this category. Asbestos mines in South Africa and Finland present serious pollutant problems for the surrounding areas. Asbestos dumps and piles of tailings are sources for blowing dust, as are roads which are paved with mine tailings. In a number of mesothelioma cases, the source of asbestos exposure seems to have been a family member working in an asbestos mine or plant. Apparently enough asbestos can be carried home on clothing or other articles to constitute a household hazard.

The second group consists of people exposed only to a general, background type of asbestos air pollution. The main problem encountered in defining the hazard to this group is that no method currently exists to measure the *concentration* of asbestos fibers in "normal" air. The *weight* of asbestos in the air can be estimated by sucking air samples through a membrane filter, burning the filter to destroy non-heat-resistant fibers, and then weighing the material remaining. Reported values are generally about 1 nanogram per cubic meter. This procedure, however, destroys the individuality of asbestos fibers and so gives no information on the number of fibers in a given sample, and no indication of their size. Asbestos fibers can be differentiated from other dust particles by instruments such as the electron microscope, but the concentration of asbestos in urban air, compared to the concentration of other particles, is too small for this method to be useful.

Although the concentration of asbestos fibers in ambient air may be unknown, the sources of the fibers are recognized. A small amount of asbestos is released into the atmosphere by natural weathering of ore deposits, by erosion, and landslides. Farming and road building in areas where the soil contains asbestos will result in the release of fibers to the air. Demolition operations contribute asbestos from insulation and building materials. Mining, milling, and the transportation of ore contribute to atmospheric concentrations of asbestos, as do factories which fabricate asbestos-containing products. A

[11] It is generally believed that fibers longer than 5 microns cause greater fibrosis of lung tissue than shorter fibers. There is, however, nothing known about the relationship of fiber length to cancer production.

number of materials and items contain asbestos, including cement, floor tile, and asbestos paper, and often paints, caulking materials, roof coatings, and certain plastics and textiles.

It is estimated that 85 percent of such products contain tightly bound asbestos, that is, asbestos which will not be released to the air during normal use of the product. Asbestos cement and floor tiles are examples of products containing bound asbestos and are presumably nonhazardous. Asbestos in cloth, paper, and especially in sprayed fireproofing, however, is not tightly bound and thus can contribute significantly to the concentration of airborne asbestos fibers. Asbestos paper products such as asbestos boards and sheets, used in building safes, protective walls, furnaces, and ovens, are almost 100 percent asbestos. They must often be cut during construction, and the cutting releases asbestos fibers. Concern has been voiced about the fate of the asbestos used in brake linings, since it involves 5 percent of the asbestos used in the United States. Conceivably brake linings could release asbestos dust as they are worn away. Investigations have indicated that the asbestos linings of brakes are converted to a nonfibrous form of asbestos by the heat of friction generated on braking. If this is the case, brake linings would not contribute to the air burden of asbestos fibers.

Talcum powder may also be a source of asbestos fibers. Although there is a mineral called talc, commercial talcum powders are in fact mixtures of various substances, including talc. They often contain significant amounts of tremolite asbestos. In one study of 51 samples of commercial talc, 3 were found to contain 80 percent tremolite asbestos, while 34 had less than trace amounts. Large amounts of commercial talc are used as a dilutant and carrier in pesticide formulations and toilet products.

Whether cosmetic talcum powder presents a serious asbestos hazard is not clear at the present time. The hazard may vary with the particular brand. A *New York Times* article in June 1972 reported the city's Environmental Protection Agency statement that Landers and Johnson & Johnson's baby powders had been found to contain between 5 and 25 percent asbestos fibers. The Johnson & Johnson Company denied that its baby powder contained asbestos. Three brands of textured ceiling paint were mentioned in the same article as containing from 5 to 80 percent asbestos.

A recent study [9] has found chrysotile asbestos in one-third of the samples examined of 17 drugs. These drugs are ones usually administered by injection. The source of contamination is believed to be chrysotile used to filter the drugs during preparation. The authors express concern, since animal experiments have shown that injected asbestos fibers can cause cancerous growths at the injection site. Injected asbestos fibers can also be transported to other parts of the body where they could cause cancers.

Asbestos in the air is washed down by rain and snow, thus finding its way into natural waters. Asbestos has been found in trace amounts in almost all waters examined.

THE HAZARD TO THE GENERAL PUBLIC

Two factors combine to obscure the nature of the hazard to the general public from airborne asbestos. First, there are no adequate methods available to detect asbestos fibers in the low concentrations found as normal background. Second, the diseases caused by asbestos inhalation require a long period of time before they are clinically diagnosable. Attention has therefore been focused on the prevalence of asbestos bodies in the lung tissue of the general population in hopes that information can be gained on the extent of exposure to asbestos fibers and the future incidence of asbestos-caused diseases. Asbestos bodies were first identified in individuals suffering from asbestosis and are invariably found in the lung tissue of asbestos workers.

Various studies of the general population have shown that 25 to 100 percent of the people examined have particles resembling asbestos bodies in their lung tissue. For instance, in a 1942 study [8], 40 percent of the housewives, 50 percent of the white-collar males, and 50 percent of the blue-collar males showed evidence of asbestos bodies in their lungs. Among workers who had a history of construction or shipyard work, however, the figure was 70 percent. Whether the golden brown, elongated particles found in lung tissues actually contain an asbestos fiber core is not easy to determine. Similar bodies can be formed in response to dust from graphite, coal, hornblende, rutile, diatomaceous earth, and carborundum. In these cases, the term *ferruginous bodies,* referring to the iron-protein coating, has been suggested as a more appropriate name. Recent electron microscopic studies have indicated that 80 percent or more of the ferruginous bodies in the general population are, in fact, asbestos cored. Investigations designed to show whether the incidence of asbestos bodies is increasing, perhaps in a parallel fashion to the incidence of mesotheliomas, have yielded only conflicting results so far.

Thus, there exists a situation in which a known carcinogen is demonstrably present in the lungs of the general population. Yet no firm information is at hand to determine how much of an actual risk of carcinogenesis this presents, and no data are available on the concentration of the carcinogen in the air which the public is breathing. Given such a situation and the additional fact that the results of breathing this unknown concentration of asbestos dust will not become evident for 20 to 40 years, the only prudent course is to reduce, to an absolute minimum, the atmospheric asbestos concentrations.

STANDARDS

The Environmental Protection Agency set final emission standards for asbestos in March 1973. These standards were not given as numerical values because of the lack of detection methods but are in the form of required control measures. Here are examples of the controls:

1. Filters are specified for use in mining, milling, and manufacturing operations to prevent the release of fibers to the outside air.
2. Methods of demolition which will reduce the release of asbestos fibers to the atmosphere are described. For example, asbestos debris may not be dropped or thrown to the ground but must be carefully lowered. Asbestos insulating materials are to be wetted and removed from buildings before demolition begins.
3. Spray-on asbestos materials, used to insulate structures, may not contain more than 1 percent of asbestos (dry weight).

The standards are to be enforced by the states on the basis of plans submitted to the EPA for approval.

REFERENCES

1. *Asbestos: The Need for and Feasibility of Air Pollution Controls,* Committee on Biological Effects of Atmospheric Pollutants, National Research Council, National Academy of Sciences, Washington, D.C., 1971.
2. B. Castleman, "Asbestos—Effects on Health," Master's thesis, Johns Hopkins University, Baltimore, 1971.

3. R. J. Sullivan and Y. C. Athanassiadis, *Preliminary Air Pollution Survey of Asbestos, A Literature Review,* U.S. Dept. of Health, Education and Welfare, Public Health Service, Consumer Protection and Environmental Health Service, National Air Pollution Control Administration, Raleigh, N.C., 1969.
4. G. W. Wright, "Asbestos and Health in 1969," *American Review of Respiratory Disease* 100 (1969): 467.
5. W. M. Gafafer, ed., *Occupational Diseases,* U.S. Dept. of Health, Education and Welfare, Public Health Service, 1966.
6. *Federal Register* 36, no. 234 (December 7, 1971): 23242.
7. H. E. Whipple, ed., "Biological Effects of Asbestos," *Annals of the New York Academy of Sciences* 132 (1965).
8. *Occupational Exposure to Asbestos: Criteria for a Recommended Standard,* U.S. Dept. of Health, Education and Welfare, Public Health Service, Health Services and Mental Health Administration, National Institute for Occupational Safety and Health, 1972.
9. W. J. Nicholson, C. J. Maggiore, and I. J. Selikoff, "Asbestos Contamination of Parenteral Drugs," *Science* (July 14, 1972): 171.
10. *Federal Register* 36, no. 234 (December 7, 1971): 23207.
11. *Federal Register* 37, no. 110 (June 7, 1972): 11318.

The Automobile and Pollution

The association in the public mind between the automobile and pollution is definite and persistent. This link is supported by numerous statistics. On the other hand, one idea circulating that is unsupported by fact concerns the widespread use of electric automobiles and a consequent reduction of pollution. That **Electric Power** is clean is a fiction; the pollutants associated with electric power are merely different, in part, from those that emerge from the automobile. Our intent in this section is to highlight the principal waste substances which stem from automotive use and point out how these are being and will be controlled. Also, note is made of the trade-off, in terms of pollution, between electric- and gasoline-powered vehicles. This section is brief, not because there is little to say, but because the material is included elsewhere in the book.

The relation of the automobile to elevated levels of **Air Pollution** is not in doubt. Nitrogen oxides and hydrocarbons, both of them constituents of automotive exhaust, react in the presence of sunlight to create the noxious components of **Photochemical Air Pollution.** The hydrocarbons arise from the evaporation and from the incomplete cumbustion of motor fuels. The oxides of nitrogen arise as a by-product of combustion, the oxygen being drawn from the air and the nitrogen being taken from both the air and, to an uncertain extent, the fuel. In the late 1960's automobiles and other motor vehicles contributed about 35 percent, nationwide, of the annual emissions of nitrogen oxides and 50 percent of the annual emissions of hydrocarbons.

Carbon Monoxide, another constituent of automotive exhaust, is a product of the incomplete combustion of fuels. In many of our major cities, Washington and Los Angeles, for example, over 90 percent of the vast quantities of carbon monoxide in the air is attributable to motor vehicles. The pervasiveness of this pollutant in urban atmospheres is illustrated by the fact that in the period of the late 1960's the amount of carbon monoxide produced yearly essentially equaled the annual tonnage of all other air pollutants combined.

The automobile is not usually thought of as a major contributor to particles in the air, but droplets of unburned fuel and of oil and particles of carbon may appreciably raise levels of **Particulate Matter** in certain cities. With new controls on vehicle emissions, however, the picture may change.

Lead is another pollutant in urban air, and virtually all of it is due to motor vehicle operation. Lead antiknock compounds average about 2.5 grams of lead per gallon of gasoline, and about 70 percent of the lead in gasoline is ultimately exhausted to the atmosphere in the form of particles. New lead-free gasolines were appearing on the market in the early 1970's. Surprisingly, the trend was due not to lead's hazards to health but to the fact that lead poisons the catalyst which will be used to remove the hydrocarbons and nitrogen oxides from exhaust streams. In 1973, the Environmental Protection Agency issued regulations on lead in gasoline. The regulations *require* lead-free gasoline to be generally available by July 1, 1974. The EPA also *proposed* a schedule for the reduction of the maximum lead content of gasoline to 1.25 grams per gallon. The reduction would take place over a several-year period. EPA cited both catalyst poisoning and health hazards as its reasons for proposing the regulations. As of March 1973 the schedule for reduction has not been finally issued.

The motor vehicle is, in addition, a major source of urban **Noise Pollution.** In most urban and suburban communities, highway noises contribute to a steady background upon which the sounds of construction equipment, airplanes, power mowers, etc., are superimposed. The only class of air pollution for which motor vehicles may not be listed as a principal source is **Sulfur Oxides.**

Automotive use contributes to **Water Pollution** as well. The extraction and importation of petroleum opens the way for **Oil Pollution** of the seas. Blowouts at offshore stations and tanker spills have in recent years attracted attention for their damage to wildlife and recreational resources. Beneath these publicized events, however, is a background of constant addition of vast quantities of waste motor oil to the water. Little account of the effects of this addition has yet been taken.

At the turn of the century, there was wide concern in New York City over the growing quantity of wastes from horses, which were then the principal mode of transportation. Seventy years later there is alarm that the wastes from the principal mode of transportation, this time automobiles, are again endangering the health of the city. Man may learn, but he seems to require a great deal of time. One is tempted to suggest that perhaps there is an odor threshold above which we become concerned.

What is being done and what may be done? Much. Auto manufacturers have responded to Congressional pressure with real accomplishments, although clean air laws appear to cause them great anguish. Hydrocarbons which used to "blow by" the pistons during compression would escape from the crankcase into the atmosphere. Crankcase blow-by is now being prevented by positive crankcase ventilation, which recycles the escaped fuel back to the combustion chamber. This eliminates losses from a source which formerly accounted for 25 percent of the hydrocarbon emissions from the auto.

Carbon monoxide and hydrocarbons are being reduced by numerous engine modifications, including timing adjustments, more exacting tolerances for carburetors, and an upward adjustment in the air-to-fuel ratio. One result of these changes is increased gasoline mileage, but future changes such as the catalytic converter may have the opposite effect. Another feature added to some manual transmission vehicles is air injection in which the exhaust gases are mixed with air for a final combustion just prior to their ejection to the atmosphere. Evaporative controls are on all 1971 cars. Gasoline vapors that are lost from the carburetor or fuel tank when the engine is turned off are stored and recycled to the combustion chambers on start-up.

All these changes have had a marked effect on vehicle emissions (see Table A.2). The first year of vehicle certification was 1971, and the results were encouraging. The regulations to which manufacturers were conforming were set in 1968; they called for no crankcase emissions from 1971 vehicles and established limits on exhaust emissions and evaporative emissions.

Autos built in the 1960's without positive crankcase ventilation released crankcase blow-by gases at an average rate of about 3.7 grams of hydrocarbons per mile. This source is now virtually eliminated.

The auto without emission controls also exhausted about 11.2 grams of hydrocarbons per mile. The standard for 1971 models is 2.2 grams of hydrocarbons per mile. Compliance by all manufacturers on 1971 cars sold in the United States is reported.

That same old car without emission controls spewed about 73 grams of carbon monoxide from the exhaust during each mile of travel. The carbon monoxide standard for 1971 models is 23 grams per mile. This too was met, and many vehicles produced only about half that amount.

The polluting monster of the 1960's also lost vapors from the carburetor and fuel tank at the rate of 2.8 grams of hydrocarbons per mile. The 1971 standard called for a reduction of this emission to about 0.5 gram per mile. Results of testing showed most vehicles at one-fifth this level and less.

The apparent technology of choice for future control is a catalytic converter. Once thought to have low potential for long-term control of automotive emissions, it now appears that auto manufacturers will rely on the device at least initially to meet the hydrocarbon and nitrogen oxide reduction requirements set by the Clean Air Act of 1970. The converter operates on the exhaust stream and will have two compartments. In the first, nitrogen oxides will be

TABLE A.2

TIME PATTERN OF EMISSION CONTROLS ON LIGHT-DUTY VEHICLES

	Hydrocarbons (g/mi)			Carbon Monoxide (g/mi)
	Blow-by	*Exhaust*	*Evaporative*	
Average emissions from vehicles of the mid-1960's	3.7	11.2	2.8	73.0
Standards for 1970 vehicles	3.7	2.2	2.8	23.0
Standards for 1971 vehicles	[a]	2.2	0.5	23.0
Standards for 1975 vehicles[b]	[a]	0.41	[a]	3.4

[a] Emissions eliminated.

[b] The law states that a 90 percent reduction is to be achieved from 1970 levels. The figures in the fourth row do not correspond to a 90 percent reduction from the second row because the testing procedure has changed several times since that period. The figures do represent a 90 percent reduction from the 1970 standards which correspond to the current method of measurement.

reduced to nitrogen; in the second, after the introduction of more air, carbon monoxide and hydrocarbons are to be oxidized to water and carbon dioxide. Low-lead and lead-free gasolines are gradually entering the market in anticipation of the day when the converters will come into use. Lead-free gasolines are essential for the converter's operation because **Lead** fouls the catalysts and makes them ineffective.

The Clean Air Act of 1970 set stringent pollution control standards for model years 1975 and 1976. Carbon monoxide and hydrocarbons were to be reduced by 90 percent from 1970 levels in time for the 1975 model year. Nitrogen oxide emissions are to be reduced by 90 percent from 1971 levels in time for the 1976 model year. The act gave manufacturers the right to appeal for one-year extensions of these standards. Requests for extension of the hydrocarbon standards for 1975 models were filed with the Administrator of the EPA in 1972 and were initially refused. Auto manufacturers, insisting that the appropriate technology for the 1975 model would not exist in time, took the matter to the courts, which directed the Administrator of the EPA to reconsider his decision.

In April 1973 the Administrator chose to modify his decision and issued interim standards for 1975. The interim hydrocarbon and carbon monoxide standards represented about a 60 percent reduction from 1970 standards. California, however, was singled out for tougher interim standards. According to the EPA, the catalytic converter would be needed to achieve California's interim standards but probably would not be necessary on more than 10 percent of the vehicles sold outside of California. The decision is thought to give auto manufacturers an opportunity to investigate the stratified-charge, dual carburetor engine system. Honda, a Japanese manufacturer, has demonstrated such a system which met 1975 standards.

The Clean Air Act of 1970 also called for studies by the National Academy of Sciences to investigate the feasibility of meeting the standards. In a report issued by a committee of the Academy in January 1972 the development of adequate technology for 1975 models was termed "possible." However, it was noted that the technology for such emission control could add $200 to the price of a new car. Fuel consumption could increase by 5 to 15 percent.

As of early 1973, the control of nitrogen oxide emissions had not been accomplished. Several systems besides the converter appeared promising, including steam injection to the combustion chamber and recirculation of exhaust gases.

The foregoing discussion has set forth advances in the technology of the gasoline-powered automobile. Are electric autos or steam cars possible? Technologically, American enterprise, given the proper stimulus, seems capable of remarkable achievements. Should the stimulus be applied? Will the use of electrical energy for the car solve pollution problems? It is not certain; both **Fossil Fuel Power Plants** and **Nuclear Power Plants** exert environmental effects on the atmosphere and on water.

In the late 1960's, fossil fuel power plants accounted for more than 40 percent of the annual **Sulfur Oxide** emissions in the United States, about 25 percent of the annual emissions of **Particulate Matter,** and about 20 percent of the annual emissions of nitrogen oxides. If, as anticipated, fossil fuel plants continue to command a large share of the **Electric Power** market, a switch to electric autos could have a marked and nonbeneficial effect on pollution. It is not clear that the electric auto is an answer to air pollution control at this time. Its widespread use might lead only to a substitution of one pollutant for another. For the time being, technological changes in the gasoline-powered automobile will make significant improvements. Probably before the decade is over, we will know our future needs and directions in new technology. Beginning now, however, research in new technology ought to be initiated and stimulated.

But trading one form of energy for another is only the technological side; society entrenched in its riding habits is on the opposite side of the coin. Investment in and utilization of large-scale transportation systems—fast rail systems, subways, buses, and elevated vehicles—is certainly a viable alternative to individual transportation with its associated pollutants. Mass transit systems use energy more efficiently and thus provide less by-product pollution for each individual per mile of travel. This factor leads us to a recommendation; it would be useful for the federal government to provide an accelerated program of investment in large-scale transportation systems for the cities. A logical source of the investment would be the Highway Trust Fund, whose coffers are fed by the federal gasoline tax. The need is to create comfortable, accessible, and desirable alternatives to the automobile. By means of revenue sharing with the cities, these modes of transport might be made financially self-sustaining even though fares alone could not support the systems. This is not a subsidy; it is a way of financing cleaner air. Finally, it would be useful for such transportation services to exist, if only to avoid some of the 50,000 annual deaths from automobile accidents.

REFERENCES

In addition to the companion entries, the reader may wish to see the certification results for particular vehicles. These are published in the *Federal Register* 36, no. 70 (April 10, 1971). The postponement decision is described in the *New York Times*, April 12, 1973.

B

Bacterial and Viral Pollution of Water

Today nearly everyone in the United States drinks water with the reasonable assurance that he will not become ill from bacterial contamination. It was not always so. Nor are the individuals in developing nations assured of water safe to drink. Bacterial pollution of drinking water is prevented by water treatment, particularly by filtration through beds of sand and by chlorination (see **Water Treatment and Water Pollution Control**).

The bacteria and viruses which cause illness, and which can survive long enough in water to infect people, form a large and varied group. Examples of waterborne diseases are: typhoid, paratyphoid, and dysentery, which are caused by bacteria in the *Salmonella* group; and poliomyelitis and infectious hepatitis, which are attributed to viral agents.

We did not arrive easily at the understanding that diseases may be spread through water. Much suffering intervened. We here recount the early cases of polluted water supplies to show how our current knowledge was gained, summarize current bacterial standards for a safe water supply, discuss the potential for viral contamination of water supplies, and describe quality standards for recreational waters.

WATERBORNE EPIDEMICS OF BACTERIAL DISEASES

Typhoid and Asiatic cholera are diseases that affect the intestinal tract. The pathogens (disease-producing microorganisms) causing them are found in the bowel discharges of infected individuals, and if these discharges find their way into water which is ultimately consumed by man, the results can be disastrous.

Typhoid fever is characterized by headache, backache, nosebleed, diarrhea, and loss of strength in the roughly two-week period which precedes its active state. The period of active illness may last four to eight weeks. The weakened individual is afflicted with rising and falling fevers ranging from 100° to 105°F, and the intestine may be perforated with ulcers. The disease is spread throughout the body by way of the lymphatic system and bloodstream. Besides being communicated via water polluted with the specific pathogens, typhoid may be spread by carriers of the disease who handle food.

Asiatic cholera, in contrast to typhoid fever, strikes more quickly and with greater intensity, and results in a higher percentage of fatalities. An extreme diarrhea brings dehydration and shock. Individuals may contract the disease and not exhibit symptoms but may still be capable of transmitting the offending bacteria during their period of infection. While such contact-carriers are common in epidemics, the occurrence of chronic carriers is infrequent.

The realization that these diseases were associated with polluted water came gradually over the period from 1850 to 1900. While today the connection is taken as obvious, the residents of St. James Parish, London, in 1854 at first saw no such connection.

The epidemic of Asiatic cholera that year in London resulted in the death of at least 700 of the 36,000 residents in the district. The death rate was 220 per 10,000 population in St. James Parish and only 9 and 33 per 10,000 in adjacent districts. An Inquiry Committee was created to uncover the cause of the outbreak. Such aspects of the community as population density, ventilation and cleanliness of housing, cesspools, sewers, and water supply received attention.

The investigation of one of the members of the Inquiry Committee, John Snow, revealed that, in the epidemic in St. James Parish, 69 of the first 83 deceased individuals had been known to drink from the Broad Street well, either constantly or occasionally. The evidence presented by Dr. Snow to the Board of Guardians of St. James Parish quickly convinced them to remove the handle from the pump on Broad Street.

In the spring of 1855 the Reverend Mr. Whitehead, a member of the committee, discovered a cesspool very near to the well. This cesspool had received dejecta from individuals who had suffered from an unidentified disease prior to the outbreak of cholera. A detailed survey of the well and surroundings revealed a drain from the cesspool at a level 9 feet above the water in the well. The drain was in a state of decay which allowed fluids originating at the privy above the cesspool to percolate through the drain floor and the soil into the adjacent well.

This historic incident in St. James was followed by a succession of waterborne epidemics in the 50 years leading to the twentieth century. The epidemics illustrated clearly the dangers of infected water supplies.

In the years 1890–1891 there were typhoid epidemics in at least two areas of the United States. A recounting of the circumstances in the Mohawk-Hudson Valley region of New York provides convincing evidence that the polluted river was a route for the spread of typhoid infection. The communities of Schenectady, Cohoes, West Troy, and Albany are situated near the confluence of the Mohawk and Hudson Rivers, and all were suffering from typhoid (see Figure B.1). Schenectady, on the Mohawk about 18 miles from the point where it joins the Hudson, was the first to be struck, in July 1890. Cohoes, at the junction of the Mohawk and Hudson, was drawing its drinking water from the Mohawk, into which Schenectady dis-

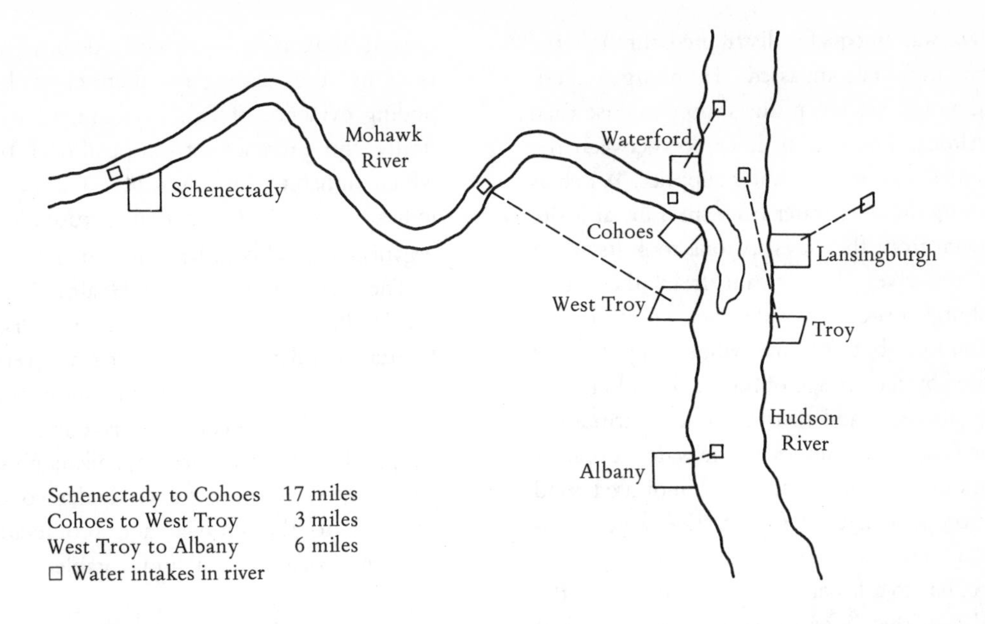

FIGURE B.1
SOURCES OF DRINKING WATER IN THE MOHAWK-HUDSON VALLEY OF NEW YORK, 1890–1891
From F. Turneaure and H. Russell, *Public Water Supplies,* Wiley, New York, 1911.

charged its wastes. The typhoid epidemic in Cohoes began in October 1890. West Troy, though located on the Hudson, took its water from the Mohawk above Cohoes. Its epidemic arrived in November 1890. Albany's source of supply was the Hudson River below the point of confluence with the Mohawk. Its water was thus partly from the Hudson, partly from the Mohawk. The outbreak occurred there in December 1890.

This wave of epidemics is contrasted to the absence of typhoid in other communities of the region. Waterford, which is across the Mohawk from Cohoes, escaped entirely, as did Troy, which is across the Hudson from West Troy. Both of these communities drew their water from the Hudson above its confluence with the Mohawk.

At the same time, in the Lowell-Lawrence area of Massachusetts there was a similar situation. Two communities, each drawing water from below the discharge of an infected community, were subsequently struck by a typhoid epidemic.

From the experience of these two regions, public health scientists learned that the sewage-polluted river could act as the transmitting agent of typhoid fever. If these events are little remembered today, it is a tribute to the efficacy of the treatment processes which were shortly incorporated into the water supplies of cities and towns.

One epidemic in particular was destined to spur the movement toward treated water supplies. It was 1892, and the city of Hamburg in Germany was struck by an epidemic of cholera. Located on the Elbe River in Germany are the adjoining cities of Altona and Hamburg. Although they had a common boundary, the cholera epidemic which fell on them in the fall of 1892 was unequally distributed through the region they encompassed. Hamburg experienced a case rate seven times more intense than did Altona. The crucial difference between the two cities was in their water supplies. Whereas Hamburg drew its water from the Elbe at a site upstream from the cities, Altona took its water from the river downstream from the cities. Hamburg treated its water not at all before distribution, but Altona, whose supply was polluted by the sewage of both cities, filtered its water through sand beds prior to distribution. These facts alone implicated the water supply as the means of transmission and indicated sand filtration as a method for effective water purification. The fact that cholera did occur in Altona but to a lesser extent is attributed to the population that lived in Altona but worked in Hamburg.

These are some of the great historical cases which brought enlightenment on how diseases spread through polluted water. Water-borne epidemics have occurred since then, but the process of learning to protect public water supplies had begun.

BACTERIAL STANDARDS

It is difficult to catalogue the presence or absence of the myriad types of disease-causing organisms in water. In addition, adequate methods are not at present available to detect all of the possible human pathogens. As long ago as 1880, scientists, recognizing these facts, suggested that if evidence of human fecal contamination of waters could be obtained, those waters could be presumed unfit to drink. On this basis, tests have been developed to determine the presence of the coliform group of bacteria. These organisms are normal inhabitants of the gut of warm-blooded animals and are generally harmless. All humans have such bacteria in their intestines. Thus, pollution with disease-causing organisms is currently detected not by isolating the pathogens themselves but by finding evidence of fecal contamination, specifically the presence of the coliform bacteria which inhabit the intestine. That evidence indicates the likelihood that disease-producing organisms are also to be found in the water.

The U.S. Department of Health, Education and Welfare, in its booklet defining **Drinking Water Standards,** recommends a several-step procedure for determining the most probable number (MPN) of coliform bacteria in a water sample. In this procedure, organisms present in the water sample are grown under conditions favorable to the survival and proliferation of coliform bacteria and unfavorable to other organisms.

Water samples are to be taken at specified time intervals depending on the number of people served by the water supply. The more people served, the more often their water must be sampled, although fewer samples per unit of population are required. For instance, a population of 1,000 requires 2 samples a month, or 1 per 500 people; a population of 1 million requires about 300 samples per month, or 0.15 per 500 people.

The sample is fermented in a special broth containing the sugar lactose. Coliform bacteria and some others are able to produce gas in this medium. Samples in which gas is produced are called presumptive positive.

Presumptive positive samples are then grown in another broth, to which chemicals have been added that will inhibit the growth of all but the coliform bacteria. Alternatively, they are grown on agar (a gelatin-like substance) to which chemicals are added that cause coliform bacteria to grow in a characteristic form. If growth is again positive, the sample is referred to as confirmed positive.

The final step involves isolation of coliform bacteria from the sample and demonstration of the characteristic appearance and behavior of the organisms.

A somewhat simpler procedure involves filtering the original sample through a membrane filter (which has such tiny holes or pores that even bacteria are held back). The bacteria are then nurtured directly on the filter. This method allows the use of larger samples and gives results more quickly. Both methods allow the estimation of the number of coliform bacteria in the water supply.

Allowable limits for coliforms are based on the percentage of times positive samples are obtained from a water source, and on the appearance of specific patterns of positive tests. For example, the absence of gas in all tubes when 5- to 10-milliliter portions of a sample are examined indicates that less than 2.2 coliforms exist in 100 milliliters of the sample. Such waters are considered safe to drink. On the other hand, a positive confirmed test in three or more 10-milliliter portions of a sample or the calculation of 4 or more coliforms in 100 milliliters of the sample indicates bacterial pollution at a level such that the water is not considered safe to drink.

Some criticism of the above procedures has been voiced, since they measure coliform bacteria from many sources other than human feces. Coliforms are found in soil, on plants, and in insects, as well as in warm-blooded animals. More discriminatory tests are based on the fact that coliforms from warm-blooded animals can grow at higher temperatures than those from nonfecal sources. Thus the appearance of coliforms able to grow at 44°C (113°F) is recognized as a more serious hazard than detection of coliforms able to grow only at 37°C (98.6°F). Laboratory experiments have established that 96 percent of coliform strains isolated from warm-blooded animal feces grow at the elevated temperature while 91 percent of coliforms isolated from unpolluted soil do not. Special broths (called EC medium) have been developed for use at the higher temperature. The membrane filter can also be used with special medium (MFC) at 44°C.

The Delaware River Basin Commission and the Pennsylvania Department of Health have adopted the use of *fecal* coliform tests in regulation of sewage disposal into the Delaware and Ohio rivers.

In some studies fecal streptococci are also determined. This is another group of bacteria indigenous to the intestinal tract of warm-blooded animals. Their presence also indicates pollution with feces. They are sometimes valuable in determining the source of pollution. Thus ratios of fecal coliform to fecal streptococci greater than 4 to 1 indicate domestic waste water as a pollution source while ratios less than 0.6 to 1 are indicative of wastes from nonhuman, warm-blooded animals.

VIRAL CONTAMINATION OF WATER SUPPLIES

Just as some bacterial diseases—typhoid and cholera, for instance—are spread by contaminated water, we have learned in the past 20 years that some viral diseases may spread via water. Although the viruses of smallpox, chicken pox, and measles are spread by droplets in the air and those of encephalitis and yellow fever are carried by mosquitoes, such well-known virus diseases as conjunctivitis and poliomyelitis have the potential to spread via water.

CHARACTERISTICS OF VIRUSES

Viruses are believed to be the simplest of living organisms. In fact, they are so simple in their chemical makeup that they are sometimes described as being on the borderline between living and nonliving substances. Viruses cannot reproduce themselves. Instead they infect plant, animal, or bacterial cells and cause the infected cell to manufacture more viruses. In the process the host cell may be destroyed.

Since viruses cannot grow and multiply by themselves, they are usually cultured, in the laboratory, in animals like chicks or mice, in fertilized eggs, or in isolated animal cells grown in various fluids. The last procedure is called tissue culture. The number of viruses in a given sample can be estimated by counting the clear areas on a plate covered by a sheet of host cells. These clear areas are called plaques and represent the places where host cells were destroyed by viral infection. The agent causing infectious hepatitis, although it is suspected to be a virus, cannot be cultured in this way, since it has not yet been grown outside the human body.

Viruses are rarely visible directly, even with an ordinary microscope, but pictures have been taken, magnified many thousands of times, with electron microscopes. Because of their tiny size, viruses pass through most filters, even those which trap bacteria.

These simple life forms are more resistant to many chemical treatments (e.g., chlorination) than are bacteria. In the waterborne stage of a virus life processes are relatively quiescent, making the package very stable. Only when presented with an appropriate host does the virus become active, injecting itself into the cells of the host to begin its life cycle.

TABLE B.1

VIRUSES AND THE DISEASES THEY CAUSE IN HUMANS

Virus	Disease
Poliovirus	Poliomyelitis
Coxsackie viruses, groups A and B	Aseptic meningitis, herpangina
ECHO[a] viruses	Aseptic meningitis, diarrhea
Adenoviruses	Eye and respiratory infections
Infectious hepatitis[b]	Infectious hepatitis

[a] Enteric cytopathic human orphan viruses.

[b] The consensus is that the causative agent in infectious hepatitis is a virus, although this has yet to be conclusively demonstrated.

VIRUSES OF PUBLIC HEALTH SIGNIFICANCE

The viruses of concern in regard to water pollution include the human enteric viruses, found in large numbers in the feces of diseased humans, and several livestock viruses. In addition, only those viruses which can survive in water for a reasonable length of time are a problem. Table B.1 lists the viruses that affect humans and the diseases they cause.

Of these viruses only two groups have been causally linked to outbreaks of disease from drinking waters. They are infectious hepatitis and polio viruses. The most extensive epidemic of waterborne infectious hepatitis occurred in India in 1955–1956. Between 20,000 and 40,000 cases resulted from pollution of a municipal water supply. Since the water supply was chlorinated to some extent and since no concurrent increase in bacterial diseases was observed, the incident also illustrates that viruses are more resistant to chlorination than are pathogenic bacteria. A number of waterborne infectious hepatitis epidemics have been reported in this country. In several cases, as in 1961 in Alabama and Mississippi, the cause is believed to have been contaminated oysters. Shellfish have been shown to take up polio, infectious hepatitis, Coxsackie, and ECHO viruses from sewage-polluted seawater.

Several polio epidemics have been traced to water contamination. For example, one in Edmonton, Canada, was attributed to a failure of the chlorination system for the public water supply.

Adenoviruses are actually respiratory viruses but are also found in human feces. No general water supply epidemics have been attributed to this group. However, outbreaks of adenovirus diseases from swimming in contaminated swimming pools are known.

WATER AND SEWAGE TREATMENTS FOR CONTROL OF VIRUSES

As normally practiced, primary and secondary sewage treatments do not kill all the pathogenic viruses in sewage. This fact was demonstrated in a study conducted on water samples from the upper Illinois River in 1962. All three sewage treatment plants in the area have activated sludge procedures; yet viruses from all the human enteric groups were isolated from treated sewage and from the river.[1] Once released into waters, viruses have been shown to remain infective for distances of at least one to four miles from the point of entry. Chlorination has been shown to be an effective sewage treatment method for elimination of viruses if the amount of chlorine added and the time allowed for it to act are carefully controlled (see [12] or [13]).

Water for drinking can be filtered through sand after coagulation with aluminum or ferric sulfate and then chlorinated. These methods should produce water of acceptable quality for drinking.

BATHING WATER QUALITY

Two general criteria are applied in judging the acceptability of waters for bathing. First, bathing waters should be sanitary; that is, the water should be free of organisms which produce disease. Second, the water should be aesthetically pleasing; that is, it should look clear and be free of unpleasant odors or large numbers of aquatic plants.

[1] See **Water Treatment and Water Pollution Control,** for descriptions of primary and secondary sewage treatment and the activated sludge process.

There is a significant number of waterborne diseases. Besides the diseases mentioned previously, one might add swimming pool granuloma or elbow lesions caused by *Mycobacterium balnei,* fungal illnesses such as ringworm, the parasitic worm disease, swimmer's itch or schistosome dermatitis.

Designing prevention standards for the transmission of diseases among bathers has its problems. The difficulty lies in the present lack of solid evidence relating the results of standard tests such as fecal coliform counts or fecal streptococcal counts to the actual incidence of diseases among swimmers. The Federal Water Pollution Control Administration Water Quality Criteria Committee has assessed the available data and recommends fecal coliform counts, which are a reflection of contamination by the feces and urine of warm-blooded creatures, including man, as the best available test.

Several workers have reported that the ratio of fecal to nonfecal coliforms in polluted waters is about 1 to 5. The U.S. Public Health Service has determined that a total coliform count over 2,300 per 100 milliliters in bathing waters constitutes a health hazard. Thus 400 fecal coliforms per 100 milliliters of water would be considered dangerous. Waters restricted to boating and fishing would still be regarded as safe for that use at counts up to 2,000 fecal coliforms per 100 milliliters. Chlorination of sewage effluents entering natural waters is the principal means of removing disease-producing organisms.

Since pH, or the acidity of water, has a relationship to eye irritation, the Federal Water Pollution Control Administration Water Quality Criteria Committee concludes that the pH should be maintained in the range 6.5 to 8.3 (tears have a pH of 7.4).

Finally, the same group notes that water warmer than 85°F (30°C) cannot cool the body during exercise (swimming) and thus may cause physiological disturbances. For this reason, 85°F is probably an upper limit for the temperature of bathing waters.

REFERENCES

EPIDEMICS

1. *Sedgwick's Principles of Sanitary Science and Public Health*, rewritten and enlarged by S. Prescott and M. Horwood, Macmillan, New York, 1935.
2. F. Turneaure and H. Russell, *Public Water Supplies*, Wiley, New York, 1911.

BACTERIAL STANDARDS AND BATHING WATER QUALITY

3. Orsanco Water Users Committee, "Total Coliform: Fecal Coliform Ratio for Evaluation of Raw Water Bacterial Quality," *Journal of the Water Pollution Control Federation* 43 (1971): 630.
4. *Standard Methods for the Examination of Water and Waste Water*, 12th ed., American Public Health Association, New York, 1965, pp. 567–626.
5. *Public Health Service Drinking Water Standards*, U.S. Dept. of Health, Education and Welfare, Public Health Service, 1962, pp. 3–6, 13–18.
6. *Sanitary Significance of Fecal Coliforms in the Environment*, U.S. Dept. of the Interior, Federal Water Pollution Control Administration, 1966.
7. H. F. Clark and P. W. Kabler, "Reevaluation of the Significance of the Coliform Bacteria," *Journal of the American Water Works Association* 56 (1964): 931.
8. *Water Quality Criteria*, U.S. Dept. of the Interior, Federal Water Pollution Control Administration, 1968.
9. H. Hawkes, "Disposal by Dilution? An Ecologist's Viewpoint," in *Microbial Aspects of Pollution*, ed. G. Sykes and F. Skinner, Academic Press, New York, 1971.

VIRAL CONTAMINATION

10. *Public Health Service Drinking Water Standards*, U.S. Dept. of Health, Education and Welfare, 1962, Public Health Service, pp. 18–19.
11. Report of the Committee on Water Quality Criteria, U.S. Dept. of the Interior, Federal Water Pollution Control Administration, 1968, pp. 140–141.
12. C. W. Chambers, "Chlorination for Control of Bacteria and Viruses in Treatment Plant Effluents," *Journal of the Water Pollution Control Federation* 43 (1971): 228.
13. R. W. Burns and O. J. Sproul, "Virucidal Effects of Chlorine in Wastewater," *Journal of the Water Pollution Control Federation* 39 (1967): 1834.
14. T. G. Metcalf and W. C. Stiles, "Enteroviruses Within an Estuarine Environment," *American Journal of Epidemiology* 88 (1968): 379.
15. G. A. Lamb et al., "Isolation of Enteric Viruses from Sewage and River Water in a Metropolitan Area," *American Journal of Hygiene* 80 (1964): 320.
16. N. A. Clark and S. L. Chang, "Enteric Viruses in Water," *Journal of the American Water Works Association* 51 (1959): 1299.
17. G. M. Little, "Poliomyelitis and Water Supply," *Canadian Journal of Public Health* 45 (1954): 100.
18. S. L. Chang et al., "Removal of Coxsackie and Bacterial Viruses in Water by Flocculation," *American Journal of Public Health* 48 (1958): 51.

Barium

Barium exists both as a stable substance and as a radioactive substance. The stable element is dealt with first. Concern here is mainly with barium as a water pollutant.

Barium carbonate and sulfate are relatively insoluble, and carbonates and sulfates are naturally present in surface water. Nevertheless, one may find barium in the microgram-per-liter range of concentration in surface waters.

Barium has toxic effects on the nervous and circulatory systems and is thus specifically mentioned in the **Drinking Water Standards.** The maximum concentration of barium tolerable in drinking water is set at 1 milligram per liter by the Public Health Service, but there is no barium standard for irrigation water.

Radioactive barium (barium-140) is of special concern as a fission product which may occur in cows' milk after radioactive fallout. As such, its hazardous nature is due to the radiation dose which would accompany ingestion.

Beryllium

Beryllium is one of the most hazardous substances being used in industry today. Prolonged inhalation of dusts of beryllium or its salts may be responsible for a severe respiratory disease known as berylliosis. The symptoms are chest pain, dyspnea (shortness of breath), cough, fever, and cyanosis (insufficient aeration of the blood causing the surface of the body to become blue). The symptoms may be delayed from one month to 20 years depending on degree of resistance and extent of exposure.

Beryllium dust in air is also responsible for a temporary dermatitis (skin disease). In addition, short exposure to high concentrations of the dust may result in a chemical pneumonitis (a chemically induced pneumonia), which is a curable disease. Over 600 cases of beryllium poisoning have been reported since the metal was recognized as toxic some 30 to 40 years ago.

Beryllium may be found in trace amounts in water. It has been isolated in the nanogram-per-liter range of concentration in the Delaware, the Monongahela, and the Allegheny rivers. Its presence in the Monongahela is believed to be due to **Acid Mine Drainage** in the surrounding region.

Although beryllium is highly toxic as an air pollutant, there is no evidence to date that it is harmful in trace quantities in drinking water, and the element is not listed in the **Drinking Water Standards** of the U.S. Public Health Service. The Committee on Water Quality Criteria of the Federal Water Pollution Control Administration recommended a maximum permissible level of 0.5 milligram per liter in irrigation water which is being applied continuously to soil.

Black Lung Disease

Black lung disease (coal worker's pneumoconiosis) is caused by inhalation of the dust of bituminous and anthracite coals over an extended period. In anthracite coal workers, the disease is referred to as anthracosilicosis (miner's asthma), and it was recognized over 40 years ago. However, awareness of the disease among bituminous coal workers in the United States did not occur until the late 1940's.

The number afflicted with the disease among the retired and active mining populations has been estimated to be as low as 38,000 and as high as 125,000. In 1969, 28,000 Pennsylvania coal miners were being compensated for disability due to pneumoconiosis. As to the current prevalence of the disease, a study of bituminous coal miners conducted by the Public Health Service between 1963 and 1965 indicated that 10 percent of the working miners and 20 percent of the nonworking miners showed evidence of the disease on X-ray.

Pneumoconiosis may begin to show on X-ray after 10 or more years of exposure to the dust. However, even though the disease may be detected by X-ray, outward symptoms may still be lacking and the individual may be unaware of his disease. Not only is the disease related to time spent underground; the specific job seems to be a factor in its development. Miners working at the coal face and those engaged in transport of the coal seem to be most at risk.

Radiographic evidence of the disease is seen as black spots or nodules in the lung. The longer the exposure, the larger the nodules. The two forms of the disease are referred to as "simple" and "complicated." While simple pneumoconiosis may only be associated with chronic bronchitis, in the advanced form there may be shortness of breath, upper respiratory infections, lung collapse, and eventual heart failure.

There appears to be no therapy for the disease. Control of coal dust or removal of workers from the mines on early evidence of disease seem to be the only preventive measures possible. Although there have been some guidelines for the limits of coal dust exposure, the values of the guidelines have trended steadily downward. The Federal Coal Mine Health and

At left, Gough section of lung from an 86-year-old man who never was a coal miner.
At right, Gough section of lung from a coal miner afflicted with black lung disease.

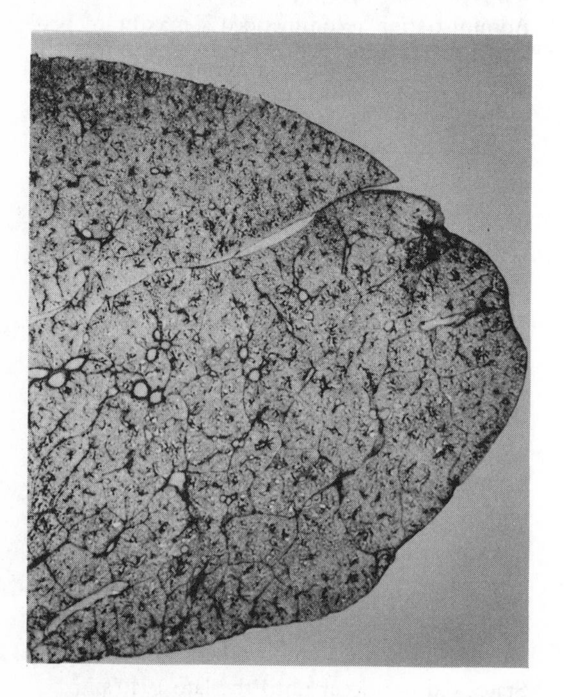

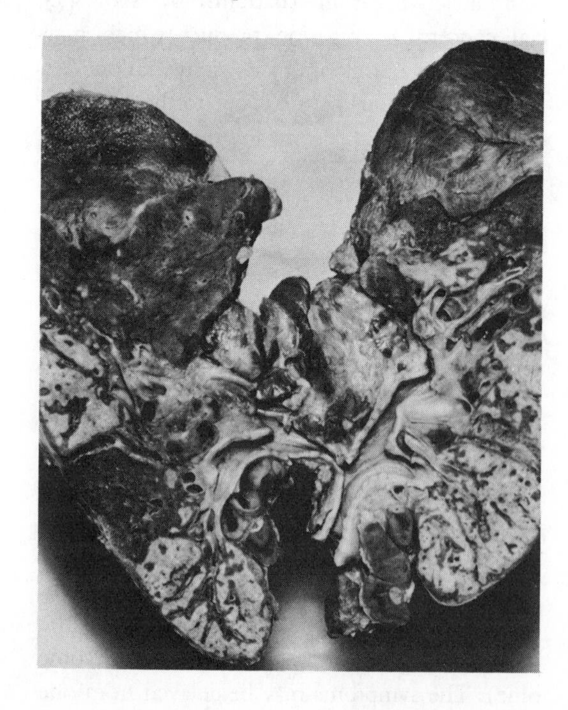

Safety Act of 1969 prescribed a standard for respirable dust of no more than 2 milligrams per cubic meter after December 31, 1972. However, temporary noncompliance at 3 milligrams per cubic meter is apparently allowed if the mine operator is engaged in achieving the primary standard.

REFERENCES

1. W. Lainhart, H. Doyle, P. Enterline, A. Henschel, and M. Hendrick, *Pneumoconiosis in Appalachian Bituminous Coal Workers*, U.S. Public Health Service, Bureau of Occupational Safety and Health, 1969.
2. *Federal Register* 37, no. 97 (May 18, 1972): 10042.

C

Cadmium

Cadmium is a trace metal which presents a little-recognized water pollution hazard. It is not known to have any beneficial effects on humans. Toxic effects include kidney damage and associated high blood pressure.

SOURCES

The element itself is insoluble. It occurs in water as various salts, particularly cadmium sulfide.

Cadmium has a number of industrial uses. It is involved in the manufacture of batteries, paints, and parts for nuclear reactors. It is used to make several alloys and is also a waste from electroplating processes. Some pesticides have cadmium as an ingredient.

Since only minute amounts of cadmium are found naturally in waters, detectable concentrations are indicative of contamination from industrial or mining operations. A whole village in Japan was poisoned by heavy metal contamination (probably cadmium) of its water supply. The contamination resulted from extensive mining. Seepage of cadmium from electroplating plants into groundwater has been reported in this country. The zinc in galvanized iron contains trace amounts of cadmium which could contaminate waters.

In a study of trace metals found in United States waters from 1962 to 1967, cadmium was detected in something less than 7 percent of the samples taken from 130 different sampling points. Of these samples, six were above the recommended limit of 0.01 milligram per liter. The highest concentration found was 0.12

milligram per liter in the Cuyahoga River at Cleveland, Ohio. The river feeds into Lake Erie.

EFFECTS

Cadmium has been shown to accumulate in soft tissues; for instance, in the kidneys, both in experimental animals and in industrial workers exposed to cadmium. Some researchers have shown that cadmium levels in the kidneys and urine of patients with high blood pressure are higher than levels in normal individuals. Animal experiments have demonstrated that cadmium interferes with a vital energy-producing biochemical process (it uncouples oxidative phosphorylation). Aquatic organisms vary in their susceptibility to cadmium. Lethal levels may range from 0.01 milligram per liter to 10 milligrams per liter. Other toxic substances may become more harmful in combination with cadmium. (See the discussion of NTA in **Eutrophication, the detergent controversy.**)

REGULATIONS

In 1962 the Public Health Service recommended that *drinking water* supplies be rejected if they contain more than 0.01 milligram per liter of cadmium. Despite the high toxicity of cadmium and the demonstrated possibility of its incorporation into water supplies, however, very few states have legal limits for the concentration of cadmium in drinking water.

REFERENCES

1. *Public Health Service Drinking Water Standards*, U.S. Dept. of Health, Education and Welfare, Public Health Service, 1962.
2. J. F. Kopp and R. C. Kroner, *A Five Year Summary of Trace Metals in Rivers and Lakes of the United States*, U.S. Dept. of the Interior, Federal Water Pollution Control Administration, 1967.
3. Report of the Committee on Water Quality Criteria, U.S. Dept. of the Interior, Federal Water Pollution Control Administration, 1968.
4. J. Kobayoshi, Report at the Fifth International Water Pollution Research Conference, San Francisco, 1970, reviewed in *Medical World News*, August 21, 1970.
5. H. M. Perry, "Hypertension and Trace Metals with Particular Emphasis on Cadmium," in D. D. Hemphill, *Trace Substances in Environmental Health*, University of Missouri, Columbia, 1969.
6. H. A. Schroeder et al., *Archives of Environmental Health* 13 (1966): 788.

Carbamates

Carbamates are chemical compounds related to carbamic acid (NH_2COOH) and are finding wide use as insecticides, fungicides, and **Herbicides.** Some of the impetus for their development comes from the growing number of instances in which insects have become resistant to the **Chlorinated Hydrocarbons** such as DDT and the **Organophosphorus Compounds.** In addition, concern over the persistence of chlorinated hydrocarbons in the environment is leading in many cases to banning or severe curtailment of their use—for instance, in Sweden, Great Britain, and the United States. Thus the search for less long-lived pesticides is economically important.

The ideal pesticide, besides being effective against pests, should present as small a hazard as possible to the workers who produce and distribute it and to those who apply it. Carba-

TABLE C.1
CARBAMATE PESTICIDES

Chemical Group	Examples	Comments
Aryl esters of methylcarbamic acid	Baygon, Zectran, Carbaryl (Sevin)	Insecticides
Alkyl esters of N-aryl carbamic acid	Chloro-IPC	Herbicides, control grasses in cotton, carrots, and sugar beets
Thiocarbamic acid derivatives	Diallate	Herbicides, used to kill annual grasses
Dithiocarbamic acid derivatives		Herbicides, fungicides, and nematocides
Metal salts	Ziram	
Salts of ethylene bis(dithiocarbamic) acid	Zineb	Fungicides; Zineb may act as a fertilizer as well since it degrades to the plant nutrient zinc sulfite
Esters of dithiocarbamic acid	Vegedex	Herbicides, nematocides

mates are not this hypothetical ideal, but they are generally a good deal less hazardous to apply than the organophosphates and are not stored in the body by animals. In common with the organophosphates, chemicals in the carbamate group combine with an enzyme in the body (cholinesterase) which is vital to proper functioning of the nervous system. However, the organophosphorus compounds make a long-lasting combination and thus can poison a person by using up his supply of the enzyme during repeated nontoxic exposures. The carbamates and cholinesterase, on the other hand, form a combination which is rapidly broken down. For this reason poisonings by carbamates are rare, although they are certainly possible if a dose is sufficiently large. Some ability to stand higher than normal doses after repeated exposure has been reported.

Symptoms of poisoning are similar to those of organophosphate poisoning. Death is due to respiratory failure.

Carbamic acid derivatives can be generally grouped into the classes noted in Table C.1. For a discussion of the environmental effects of carbamates see **Pesticides.**

REFERENCES

1. Report of the Secretary's Commission on Pesticides and Their Relationship to Environmental Health, U.S. Dept. of Health, Education and Welfare, December 1969.
2. N. N. Melnikov, "Chemistry of Pesticides," *Residue Reviews* 36 (1971).

Carbon Dioxide

The products of the combustion of fossil fuels, coal, petroleum, and natural gas are numerous. Many are discussed under **Air Pollution,** but two substances are not mentioned there. These are water vapor and carbon dioxide gas. In the last decade scientists have foreseen a possible hazard associated with the release of carbon dioxide to the atmosphere.

Carbon dioxide is a normal constituent of the atmosphere at a concentration of about 0.05 percent (on a weight basis). It is also vital to plant and animal life. Green plants utilize carbon dioxide and water to make carbohydrates by the process of photosynthesis, and oxygen is produced as a by-product. This fixation of carbon dioxide and water is both the source of our food and the source of oxygen replenishment to the atmosphere. Both plants and animals consume oxygen in respiration, producing carbon dioxide and water vapor as end products. Only plants, however, are able to utilize carbon dioxide and *produce* oxygen. The central point is that carbon dioxide is naturally present in the air and necessary for life.

Suddenly, however, in the last several hundred years (it is sudden if you think that the earth is two billion years old), we have begun to release enormous quantities of carbon dioxide into the air by the combustion of fossil fuels. In addition, these releases are increasing at a steady pace. (See **Electric Power** for projections on fossil fuel power for the year 2000.) In the late 1960's the amount of carbon dioxide being produced by combustion probably amounted to between 5 and 7 percent of the total production by green plants.

Apart from the fact that green plants use carbon dioxide in photosynthesis, there are other sinks or reservoirs for the gas. Carbon dioxide is soluble in water, so that the vast oceans are sinks. Calcium and magnesium from the weathering of rock are in solution in the ocean, too, and these substances combine with carbon dioxide to give limestone and dolomite sediments, making room for more carbon dioxide to dissolve. Before our "burning" began, release and capture of carbon dioxide were in rough equilibrium. Now the situation may be altered; the natural processes of carbon dioxide removal are not keeping pace with processes which produce it (see Figure C.1; the horizontal

FIGURE C.1

MEAN MONTHLY CONCENTRATIONS OF CARBON DIOXIDE AT MAUNA LOA, 1958–1972

From Lester Machta, "Prediction of Carbon Dioxide in the Atmosphere," Paper presented at the Brookhaven Symposium in Biology, Brookhaven, N.Y., May 1972. Data provided by C. D. Keeling, Scripps Institution of Oceanography, sponsored by the National Science Foundation. Used by permission of Lester Machta.

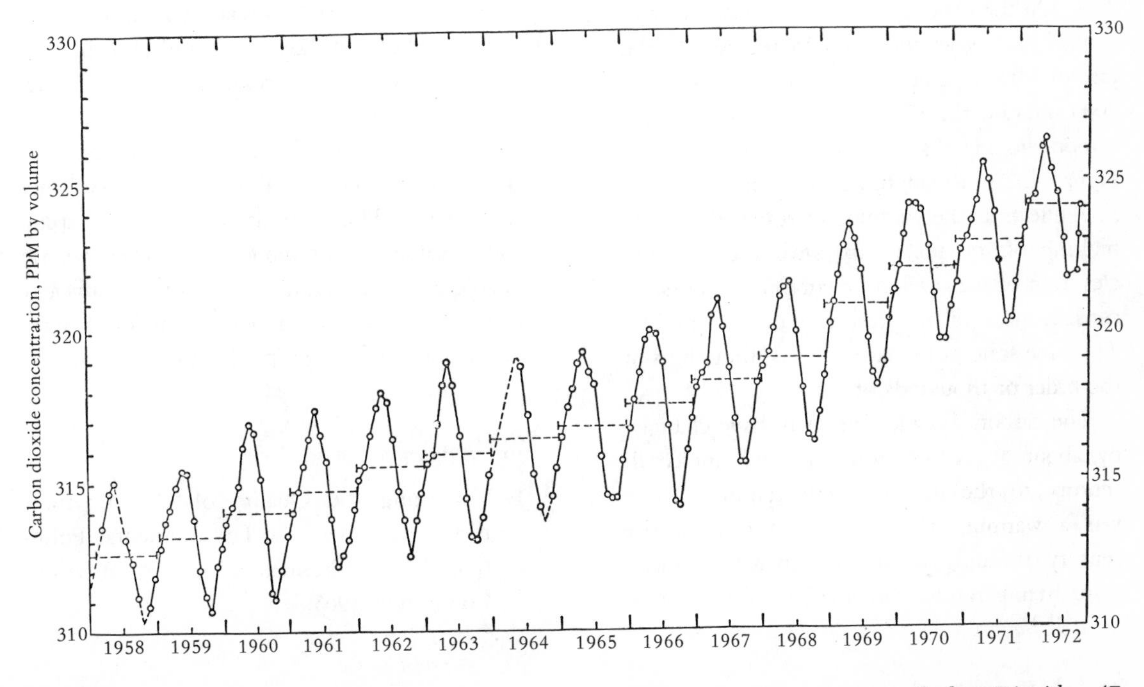

dashed lines indicate average yearly value). The resulting buildup could have important consequences.

While the gas does not block short-wavelength solar radiation from reaching the earth, it does absorb the long-wavelength radiation which is back-radiated from the earth. In the same way that the glass panes of a greenhouse prevent heat from being radiated out of the greenhouse, the carbon dioxide may lessen heat escape from the earth below. This is the origin of the term *greenhouse effect.* Scientists have speculated on the potential effect on climate since the turn of the century.

Measurements of atmospheric levels of the gas which were made during the period 1958–1963 showed the concentration increasing at about 0.25 percent each year. It appeared from these increases that about half of the annual carbon dioxide production from fossil fuel was accumulating. Assuming that civilization continues to combust fossil fuels in the present quantities, scientists have estimated that the concentration of carbon dioxide will increase to 14 percent above the 1950 level by the year 2000. On the other hand, if the quantity of fuel burned each year continues to increase at the rate of 3 to 4 percent, projections for the year 2000 indicate the presence of 25 percent more carbon dioxide than was present in 1950 (see Figure C.2.). In addition to warming of air, such effects as the warming of seawater and the melting of the polar caps with the resulting elevation of sea level are mentioned as possible consequences of the carbon dioxide addition. The time scale of the latter effects, though, is of the order of thousands of years.

The carbon dioxide level may have changed by about 5 percent from the mid-nineteenth century to the mid-twentieth century. There was a warming trend from the turn of the century through about 1940 in which mean annual temperatures rose a degree or two. Since 1940, however, a modest cooling trend has been evident. The cooling trend would not be expected were a greenhouse effect actually operating. Yet the cooling trend may be explained by the same mechanism as the warming trend: combustion. Combustion produces particles; coal especially is responsible for **Particulate Matter.** Particles in the atmosphere reflect solar energy away from the earth. We know that following a volcanic eruption in the 1880's in what is now Indonesia two severe winters were experienced. As part of a program to abate **Air Pollution, Particulate Matter** is coming under increasing control. The effect on climate of this action is not clear.

Still other possibilities limit our capacity to predict the climate in the year 2000. For instance, there is a mechanism which could counteract the warming trend. Evaporative losses from the ocean will increase as the earth's temperature rises. The water vapor added to the atmosphere may contribute to an increased cloud cover, which in turn could diminish the quantity of the sun's radiation reaching the earth. On the other hand, the warming trend could accelerate as the oceans warm, since the quantity of carbon dioxide (or any gas) that can dissolve in water decreases as the water temperature rises.

In summary, we know enough to be concerned about the global effects of fossil fuel combustion. This commits us to further study and research. At the moment, however, we are unable to forecast climatic effects with sufficient certainty to make recommendations for changes in our mode of energy production.

REFERENCES

1. "Restoring the Quality of Our Environment," Report of the Environmental Pollution Panel, President's Science Advisory Committee, 1965.

FIGURE C.2

PREDICTION OF ATMOSPHERIC CARBON DIOXIDE FROM THE COMBUSTION OF FOSSIL FUELS, 1860–2000

The addition of carbon dioxide by fossil fuel combustion is assumed to increase at 4 percent per year from 1969 to 1979 and at 3.5 percent per year from 1980 to 1999.

From *Changing Chemistry of the Oceans,* ed. D. Dyrssen and D. Jagner, Wiley, New York, 1972. Originally appeared in Lester Machta, "Prediction of Carbon Dioxide in the Atmosphere," Paper presented at the Brookhaven Symposium in Biology, Brookhaven, N.Y., May 1972. Used by permission of Lester Machta.

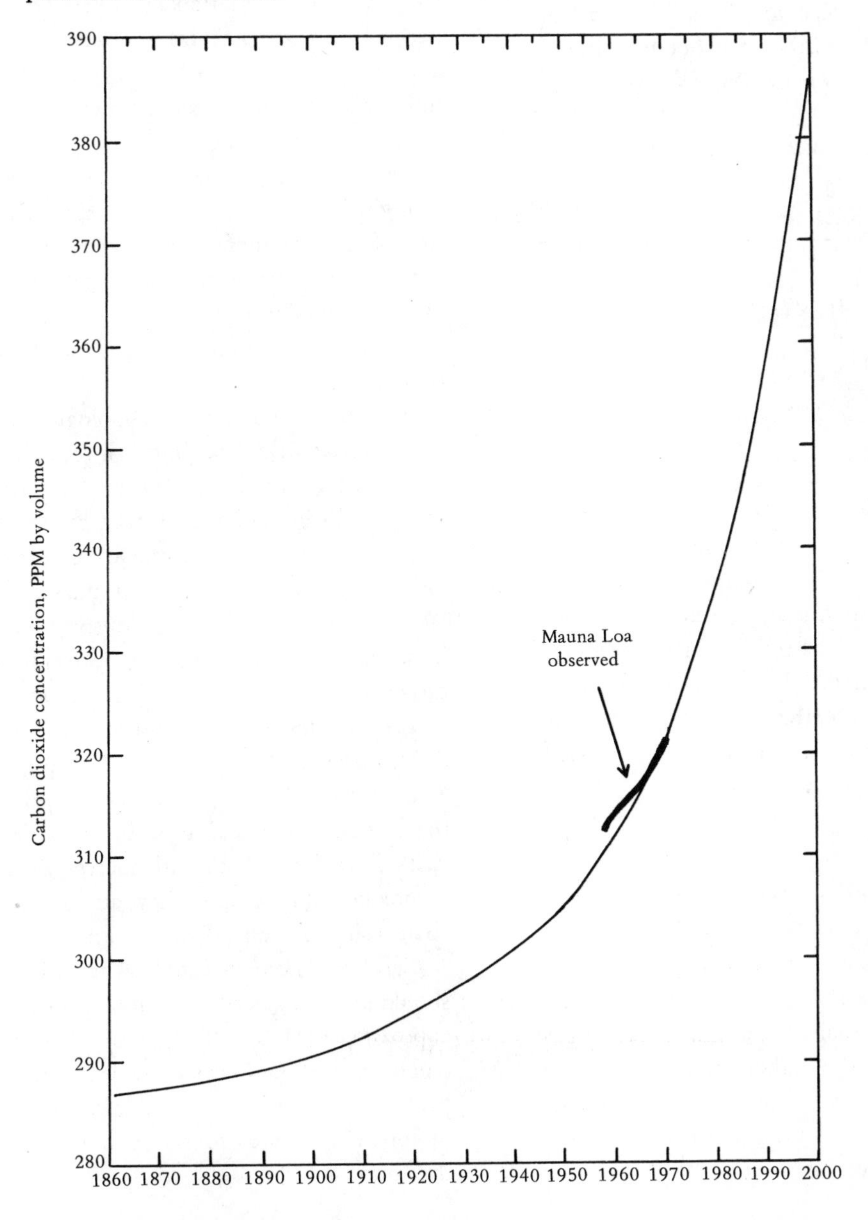

2. *Man's Impact on the Global Environment*, M.I.T. Press, Cambridge, Mass., 1971.
3. Lester Machta, "Prediction of Carbon Dioxide in the Atmosphere," paper presented at the Brookhaven Symposium in Biology, Brookhaven, N.Y., May 1972.

The data on carbon dioxide concentrations at Mauna Loa derive from:

4. J. Pales, and C. Keeling, "The Concentration of Atmospheric Carbon Dioxide in Hawaii," *Journal of Geophysical Research* 70 (1965): 6053–76.
5. Personal communication to Lester Machta from C. Keeling and A. Bainbridge, 1971, and from C. Keeling, 1972.

Carbon Monoxide

Carbon monoxide has no odor, no color, and no taste. It pervades our urban atmospheres. Stemming largely from the incomplete combustion of carbon-containing fuels, carbon monoxide reduces the ability of hemoglobin in the blood to carry oxygen. It has long been known to be lethal in high concentrations, but its effects at lower concentrations such as are found in urban atmospheres are just being explored.

ORIGIN AND CONCENTRATIONS

Carbon monoxide is both a pollutant and a mystery. The compound arises from the incomplete combustion of carbon in fossil fuels and other combustible substances. This is the origin of nearly all of the carbon monoxide which pollutes the air in the cities. Carbon monoxide is, however, produced from many other natural sources and in vast quantities: from volcanic gases and forest fires, from the dissociation of carbon dioxide in the upper atmosphere, from seaweed and plankton in the ocean, from marshes, and from other sources. So much is produced from these natural sources that the background concentration might be expected to rise steadily, but it does not. Background levels average around 0.1 milligram per cubic meter, although a concentration just over 1 milligram per cubic meter has been observed in what appeared to be "clean" air. Somehow carbon monoxide is being removed from the air. Is there a movement of carbon monoxide to the upper atmosphere where in the presence of sunlight it is oxidized to the dioxide? Does the use of carbon monoxide by certain bacteria account for its removal? Is the ocean a sink for carbon monoxide as it is for carbon dioxide? All these fates have been suggested but none has been proved.

From the standpoint of technological sources, the motor vehicle contributes an overwhelming portion of the total quantity. In New York, as in Boston, Chicago, and Denver, transportation activities accounted for over 90 percent of the carbon monoxide in the air in 1968. What of Washington, D.C., and Los Angeles? They are in a class by themselves; 99 percent of the carbon monoxide in the capital and 98 percent in Los Angeles is from transportation. Why in St. Louis and Philadelphia did transportation account for only 77 percent and 60 percent of the carbon monoxide? Probably because these cities have large industrial sources of carbon monoxide, not because they are more clever about using the auto than the others.

Gasoline-powered vehicles furnished (we should say "exhausted") about 60 percent of the approximately 100 million tons of carbon monoxide emitted nationwide in 1968. About 80 percent of the carbon monoxide emissions from motor vehicles are from light-duty vehicles—the auto and the light truck. On a national level, the incineration or other burning of solid wastes accounted for about 8 percent of the total carbon monoxide emissions, and industrial activities, such as iron and steel making and petroleum refining, produced about 11 percent. Forest fires are thought to have accounted for nearly 8 percent of the total. Home heating and power generation are innocents in the big picture.

Levels of carbon monoxide in urban atmospheres typically rise during the morning and evening periods of heavy traffic volume. Mornings in the fall may see the most significant peaks when inversions (see **Air Pollution, meteorological aspects**) trap pollutants in the lower layers of the atmosphere. Although urban areas do not appear to show definitive trends in carbon monoxide levels, areas which are "urbanizing," attracting industry, commerce, and traffic, are likely to be experiencing increased concentrations.

The method by which atmospheric levels are reported takes into account the time required for a carbon monoxide buildup in human blood. The concentration in human blood reaches a relatively stable value within 8 to 12 hours. Thus, a reporting of the average atmospheric concentration over 8-hour periods gives a relative indication of the levels that could be reached in the blood of exposed individuals.

TABLE C.2

CARBON MONOXIDE LEVELS IN SEVERAL CITIES, 1962–1967 (IN MILLIGRAMS PER CUBIC METER)

	Average Levels Exceeded	
City	*50% of the 8-hour periods*	*10% of the 8-hour periods*
Chicago	14	24
Philadelphia	8	14
St. Louis	7	12
Washington, D.C.	5	9

In sketching pollution levels observed in the past in major cities, it is useful to notice the fraction of the total number of 8-hour periods in which the average pollutant concentration exceeded a stated level. In Chicago, during the period 1962–1967, 50 percent of the 8-hour periods exhibited average values greater than 14 milligrams per cubic meter, and 10 percent of the 8-hour average values exceeded 24 milligrams per cubic meter. The experiences of several other major cities in that period are described in Table C.2.

Certain occupational groups are more exposed than the general population. Cab drivers, traffic policemen, delivery men, workers at loading platforms, guards in tunnels, and parking attendants in underground garages are among the most highly exposed. Concentrations of greater than 100 milligrams per cubic meter have been observed in the work environments of the last two groups. A study of New York's Holland tunnel estimated an average daily carbon monoxide concentration of 80 milligrams per cubic meter. The exposure of the ordinary driver and his passengers depends on the average speed of vehicles on the route he drives and the traffic intensity. *Intensity* refers to number of cars passing a particular point per unit of time. Frequent acceleration and deceleration increase carbon monoxide emissions. Therefore, a low-speed, stop-and-go, heavily traveled route provides the worst exposures. Studies conducted on center city routes in a number of major cities in 1966–1967 showed carbon monoxide levels in traffic ranging between about 24 and 44 milligrams per cubic meter.

EFFECTS ON MAN

Carbon monoxide in high concentration over sufficient time is known to cause death. At lower concentrations, such as those experienced in urban atmospheres and in workplaces, the effects are (fortunately) less well defined. We can be relatively precise in stating the mechanism of carbon monoxide interference with body chemistry; the compound is one of the few contaminants in air for which this can be done.

Oxygen and carbon dioxide are carried within the human body by hemoglobin, a molecule found in the red cells of the blood. Carbon dioxide is produced during cell metabolism in all parts of the body, and blood carries this waste product from the cells. The carbon dioxide may be dissolved in the blood as the bicarbonate ion or it may be attached to the hemoglobin molecule. At the lung, the carbon dioxide is traded for oxygen; the oxygen attaches to the hemoglobin and is ultimately carried to the cells.

Carbon monoxide interferes with the oxygen-carrying capacity of hemoglobin; the gas has a 200 times greater affinity for the hemoglobin molecule than does oxygen. A hemoglobin molecule to which carbon monoxide has become attached is called carboxyhemoglobin, and it is not available to carry oxygen to the tissues. Because of carbon monoxide's great affinity for hemoglobin, it exerts an inordinately large effect on the ability of the blood to transfer oxygen—even at low atmospheric levels of the pollutant. Concentrations of about 70 milligrams per cubic meter for 8 to 12 hours will elevate the fraction of hemoglobin in the blood which has been converted to carboxyhemoglobin to about 10 percent.

It is true that carbon monoxide is already present in the blood and keeps about 0.5 percent of the hemoglobin out of service at any one time. The source of this background carbon monoxide is apparently the natural chemical breakdown of aged or diseased heme molecules. It is also true that smokers regularly tolerate higher levels of unusable hemoglobin due to the carbon monoxide they inhale; about 2.5 percent of their hemoglobin may be out of service because of their habit.

Studies have tried to correlate test-taking ability with levels of carboxyhemoglobin, and the results yield something for everyone. Those who dispute the effects of carbon monoxide on performance can argue that tests indicating such effects were complicated by boredom and fatigue. Yet those who accept test results indicating no impairment of performance can be challenged by citing the individual who was scoring well, stood up, walked several steps, and fainted.

One test did show decreased driving ability at blood levels of 10 and 20 percent carboxyhemoglobin. Recall that the 10 percent level corresponds to exposure to about 70 milligrams per cubic meter for 8 to 12 hours. (By this length of time, blood levels have usually stabilized at an "equilibrium" value.) While the 70-milligram level may be uncommon on the street, it does occur in workplaces like tunnels and loading platforms. The ability to respond to brake lights and to changes in speed of the car ahead was diminished, as was the ability to maintain a constant speed and constant distance from the car ahead.

At levels as low as 5 percent of carbon monoxide–bound hemoglobin, certain performance test scores have deteriorated. Even at the 3 percent level, the ability to distinguish the difference between light intensities has been seen to decrease. The 3 percent level corresponds to an exposure of about 18 milligrams per cubic meter for 8 to 12 hours. It could also occur after only a 4-hour exposure to a level of about 35 milligrams per cubic meter.

A number of studies have sought carbon monoxide effects among individuals exposed to low levels of the gas in their occupations. Attempts have been made to link the quantity of carboxyhemoglobin with exposure, but the effect of cigarette smoking has clouded the picture. Smoking alone brings about levels comparable to low occupational exposures. It would not be unusual to find a smoker who is not occupationally exposed with 2.5 percent carboxyhemoglobin; a nonsmoker would need to be exposed to about 14 milligrams per cubic meter for 8 to 12 hours to achieve the same level.[1]

Studies have been made which attempt to link carbon monoxide with deaths, illnesses, absences, and accidents. One study found that the number of individuals who die of heart attacks was elevated during periods of high carbon monoxide concentrations,[2] but no link of illnesses and absences to carbon monoxide levels is yet established. One investigator did find higher carbon monoxide–bound hemoglobin among individuals involved in auto accidents than in other groups, but the suggested link is still tenuous.

CONTROL OF EMISSIONS

This discussion centers on the control of motor vehicle exhausts, since vehicles are the single most important source of carbon monoxide in urban atmospheres (see also **The Automobile and Pollution**). As with hydrocarbons and oxides of nitrogen (see **Photochemical Air Pollution**), one can envisage societal changes which could diminish carbon monoxide emissions. Clearly, we are referring to more extensive use of large transportation systems, such as railroad, subway, or buses in place of private vehicles. For the present, however, technological changes in the automobile appear to be a feasible avenue of change, although in the long run transportation systems will probably become essential.

Another technique suggested to reduce carbon monoxide levels in the centers of cities is to enable traffic to move more rapidly. Low-speed driving in the city tends to produce higher emissions because of frequent changes of gear. One-way streets or light timing might be used to speed traffic. This suggestion, however, requires careful scrutiny. While individual cars might decrease their emissions, an increase in traffic flow in response to improved conditions could easily offset the reductions. Transportation planners have frequently seen vehicle flow increase in response to lower travel times.

There is still another reason that higher speeds may not be desirable. The **Noise** from automobile tires at higher speeds and greater traffic volumes could add significantly to the noise burden in an urban neighborhood. A doubling of traffic flow may produce a 3-decibel increase in noise levels, and a doubling of traffic speed could induce a 6-decibel increase. If the effects are independent, a 9-decibel increase could occur for a doubling of vehicle speed and traffic flow. A 10-decibel increase would be perceived by the human ear as about a doubling of noise volume. Since urban areas already tolerate among the highest average noise levels, increasing velocities and the flow of vehicles could potentially degrade the urban environment still further.

Technological improvements based on knowledge of the combustion process have already been made. In the gasoline-powered engine, the amount of air in the air-fuel mixture has been increased over earlier levels, diminishing the amount of carbon monoxide in the exhaust by making more oxygen available for combustion. The ratio of 14.5 pounds of air per

---

[1] This is an interesting figure in light of the recent **Air Quality Standards** that have been set. It illustrates the description of smoking, attributed to Professor Charles D. Gates of Cornell University, as a "personal form of air pollution."

[2] This statement does *not* say that elevated carbon monoxide levels *cause* heart attacks or deaths. It merely establishes the presence of the compound at the scene.

pound of fuel furnishes almost exactly the amount of oxygen needed to oxidize all the carbon in the fuel to carbon dioxide. It is called "the stoichiometric quantity." Increasing the amount of air still further accomplishes very little in the way of further carbon monoxide reduction. A number of the recent engine modification systems incorporate a larger amount of air (approaching the "stoichiometric quantity") and better fuel metering; the packages have been on autos since 1968.

Another approach to reducing carbon monoxide emissions is to inject air into the hot exhaust stream to combust remaining hydrocarbons and carbon monoxide. American Motors, Ford, and General Motors use the technique on manual transmission vehicles. Chrysler, which pioneered the principles of the engine modification systems now in use on vehicles from all manufacturers, does not use air injection.

The future mode of control appears at this time to be the catalytic converter, which may be on cars by 1976. The device operates on the exhaust stream in two stages; it reduces nitrogen oxides in a first chamber and promotes oxidation of hydrocarbons and carbon monoxide in a second stage. Additional air is drawn in for the second reaction stage. Originally counted out of the clean air developments because of poor catalyst life, the device now offers more promise because of the gradual removal of **Lead** from gasoline.

In stationary combustion, higher temperatures, longer residence times, and air in close to the "stoichiometric quantity" may all be combined to diminish carbon monoxide emissions. Use of the sanitary landfill instead of burning solid wastes could reduce carbon monoxide emissions. In the iron and steel industry, as well as the petroleum industry, carbon monoxide generated may commonly be burned as a source of energy.

REFERENCES

Three recent and authoritative books from the National Air Pollution Control Administration, now the Office of Air Programs of the Environmental Protection Agency, are:

1. *Air Quality Criteria for Carbon Monoxide*, AP-62, 1970.
2. *Control Techniques for Carbon Monoxide, Nitrogen Oxide and Hydrocarbon Emissions from Mobile Sources*, AP-66, 1970.
3. *Control Techniques for Carbon Monoxide Emissions from Stationary Sources*, AP-65, 1970.

Chlorinated Hydrocarbons

The chlorinated hydrocarbons are chemicals composed mainly of carbon, hydrogen, and oxygen atoms plus one or more chlorine atoms; these were the first synthetic chemical pesticides developed. Although many compounds were used as pesticides before 1945, it was after this year, when DDT came into widespread use, that the era of chemical control of pests really began. By stopping a wartime typhus epidemic in Italy, DDT proved its effectiveness and sparked a revolution in pesticide usage. In developing nations, mosquito control for prevention of yellow fever and malaria is mainly dependent on DDT, which has the ideal economic characteristics of low initial cost and long-lasting effect. DDT sprayed on the inside of house walls will kill mosquitoes for up to a year.

Aldrin, dieldrin, chlordane and heptachlor are other commonly used chlorinated hydrocarbons. In 1967 the chlorinated hydrocarbons accounted for half of the total United States production of pesticides. However, the use of these compounds in the United States has been steadily decreasing, partly because insects have developed resistance to them. The decrease in use may also be attributed to public concern about the large-scale application of these slow-to-disappear chemicals. Some scientists have estimated that residues of DDT in the environment may have a half-life (the time necessary for half of the given quantity to disappear) of 20 years or more.

Chlorinated hydrocarbons are not natural chemicals, in that they are not synthesized by animals or plants. Produced by men in laboratories, these compounds present nature with a difficult disposal problem. This is not to say that nothing can destroy the chlorinated hydrocarbons. Some few bacteria have been found which will grow on them, and normal weathering processes will degrade them to an unknown degree. In addition, some scientists have pointed out that, considering the amount of DDT and its high fat solubility, an even greater level of DDT residues in human and animal fatty tissues might have been expected than has actually been found. They thus postulate some undiscovered and fortunate sequestering mechanisms which keep the total quantity of DDT used from being available to plants and animals.

However, it is nonetheless true that chlorinated hydrocarbons, which are highly soluble in fats, are accumulating in our adipose tissues and we can only suspect the consequences. It is this same fat solubility that allows the chemicals to easily penetrate insects' outer coverings, which contain a fatty-type substance.

The fact that these compounds are persistent makes them useful in control of termites and cockroaches; they are there when the insect comes out of its hole or its egg. But because they don't go away, they accumulate in the soil, in the air, and in the fatty tissues of fish, and birds, and people. We see, then, that the very qualities which make chlorinated hydrocarbons useful, persistence and fat solubility, are the qualities which make them long-term hazards. Many of the problems caused by the environmental accumulation of these pesticides are discussed under **Pesticides, effects on organisms other than man; effects on man.**

The acute toxicity of the chlorinated hydrocarbons varies, from endrin, which is extremely toxic, to methoxychlor, which is only slightly toxic. DDT is actually considered very toxic but does not present the hazards in application that organophosphorus compounds like parathion do. Its relative safety is probably due to the fact that it is less easily absorbed through the skin than parathion. Symptoms of chlorinated hydrocarbon poisoning are nervous system effects such as paresthesia (tingling of the skin), increased excitability, vomiting, tremor, and depression. Dermatitis due to skin contact has also been noted. No clear-cut biochemical effects, such as the **Organophosphorus Compounds** have on an enzyme, have been discovered. Long-term effects include liver and kidney damage. DDT is also known to have a stimulat-

TABLE C.3
ORGANOCHLORINE PESTICIDES

Chemical Group	Examples	Comments
Aliphatic derivatives	Nemagon	Include many nematocides
Diphenyl aliphatic derivatives	DDT, DDD, TDE, methoxychlor	
Chlorinated aryl hydrocarbons	BHC (lindane), chlordane, heptachlor, endrin, toxaphene, aldrin, dieldrin	Aldrin is believed to convert to dieldrin in the soil; thus values for the two are often combined when residue data are given

ing effect on the liver enzymes which detoxify substances in the body. This can be viewed as a helpful adaptive effect, and DDT has even been used as a drug in one form of liver malfunction. Death from organochlorine poisoning may be due to inability of the heart to pump blood effectively. Chemically, the organochlorine derivatives can be grouped into three classes, which are noted in Table C.3.

REFERENCES

1. Report of the Secretary's Commission on Pesticides and Their Relationship to Environmental Health, U.S. Dept. of Health, Education and Welfare, December 1969.
2. N. N. Melnikov, "Chemistry of Pesticides," *Residue Reviews* 36 (1971).
3. G. M. Woodwell, P. P. Craig, and H. A. Johnson, "DDT in the Biosphere: Where Does It Go?" *Science*, December 10, 1971, p. 1101.

D

DDT

See **Chlorinated Hydrocarbons**; **Pesticides**.

Detergents

The term *detergent* can refer to any washing product, but it is commonly used to mean the synthetic, nonsoap preparations intended for clothes and dishwashing. Such products have been widely criticized on two grounds within the past 10 years. It was first noted that synthetic detergents were causing unsightly foams in natural waters. After this problem was solved by changing the surfactant component, another ingredient, the phosphate builder, was indicted for contributing to the **Eutrophication** of waters.

THE PROBLEM OF BIODEGRADABILITY

Most detergents contain a surfactant, an ingredient designed to help clothes be "wetted" by water and to lift soil off the material. Detergent surfactants, or surface-active ingredients, were originally alkyl benzene sulfonates, or ABS derivatives. When dissolved in water, molecules of ABS clump together into structures called micelles. The micelles "capture" dirt particles and so remove soil from clothes.

As early as 1950, foaming groundwaters containing ABS were seen. The problem was due to the fact that ABS was not biodegradable; that is, microorganisms were not able to break

Water contaminated by nonbiodegradable detergents. The introduction of biodegradable detergents in the early 1960's has virtually eliminated such unsightly foam from groundwater.

it down into simpler compounds. It thus passed virtually unchanged through sewage systems and treatment plants into natural waters. High enough concentrations of ABS caused foam to be formed there, just as it was formed in washing machines. The foam was an aesthetic problem rather than a health hazard. It was not toxic to man, caused no fish kills, and added no odors or flavors to water.

Research uncovered the fact that the branched side chain was preventing microbes from breaking ABS down (see Figure D.1).

FIGURE D.1
ALKYL BENZENE SULFONATE (ABS)

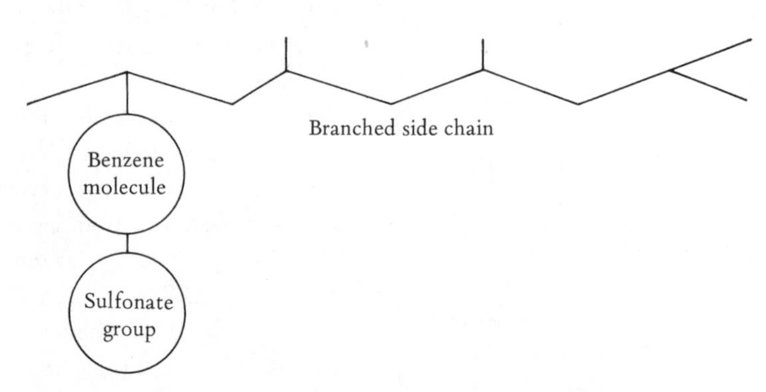

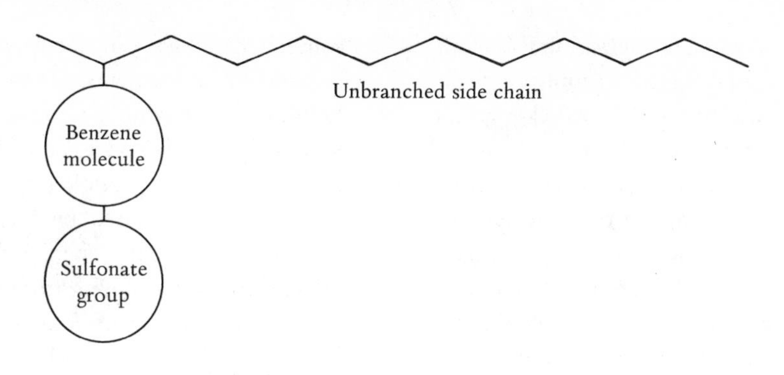

FIGURE D.2
LINEAR ALKYL SULFONATE (LAS)

Compounds with unbranched (linear) side chains were soon developed and tested (see Figure D.2). They appeared to be both biodegradable and nontoxic. Congressional pressure, in the form of deadlines for the changeover from ABS to linear alkyl sulfonate (LAS), induced the detergent industry to accomplish a complete switch to the new compounds by June 1965.

Subsequent practical experience has shown that the LAS compounds are removed reasonably well by septic tank–percolation field systems. Sewage treatment plants with the activated sludge process[1] effect about 90 percent removal of LAS. Thus, where adequate sewage treatment facilities exist, the removal of LAS is quite good. More to the point, instances of foam contamination have decreased markedly, and the concentration of surfactants in waterways has been substantially reduced.

THE PHOSPHATE PROBLEM

Although the problems caused by nonbiodegradable detergents seem to have been satisfactorily resolved, it is not possible to relate a happy ending to the problem of phosphates in detergents.[2] Typical heavy-duty detergents contain about 50 percent by weight of a phosphate compound such as sodium tripolyphosphate; the function of this compound is to act as a "builder." Builders are not soaplike themselves but function to soften water, to prevent soil from redepositing on clothes, and to maintain a proper level of alkalinity in the wash water. Approximately half of the phosphate content of domestic sewage is contributed by detergents, the remainder being derived from human wastes.

Phosphate is a nutrient required for plant growth; it has been implicated as a cause of cultural **Eutrophication** of lakes and rivers. The contribution of detergent phosphates to eutrophication is at present a matter of dispute. Conservationists and some scientists feel that the contribution is significant and, at the least, unnecessary, since there are alternative chemicals to accomplish the same job as phosphates in detergents. The soap and detergent industry, on the other hand, and other scientists claim that the contribution of detergent phosphates, while large, is not significant. They argue that

[1] See **Water Treatment and Water Pollution Control** for a description of this process.

[2] See **Eutrophication** for a detailed discussion of this problem and alternative ways of dealing with it.

human wastes, urban and agricultural runoff, and natural sources already supply enough phosphates to cause eutrophication. This group further states that no suitable replacement for phosphates has yet been found. They therefore favor a program of sewage treatment to remove phosphates from all sources. In reply, conservationists cite figures to indicate that a comprehensive system of advanced waste treatment for this country lies many years in the future. They also dispute the industry figures showing that the additional phosphate contributed by detergents is not significant.

Regrettably, there is not sufficient evidence at the present time to decide the merits of these two positions. However, one might be guided by the following rule: When there is enough evidence to prove that a chemical may irreversibly damage the environment (and there is the possibility that eutrophication is irreversible once it has reached a certain point), one should err on the side of caution in the use of the chemical. According to this rule, the phosphate content of detergents should be either reduced in quantity or eliminated.

One must, on taking such action, be prepared to grapple with the problem of suitable replacements. Soap is probably the best alternative, since it has been well tested as an environmental additive. However, hard water and the special requirements of synthetic fabrics make it less than ideal. What is clear is that we must not make the mistake of adding another harmful chemical, such as NTA, to the environment as a result of inadequate testing.

The rather homely puzzle of how to get one's clothes clean has serious environmental aspects, and it is not just the detergent industry's problem.

THE ENZYME PROBLEM

A relatively recent addition to detergents has been enzymes. These are biologically produced compounds which can, for example, speed the breakdown of proteins. They may thus be helpful in removing protein-containing stains such as blood, chocolate, or grass. Concern has been voiced about possible respiratory problems which could be caused by breathing enzyme-containing detergents. There is also a possibility of skin irritation resulting from wearing clothes with enzyme-detergent residues. A report released in November 1971 by the National Research Council and requested by the Food and Drug Administration finds that in normal use no significant increase of skin or respiratory irritation is caused by enzyme-containing detergents in comparison with regular detergents. Only a few cases of proved allergic sensitivity to the enzyme products were found. The Council did, however, strongly recommend continued study to determine the effects of the addition of enzymes to the environment.

REFERENCES

1. T. E. Brenner, "Biodegradable Detergents and Water Pollution," in *Advances in Environmental Sciences*, vol. 1, ed. J. N. Pitts and R. L. Metcalf, Wiley, New York, 1961, pp. 147–96.
2. Harold M. Schmeck, Jr., *New York Times*, November 18, 1971.

Drinking Water Standards

The U.S. Public Health Service is responsible for setting standards for public drinking water supplies. The standards are concerned with parameters such as the bacteriological, physical, and chemical characteristics of water as well as possible radioactivity. These standards were last revised in 1962. Only a portion of the total document setting forth the revisions is discussed here. Many other subjects are cross-referenced and discussed fully elsewhere in this book.

BACTERIOLOGICAL QUALITY

The section on bacteriological quality specifies the frequency of sampling of public water supplies and the scale upon which to judge water quality. The frequency of occurrence of coliform bacteria, the indicator organisms for bacterial pollution, determines whether a water supply is acceptable. More detail is found under **Bacterial and Viral Pollution of Water.**

PHYSICAL CHARACTERISTICS

The section on physical characteristics sets limits on turbidity (amount of suspended matter as measured by its effect on light transmittance), color, and odor in waters. In addition, it specifies that drinking water should not give "offense to the sense of sight, taste or smell."

RADIOACTIVITY

The section on radioactivity specifies allowable limits for various radioactive substances or combinations of substances. A discussion of how the standards are set is found under **Radiation Standards.**

CHEMICAL QUALITY

The section on chemical quality lists possible components of drinking water which are known or suspected to be deleterious to human health or which are displeasing from an aesthetic standpoint.

A number of metals are found in water in extremely small quantities. The concentrations may be so minute that detection frequently is possible only by sophisticated scientific methods. Some of the metals are beneficial to man; that is, they are required in trace amounts in human nutrition. Among these are cobalt, copper, and zinc. Other metals are very toxic to humans. Examples are **Lead** and **Arsenic.**

Both the Committee on Water Quality Criteria of the Federal Water Pollution Control Administration, formerly in the Department of the Interior, and the U.S. Public Health Service have suggested criteria for the levels of certain trace metals in public drinking water supplies.[3] There are, however, several reasons why suitable standards are not easily set. In the first place, even beneficial metals become toxic above certain levels. Thus, for any given metal, a set of concentrations might be specified whose values range from essential and favorable through chronically toxic to acutely toxic. Other substances in the water may act synergistically with a trace metal; that is, the combination of the metal and the synergistic substance may be more toxic than either alone. An example of this phenomenon is the action of nitriloacetate (NTA) with methyl-mercury or cadmium (see **Eutrophication**).

Water to be used in agriculture may have special requirements. For instance, sodium should be present in a particular ratio to other elements, and boron is toxic to many plant species. Industries may also have special requirements for the waters they use. In general, however, water which meets Public Health Service criteria for trace metals in drinking waters will also satisfy agricultural or industrial requirements.

The major portions of trace elements in natural water supplies are probably adsorbed on the surfaces of particles carried in the water. Standards for trace metal concentrations do not usually take this factor into account but rather

---

[3] See note d in Table D.1 for a comparison of the two sets of standards.

specify the allowable amount of metal in solution, a lesser quantity. The error is not serious as a rule because particles are removed during water treatment processes (see **Water Treatment and Water Pollution Control**). It should be noted, nevertheless, that standards and reported values may include only the amount of metal in solution, not the total carried in the water.

The sections of the Public Health Service drinking water standards dealing with chemical contaminants are reproduced below. One should note that there are two classifications of substances. With respect to the first list, concentrations above the given levels are acceptable if better supplies cannot be made available. With respect to the second list, the directions explicitly say that concentrations above the stated level are grounds for rejection of the water supply. More complete discussions of a number of the substances can be found under their specific headings: **Arsenic, Nitrates and Nitrites, Barium, Cadmium, Lead, Detergents** (alkyl benzene sulfonate), **Acid Mine Drainage** (manganese), and **Beryllium.**

5.2. Limits Drinking water shall not contain impurities in concentrations which may be hazardous to the health of the consumers. It should not be excessively corrosive to the water supply system. Substances used in its treatment shall not remain in the water in concentrations greater than required by good practice. Substances which may have deleterious physiological effect, or for which physiological effects are not known, shall not be introduced into the system in a manner which would permit them to reach the consumer.

5.21 The following chemical substances should not be present in a water supply in excess of the listed concentrations where, in the judgment of the Reporting Agency and the Certifying Authority, other more suitable supplies are or can be made available:

Substance	*Concentration (mg/l)*
Alkyl Benzene Sulfonate (ABS)	0.5
Arsenic (As)	0.01
Chloride (Cl)	250.0
Copper (Cu)	1.0
Carbon Chloroform Extract (CCE)	0.2
Cyanide (CN)	0.01
Fluoride (F)	See [4]
Iron (Fe)	0.3
Manganese (Mn)	0.05
Nitrate (NO_3)[4]	45.0
Phenols	0.001
Sulfate (SO_4)	250.0
Total dissolved solids	500.0
Zinc (Zn)	5.0

5.22 The presence of the following substances in excess of the concentrations listed shall constitute grounds for rejection of the supply:

Substance	*Concentration (mg/l)*
Arsenic (As)	0.05
Barium (Ba)	1.0
Cadmium (Cd)	0.01
Chromium (Hexavalent) (Cr^{+6})	0.05
Cyanide (CN)	0.2
Fluoride (F)	See [4]
Lead (Pb)	0.05
Selenium (Se)	0.01
Silver (Ag)	0.05

The standards are currently being revised to include limits for **Pesticides, Mercury**, sodium, and molybdenum. The proposed set of limits for the concentrations of pesticides in drinking waters can be found under **Pesticides, water contamination.**

The drinking water standards of the Public Health Service are not actually the law of the land but are guidelines. States and local governments set and administer the legal limits for their own water supplies. In many cases these

[4] In areas in which the nitrate content of water is known to be in excess of the listed concentration, the public should be warned of the potential dangers of using the water for infant feeding. [See **Nitrates and Nitrites.**]

are based on the Public Health Service guidelines. Two recent studies (see [1] and [2]) allow comparison of the quality of United States surface waters and waters used for public drinking supplies with the guidelines of the Public Health Service and the Committee on Water Quality Criteria.

A five-year study to determine the concentrations of trace metals in surface waters was carried out by Kopp and Kroner from 1962 to 1967. A total of 1,500 samples from across the country were examined for their trace metal content. Some of the data are summarized in Table D.1. The report itself also contains regional breakdowns of the data as well as maximum values detected. On the whole, the results were encouraging; only 0.6 percent of the determinations exceeded the Water Quality Criteria standards. On the other hand, certain geographic areas were disproportionately represented. The Cuyahoga River at Cleveland, Ohio, part of the Lake Erie basin, provided the highest concentrations of zinc (1,182 micrograms per liter), cadmium (120 micrograms per liter), and nickel (130 micrograms per liter).[5]

The region around Pittsburgh, Pennsylvania, is responsible for 32 violations from the Monongahela River and 16 violations from the Allegheny River. All the violations were due to excess manganese, which is derived from **Acid Mine Drainage.** Samples from the Ohio River basin as a whole (which includes Pittsburgh) violated trace metal standards more frequently than those from any other basin (81 violations out of a countrywide total of 179). The Lake Erie basin was second with 20 violations. The Great Basin and California basin did not yield any samples in violation of standards, and the Alaska basin provided only one.

A similar study, but one dealing specifically with public drinking waters, was made by McCabe et al. in 1970. A total of 969 water systems covering nine regions[6] of the United States were examined. The regions were chosen as examples of different types of water supplies. Thus, the survey included such water supply types as surface waters which are only disinfected, groundwater supplies serving semiarid regions, and surface waters receiving industrial municipal and agricultural wastes. Of those trace metals which have established limits, only silver was never found to exceed the limits. Iron, manganese, cadmium, and lead were each found at least once in concentrations 10 times their limits. Data for the individual metals are found in Table D.1. A calculation of the percentage of the survey population (18.2 million people) exposed to violations of the standards for iron, lead, and manganese was given. Four percent of the population were drinking water with more than the recommended concentration of iron, 5 percent with more than the standard for manganese, and 2 percent with excessive lead. The fact that arsenic, cadmium, and lead were found in some water supplies above their recommended limit is especially disturbing because of the established toxicity of these substances.

In addition to the violations of the trace metals standards which are found in the table, 2 percent of the water supplies (19 systems) exceeded the recommended limit for nitrates. The limit for radium-226 was exceeded in 6 of the 969 systems. In terms of the percentage of the population in the survey exposed to violations of the Public Health Service standards, the following were the major problems found: manganese, iron, dissolved solids, bacterial con-

[5] High trace metal concentrations are not the Cuyahoga's only problem. So much oil and other industrial wastes are disposed of in the river that, in July 1969 in the Cleveland factory area, the river actually caught fire.

[6] State of Vermont; New York, N.Y.; Charleston, W.Va.; Charleston, S.C.; Cincinnati, Ohio; Kansas City, Mo.; New Orleans, La.; Pueblo, Colo.; San Bernardino, Riverdale, and Ontario, Calif.

TABLE D.1

TRACE METALS IN U.S. WATERS

		Drinking Water Standards (μg/l)[a]		Comparison of Observed Values to Standards					
				Surface Water[b]			Public Drinking Water[c]		
Metal	Toxicity	*Permissible*[d]	*Desirable*	*Percent of samples containing element*[e]	*Mean value detected (μg/l)*[e]	*Violations of drinking water standards*[f]	*Violations of drinking water standards*[g]	*Percent of samples violating standards*	*Supplies exceeding standards*[h]
Aluminum	Currently not considered a health problem	—	—	31	74	—	—	—	—
Arsenic	Toxic at low levels because of cumulative effects	50	Absent	6	34–308	41	5	0.2	2
Barium	Toxic effects on heart, blood vessels, and nerves	1,000	Absent	Almost 100	43	0	2	0.1	1
Beryllium	Toxic to breathe, less so to drink; poorly studied	—	—	6	Generally trace	—	—	—	—

Note: A dash indicates that no standards have been recommended.

[a] Data are from *Report of the Committee on Water Quality Criteria*, U.S. Dept. of the Interior, Federal Water Pollution Control Administration, 1968.

[b] Data are from J. F. Kopp and R. C. Kroner, *Trace Metals in Waters of the United States*, U.S. Dept. of the Interior, Federal Water Pollution Control Administration, Division of Pollution Surveillance, Cincinnati, Ohio, 1962–1967.

[c] Data are from L. J. McCabe et al., "Survey of Community Water Supply Systems," *Journal of the American Water Works Association* 62 (1970): 670.

[d] These are also the U.S. Public Health Service 1962 drinking water standards, with the exceptions of boron, for which a limit was not set, and arsenic, which has a recommended limit of 10 as well as the mandatory limit of 50.

[e] These values depend on the sensitivity of the method of detection. The smallest detectable level varies from 0.1 microgram per liter for beryllium to 100 micrograms per liter for lead. A range of values indicates the data were presented separately for the various river basins.

[f] Out of a total of 1,500 samples.

[g] Out of a total of 2,595 samples.

[h] Out of a total of 969 water supply systems.

TABLE D.1 (CONTINUED)

Metal	Toxicity	Drinking Water Standards (μg/l)[a] Permissible[d]	Drinking Water Standards (μg/l)[a] Desirable	Surface Water[b]: Percent of samples containing element[e]	Surface Water[b]: Mean value detected (μg/l)[e]	Surface Water[b]: Violations of drinking water standards[f]	Public Drinking Water[c]: Violations of drinking water standards[g]	Public Drinking Water[c]: Percent of samples violating standards	Public Drinking Water[c]: Supplies exceeding standards[h]
Boron	Toxicity to man unknown; toxic to some plants	1,000	Absent	98	101	—	20	0.8	—
Cadmium	May cause high blood pressure; very toxic	10	Absent	3	10	6	4	0.2	3
Chromium (hexavalent)	Carcinogenic when breathed; ingestion hazard unknown	50	Absent	5–56	Trace to 25	4	5	0.2	4
Cobalt	Essential nutrient with low toxicity	—	—	3	1–48	—	—	—	—
Copper	Beneficial in trace amounts; unpalatable at toxic levels	1,000	Virtually absent	74	15	0	42	1.6	11
Iron	Limits set less on toxicity than on taste and color	300	Virtually absent	59–99	19–173	25	223	8.6	96
Lead	Very toxic, cumulative poison	50	Absent	20	23	27	37	1.4	14
Manganese	Standards set for taste and color rather than for health	50	Absent	51	20–600	74	211	8.1	90

Metal	Toxicity	Drinking Water Standards (μg/l)[a]		Comparison of Observed Values to Standards					
				Surface Water[b]			Public Drinking Water[c]		
		Permissible[d]	Desirable	Percent of samples containing element[e]	Mean value detected (μg/l)[e]	Violations of drinking water standards[f]	Violations of drinking water standards[g]	Percent of samples violating standards	Supplies exceeding standards[h]
Molybdenum	Currently not considered a pollution problem	—	—	33	68	—	—	—	—
Nickel	Human poisoning very rare; toxic to plants	—	—	16	19	—	—	—	—
Silver	Limits set for cosmetic reasons; causes bluish discoloration of eyes, skin, and mucus membranes	50	Absent	7	2.6	0	0	0	0
Strontium	No toxic effects known from nonradioactive strontium	—	—	99	Less than 200	—	—	—	—
Vanadium	May have some biological functions; toxic to some plants	—	—	3.4	105	—	—	—	—
Zinc	Rarely exceeds standards, for it easily adsorbs to soils and sediments	5,000	Virtually absent	76	64	0	8	0.3	1

tamination, lead, nitrate, turbidity, and organic materials.

In general, the larger the population served by a water supply, the better was the water quality. When all the parameters of water quality were examined together (including physical and bacteriological quality and chemical constituents), 59 percent of the water supplies sampled, serving 86 percent of the population, met the U.S. Public Health Service standards.

REFERENCES

1. J. F. Kopp and R. C. Kroner, *Trace Metals in Waters of the United States*, U.S. Dept. of the Interior, Federal Water Pollution Control Administration, Division of Pollution Surveillance, Cincinnati, Ohio, 1962–1967.
2. L. J. McCabe et al., "Survey of Community Water Supply Systems," *Journal of the American Water Works Association* 62 (1970): 670.
3. Report of the Committee on Water Quality Criteria, U.S. Dept. of the Interior, Federal Water Pollution Control Administration, 1968.
4. *Drinking Water Standards*, Publication No. 956, U.S. Dept. of Health, Education and Welfare, Public Health Service, 1962.

E

Electric Power

See also Fuel Resources and Energy Conservation.

In order to view electric power in the proper perspective, one needs to be aware not only of the growth in electric power demand but also of the relative importance of the principal sources of power and the environmental effects of the different power technologies. What choices may an individual make that will influence the environmental balance sheet? What are the merits of potential technologies, and how is research and development funding to be applied in these areas?

THE BALANCE OF POWER

As of 1968, about 52 percent of our electric power was being generated from coal-fired plants, 7 percent from oil-fired plants, 23 percent from gas-fired plants, 17 percent from hydroelectric stations, and less than 1 percent from nuclear plants. Since that time, however, in order to diminish emissions of **Sulfur Oxides,** many utilities have been switching from coal to oil. The Department of the Interior estimates that nuclear power will grow to supply nearly 53 percent of our total electrical requirements by the year 2000. Although Interior envisions that coal's share of the power supply will drop to 30 percent by the year 2000, this would still constitute a trebling of the amount of coal required yearly, because of the enormous increase in electrical power demand. Gas-fired plants, however, are not expected to grow appreciably in numbers because of shortages of gas and high costs; the share of electrical needs supplied by gas may dwindle to 4 percent in the year 2000. Interestingly, the Atomic Energy Commission believes that nuclear power will be supplying nearly 70 percent of our electric needs by the turn of the century.

There are two components to the phenomenal rise in electric power demand. First, energy use is growing. Second, there is a striking projected growth in the proportion of our energy needs to be satisfied by electricity. The history of electrical use in the past two decades and projections through the year 2000 are summarized in Table E.1.

The pattern of use underlying this growth in electric power demand is not expected to change substantially during the period 1970–2000. Losses in transmission will run 8 to 10 percent of the power generated. After losses are subtracted, the remainder is the actual electrical

TABLE E.1

ELECTRIC POWER GENERATION, 1950–2000

Year	Kwh Generated (Billions)[a]	Generating Capacity (Millions of Kw)[b]	Electric Energy as Percent of Total Energy
1950	389	69	16.1
1960	844	168	18.9
1970	1,638	340	25.0
1980	3,202	668	33.0
1990	5,970	1,261	41.7
2000	10,150	2,150	45.5

Adapted from tables 4 and 5 of the Report of Working Group III on Energy and Economic Growth, in *Engineering for the Resolution of the Energy-Environment Dilemma*, National Academy of Engineering, Washington, D.C., 1972.

[a] Amount of energy consumed.

[b] Rate at which energy can be made available.

Night lights in Cambridge, Massachusetts, viewed from Boston across the Charles River.

use. The percentages we quote are of actual use. Residential use will, it is anticipated, remain at about 30 percent of total use, although commercial use may rise several points to about 25 percent. Industrial use may fall slightly to 37 percent. Remember, it is the proportional use we are indicating; the total electrical use and uses in the individual categories are all rising. Large consumers in the industrial category are the primary metal industries and the chemical industries. Steelmaking alone accounted for over 12 percent of the industrial consumption of electricity in 1966. Another sector, which may be categorized as public uses, includes street lighting and rapid transit systems. This use may grow from about 4 percent in 1970 to 8 percent by the end of the century.

The category of residential use is interesting to examine. An all-electric household might be expected to consume 26,000 kilowatt-hours of electricity annually. About 40 percent would be devoted to space heating, 20 percent to water heating, and 20 percent to central air conditioning. In 1971, electric heating had not yet achieved widespread use. Nationwide, room air conditioning had reached 42 percent of the households, and electric water heating was used by 32 percent of the households. One can see that as these electrical uses spread, the growth in residential power use will be significant.

ENVIRONMENTAL IMPACT

All power plants are polluters; even the releases from hydropower plants may impair water quality. There are differences, however, in both the type and the extent of the pollutants discharged by the various kinds of plants. For purposes of comparison, Table E.2 presents the principal environmental effects which arise from the use of our major power sources. (See **Nuclear Power Plants, Reservoirs,** and **Fossil Fuel Power Plants;** each of these types of power generation is discussed from the standpoint of its actual and potential environmental effects.) Most of the special terms in the table are discussed briefly in the following paragraphs.

Thermal Pollution refers to the discharge of heated water used to condense steam in steam-powered electric plants. The resulting temperature elevation of the receiving water may be detrimental to aquatic species. Cooling towers and cooling ponds can be used to dissipate the waste heat to the atmosphere.

Strip Mining of coal is accomplished by

Land in Kentucky scarred by strip mining bordering the electric plant for which the mined coal is destined. Smokestacks discharge particulate matter, sulfur oxides, and nitrogen oxides. Cooling towers cool the condenser water from the plant.

TABLE E.2
ENVIRONMENTAL EFFECTS OF ELECTRIC POWER

<table>
<tr><th>Source of Power</th><th>Principal Contributions to Water Pollution</th><th>Principal Contributions to Air Pollution</th></tr>
<tr><td>Hydroelectric power generation</td><td>1. Sedimentation behind reservoir
2. Impairment of quality of impounded water[a]</td><td>None</td></tr>
<tr><td>Coal-fired power plants[b]</td><td>1. Thermal Pollution[c]
2. Acid Mine Drainage from coal mines and strip mining</td><td>1. Sulfur Oxides[d]
2. Particulate Matter[e]
3. Nitrogen oxides[f]
4. Mist from cooling towers, if used[g]</td></tr>
<tr><td>Oil-fired power plants</td><td>1. Thermal Pollution[c]
2. Oil Pollution from tankers and offshore drilling</td><td>Same as for coal-fired power plants, but usually less severe in all categories except mist from cooling towers</td></tr>
<tr><td>Gas-fired power plants</td><td>Thermal Pollution[c]</td><td>1. Nitrogen oxides[f]
2. Mist from cooling towers, if used</td></tr>
<tr><td>Nuclear power plants</td><td>1. Thermal Pollution[c]—more severe than from coal, oil, and gas fired plants
2. Radioactive elements in trace quantities[h]</td><td>1. Radioactive gases and particles in trace quantities
2. Mist from cooling towers, if used[g]</td></tr>
</table>

[a] See **Reservoirs.**
[b] See **Black Lung Disease; Strip Mining.**
[c] See also **Water Pollution, heat.**
[d] See also **Air Pollution, contaminants—sulfur oxides.**
[e] See also **Air Pollution, contaminants—particulate matter.**
[f] See **Air Pollution, contaminants—photochemical pollution; Photochemical Air Pollution.**
[g] See **Thermal Pollution.**
[h] See **Nuclear Power Plants.**

power shovel and auger. Surface deposits are removed without the necessity of sinking shafts. Unreclaimed land gives the appearance of a "moonscape" and may become uninhabitable for plant or animal species.

Acid Mine Drainage arises when groundwater dissolves iron pyrites—minerals which accompany coal deposits. The yellow-brown, acidic waters drain from both underground mines and lands left unreclaimed after strip mining.

Oil Pollution refers to chronic oil discharges and to the more spectacular tanker spills and blowouts of offshore wells. Harmful effects on marine animals have been widely noted. In the late 1960's, about three-quarters or more of the oil used by electric utilities was imported. Thus, it must have been transported across the ocean in tankers.

Sulfur Oxides are a class of air pollutants derived from the combustion of sulfur in coal and in oil. Sulfur dioxide damages materials and plants and has been firmly linked with the incidence of respiratory illness. Sulfur trioxide combines with water vapor to create a corrosive sulfuric acid mist. About 45 percent of the annual emissions of sulfur oxides in the United States stem from **Fossil Fuel Power Plants.**

The emission of **Particulate Matter** from fossil fuel power plants is well controlled; yet such a massive amount is generated that 20 percent of the annual particle emissions in the United States are derived from this source. Particles corrode and soil the surface of materials and have also been linked with respiratory illnesses.

Nitrogen oxides are produced during the combustion of coal, oil, and gas. Nitrogen principally from the air and oxygen from the air combine during high-temperature combustion of these fuels. While the health effects of nitric oxide and nitrogen dioxide are not well catalogued, the reactions of these substances with atmospheric hydrocarbons produce **Photochemical Air Pollution.** The pollutants in the last category are associated with eye and throat irritation.

Nuclear Power Plants discharge trace quantities of radioactive elements in the condenser water effluent and in stack releases. Gaseous wastes are discharged from a tall stack after treatment, and liquid leaks from the system are treated and diluted in the condenser water. Levels of radiation from these plants have been very low even relative to the **Radiation Standards** set for them, and they promise to be even lower in the future.

TRADE-OFF CHOICES

Because the by-products of electric power generation present a formidable challenge to the protection of environmental quality, one ought to consider the impact of prevailing and projected power uses. At the same time, one should note the alternative energy sources which are available.

Electricity is sold as a clean energy source for the home. It may be, at the home. However, if the pollutants produced by the power plant generating the electricity are taken into account, electric heating cannot be called the cleanest available heating method without numerous qualifications. There are several alternative energy sources. The following discussion focuses on residential heating for two reasons. First, the trade-off is straightforward, and second, a significant portion of the projected growth in power demand is based on the movement to electric heating. Barely 1 percent of homes still heat with coal; thus oil, gas, and electricity are the alternatives to be considered.

Suppose that a home is heated by electricity generated at a fossil fuel power plant. It is assumed that electricity used to heat a home converts entirely to heat at the home. Further, it is assumed that the best possible thermal efficiency for the current generation of coal-fired power plants prevails; that figure is 40 percent. That is, 40 percent of the heat derived from burning the coal is converted to electrical energy. The remainder of the heat is wasted to the environment either as hot stack gases or as heated water from the condenser loop (see **Thermal Pollution**).

On the other hand, suppose a home is heated by gas or oil. An accepted figure for the efficiency of a new domestic forced air heating unit which burns gas or oil is 80 percent. That is, 80 percent of the heat derived from burning the fuel is transferred to the air in the house, and 20 percent is wasted as hot stack gases. If a furnace is out of adjustment or if carbon has coated the heat exchanger, the performance may fall to 60 or 70 percent efficiency. Thus, because the heating value of oil and gas is more efficiently utilized in the home unit, the quantity of fuel which must be burned in the home is much less than the quantity which must be burned at the power plant to heat the home electrically. In fact, the power plant must produce twice as much heat energy as the home furnace and consume more fuel to do the same heating job.

The fuel at the power plant, be it coal or oil, is richer in sulfur than the oil which is used for home heating. The consequence of widespread utilization of electric heating from coal- or oil-fired plants is a significant increase in the pollution burden of sulfur oxides. There are technologies in the research and development stages to reduce sulfur oxide emissions from power plants (see **Sulfur Oxides**). How soon these processes could shift the sulfur oxide balance to favor electric heating is not known. A decade or more may elapse before the answers begin to arrive. Since natural gas has almost no sulfur, the use of electricity generated from gas-fired power plants probably does not produce significant levels of sulfur oxides.

Other air pollutants of importance associated with power generation and home heating are **Particulate Matter** and nitrogen oxides (see **Photochemical Air Pollution**). Comparing the use of electricity from a coal plant to home heating directly with oil or gas clearly favors direct heating because the coal plant is a monstrous source of particles. But comparing the use of electricity from oil- or gas-fired power plants with direct heating by these fuels does not present a clear choice in terms of particles. Probably the same is true for nitrogen oxides.

The choice between direct heating (fuel burned on premises) and indirect heating (heat generated at power station) becomes more academic with regard to air pollutants if one considers electricity from nuclear power plants. Here the pollutants are vastly different in nature from those named above. We do not propose to explore that comparison for we have another more compelling reason to recommend against indirect (electrical) heating from nuclear power plants than the problem of air pollution. The reason is **Thermal Pollution.**

We noted that the home heating unit using oil or gas may be about 70 percent efficient in transferring heat to the air in the house. Present-day nuclear plants typically have thermal efficiencies reaching no higher than 32 percent. That is, only 32 percent of the heat energy generated by the plant is converted to electricity. The remaining 68 percent is wasted entirely to the cooling water. If we compare the home heated by fuel combustion on the premises to the home heated electrically by power from a nuclear plant, it can be seen that the heat wasted to the environment by the nculear plant is about eight times greater than the heat wasted by the home unit. Furthermore, almost all of the nuclear plant's waste heat is directed to the water which cools the plant.

Coal, oil, gas, uranium—which fuel is at the power plant does not matter. Thermal pollution occurs where it need not occur if heating is by electricity. Similar comparisons may be made for the use of gas stoves and electric stoves and between the use of gas-fired hot water heaters and electrical hot water heaters. This section should not close without mention of the potential of solar energy to provide heat for homes. In the next section, we discuss the use of solar energy to generate electric power, but the technology to accomplish this efficiently may be some time off. Solar house heating, on the other hand, could become feasible soon. In Japan, solar energy has already been used to provide hot water in residences.

FUTURE SOURCES

Numerous ideas for power generation technology exist. Some may never be widely utilized. Others may eventuate in dominant sources of power at some future time. It is too early to make predictions about most systems.

A modification of existing nuclear power plants which will improve thermal efficiency may be possible in the next decade or so. Two prototypes of the modified nuclear reactors exist, and they incorporate the feature known as superheating of steam. Eventually the two prevailing types of nuclear reactors, the boiling water reactor (BWR) and the pressurized water reactor (PWR), should have the superheat feature. This addition will allow them to reach about 40 percent thermal efficiency. That is, 40 percent of the thermal energy produced will be converted to electricity by such a plant, as compared to the 40 percent thermal efficiency of the *now available* fossil fuel plant which can superheat steam. Since today's nuclear power plants have a thermal efficiency in the range of 30 to 32 percent, the increase in efficiency due to superheating does not sound like much, but it is. A 30 percent efficient plant produces about 50 percent *more* waste heat per unit of electrical energy output than a 40 percent efficient plant.

Other improvements in existing technology will be occurring in the design of **Fossil Fuel Power Plants.** Processes for removing sulfur oxide emissions are already being tested on full-scale plants. Some nitrogen oxides may also be removed by the most promising process, the limestone injection process (see **Sulfur Oxides**). Active programs of research were under way in the early 1970's. They were sponsored by the National Air Pollution Control Administration, which is now part of the Environmental Protection Agency.

Power generation via gas turbines has likewise been suggested. In this method, compressed air from the atmosphere is burned with the fuel (gas or highly refined oil) to yield gases at high temperature and pressure; these are used to turn a turbine and are expelled to the atmosphere. Some power is already generated by this method. However, because of the relatively low thermal efficiency of gas turbines (more fuel required per unit of electrical output), their use is limited at present to power production during periods of peak power demand. Nevertheless, advances in gas turbine technology in the military and civil aircraft programs promise eventual thermal efficiencies in the range of 36 to 38 percent.

Since the hot gases are discharged to the atmosphere, the gas turbine requires no cooling water. In the south central region of the United States, which is rich in natural gas, one estimate indicates that electric power from gas turbines might be obtained in the 1980's at prices lower than from steam-powered fossil fuel plants. This is especially significant because the South will have great difficulty in providing adequate cooling water for steam plants. The technology should not be overlooked in any investigation of future alternatives even though current gas shortages may dampen our view of its promise.

A power technology whose use appears to be growing is "pumped storage." As with gas turbines, power generated by this method would be used to meet peak demands rather than to provide base load power. Water is pumped during periods of slack electric usage to reservoirs at high elevations. During peak usage, the water is allowed to flow down to turn hydraulic turbines to generate additional power. About two-thirds of the electric power invested in storing the water at the high elevation is recoverable. Unfortunately, the construction of an artificial reservoir at a high elevation may destroy valuable recreational resources. One pumped storage stystem proposed in 1963, the Storm King project, was still being fought in the courts in 1972.

The other technologies are more esoteric. We shall discuss briefly the liquid metal fast breeder reactor (LMFBR), the magnetohydrodynamic (MHD) power plant, and geothermal power. Fusion power is still too distant for substantial comment.

The breeder reactor has been a priority research area of the Atomic Energy Commission AEC) since the 1950's. The light water reactors (the pressurized water reactor and the boiling water reactor) were technologically simpler, however, and came along faster. The utility industry chose to enter the nuclear age with the light water reactors, first demonstrated commercially in 1959–1960. The mid-1980's are estimated as the time of entrance of the breeder reactor to commerical power production. The reactor is expected to have a thermal efficiency equal to or slightly better than that of the best fossil fuel plant; it should also produce less radioactive effluents than present-day nuclear power plants.

The breeder reactor will use uranium-235 as the starting fissile material. Neutrons from its fission will activate uranium-238 to give plutonium-239, which will also undergo fission. Neutrons from the plutonium fission will activate more uranium-238, in essence creating more fuel. Because of the limited quantity of natural uranium-235 and much greater abundance of uranium-238, the breeder will multiply our fuel supplies by several orders of magnitude. Liquid sodium will be the coolant in a planned fast breeder reactor. Its heat will be used to boil water for the necessary steam. Many obstacles lie on the road to development, and misgivings have been expressed as to the economic merit of the LMFBR [3, 4]. Although other nuclear technologies such as the high-temperature gas-cooled reactor (HTGR) are being pursued as backup to the LMFBR, it remains the AEC's priority goal. A demonstration plant which will generate 300 to 500 megawatts of electricity is planned for the end of the decade in the United States. Other industrial nations are also engaged in active development of the breeder reactor [2, 5, 6].

Magnetohydrodynamic (MHD) power generation, which utilizes fossil fuels, holds promise as a technique for increasing the thermal efficiency of electric power generation to between 50 and 65 percent. At the 50 percent level, the waste heat production per unit of electrical energy is reduced by one-third from the output of the best conventional plant. At 65 percent, the waste heat production is reduced by two-thirds. Such high thermal efficiencies have several advantages.

First, considerably less waste heat will be rejected to the environment, and thus less thermal pollution will occur. Second, less fuel will be required to produce a unit of electrical energy because more energy is derived from the heat in a unit of fuel; there will thus be a saving in resources. Third, since less fuel will be required per unit of electrical energy, the pollutants emitted per unit of electrical energy may be significantly less. Coupled with the planned emission control techniques, the process is promising from the standpoint of pollution abatement.

The process is often thought of as a topping cycle for conventional steam power systems. In this method of power generation the combustion products from fossil fuel are seeded with an ionizable material and then pass through a magnetic field at high speeds; a high voltage is extracted by placing electrodes in the gas stream. Such generation resembles conventional generation in the sense that, rather than having an armature rotate in a magnetic field, an ionized gas is passed through it. The gas exiting from the MHD channel is then used to generate steam for conventional power generation. A small magnetohydrodynamic plant exists in the Soviet Union, but progress in the United States is slow. Estimates of time until the process could become a feasible method for commercial generation of electricity run from 10 years up to 30 years. Engineering problems include extremely high temperatures and corrosion.

Of all the *major* power technologies, the hydroelectric reservoir, destructive as it is of natural resources, has the least pollution potential. There are, however, several other concepts for power generation with low or minimal environmental impact. Solar energy, energy from the tides, and nuclear fusion are all future though distant possibilities to provide power. In addition to these, geothermal energy offers few environmental effects, and the harnessing of the earth's heat for steam generation has already been accomplished.

The use of solar energy has long been a dream of mankind. Although in a few countries water may be heated for residential use by solar energy, the commercial production of electricity from solar energy would at present require vast land areas to be covered with solar cells. A 545-square-mile area including Washington, D.C., and Prince Georges County, Maryland, might require a solar station 73 square miles in area [16]. A recent proposal for electricity production from solar radiation would use satellites carrying solar cells. The cells would capture the sun's radiation and produce electrical energy to be beamed to the earth as microwaves [10, 11].

The generation of electric power from tides is limited to areas of high tides. In these regions immense reservoirs would be built for storing water from the high tides; the water would be released through hydraulic turbines at lower tides to generate electricity. A plant of 544 megawatts (a reasonable size electric plant by modern standards) has been operating on the Rance River in France since 1967. The Bay of Fundy between the United States and Canada is often mentioned as a possible source of tidal electric power. Unfortunately, geographic factors make widespread use of tidal power unlikely. The location of a generation source distant from load centers would make transmission costs and losses prohibitive. On the other hand, electric intensive industries (such as aluminum production) might find location near such a source attractive.

Nuclear fusion would use as fuel a naturally occurring and relatively abundant isotope of water, deuterium. The enormously high temperatures required pose formidable engineering problems whose solutions are not yet at hand. Fusion power may not enter the arena of commercial power generation until the turn of the century.

Steam produced by the earth's heat already generates electricity in a number of countries. Italy, in particular, has pioneered development of geothermal resources. The geysers of Yellowstone National Park, which we do not cite as an example for commercial development, are a well-known source of geothermal steam. The steam may be used to turn turbines for the generation of electric power. The consumption of a nuclear or fossil fuel is bypassed in this generation, making the process attractive from the standpoint of air pollution control. However, the waste heat in the steam exiting from the turbines must still be dissipated to the environment. In order to prevent **Thermal Pollution,** cooling towers may be used in conjunction with the geothermal steam plant. There are several drawbacks to the development of this resource. First, geothermal areas are not widespread. In this country, California has the greatest availability of geothermal steam deposits. Additionally, problems of corrosion present themselves because of the high concentration of salts in the subterranean water. Under the Geothermal Steam Act of 1970, the United States government will soon be leasing public lands for exploration and production of electric power.

RESEARCH AND DEVELOPMENT FUNDING

Estimates of federal funding for power research for 1970 were made in 1969 by the President's Office of Science and Technology. Of the $368 million package, all but 15 percent was destined for the AEC. Thus less than $50 million went to other power-related research. Of this remainder, about 40 percent was spent on air pollution R&D and 50 percent on coal research. Coal R&D involved the areas of mining and mine safety as well as liquefaction and gasification of coal.

Of the AEC's R&D projected funding of about $310 million, close to 40 percent was devoted to the fast breeder reactor. A bit less than 10 percent went for fusion research, and 15 percent was designated for other nuclear technologies, regarded as backups to the LMFBR. The rest was to be consumed in general reactor technology and safety.

Investments by electric utilities in power R&D were about 15 percent of the federal contribution in 1968.

COMMENTARY

Two points are clear from the preceding discussion. First, potential technologies other than nuclear received almost negligible contributions from the government; MHD research was funded at less than a million dollars in 1971. Second, utility investment in R&D is inadequate. It constituted in 1968 less than 0.25 percent of the operating revenues of the utilities (which were a little less than $20 billion in that year). One explanation for the reluctance of utilities to engage in R&D is apparently related to the apprehension that research losses may not be charged against operating expenses, although there is precedent that they may be. A more likely explanation is that, while regulated, utilities are also protected in that their market is secure even though their technology does not improve substantially with time.

A more fundamental point concerns the power situation as it stands today and as it is projected for the next several decades. We have roared full steam, so to speak, into an era of burgeoning demand with technology that is inadequate to environmental needs. The enormous growth was not properly foreseen, and the impact of power on the environment was not realized. If these facts had been recognized, the apparent confrontation between energy and the environment might have been prevented.

Since the 1950's the government has relied on nuclear power development by the AEC to meet future needs. The AEC has given priority to the development of the liquid metal fast breeder reactor. The LMFBR, clean as it is expected to be, is still estimated at no more than 43 percent thermal efficiency. This is only a modest improvement over today's best fossil fuel plant.

Further, the cooling towers that are being built—and becoming necessities as plants grow larger and larger—are monstrous structures, tall as 40-story buildings. What a legacy of misty monuments to leave for the next generation! There is probably no politically acceptable action that can derail the onrush of power demands and the consequent need for cooling towers. The towers are virtually essential in many areas where plants are being erected. No dramatic action is possible now, but there are activities which will serve the future.

The first recommendation is for research and development in new power technologies. The government has placed its faith in nuclear technology up to now. MHD also deserves substantial research funding because of its promise for greater thermal efficiencies; air pollution R&D applied to fossil fuel plants ought to be pushed more vigorously. Solar power and geothermal power likewise ought to receive R&D funding. It seems unwise to rely solely on nuclear technology as the future choice for power and leave unattended so many avenues of development. Cooling technology to our knowledge is receiving negligible R&D support from the federal government. Given that we cannot escape the problem of waste heat even with LMFBR and MHD, cooling R&D is essential.

Another avenue of potentially valuable research is in coal gasification. The conversion of coal to a gas comparable to natural gas would supplement our tightly squeezed gas supplies *and* provide a fuel whose environmental impact is least among all the fossil fuels.

In 1970 the President's Office of Science and Technology recommended research in most of these areas also. The report stressed the need for nonfederal sources of funding. It does not seem likely that the utilities will meet on their own the need for research funds—for the following reason. Even if utilities could charge R&D against operating revenues, they might be reluctant to do so because higher rates discourage expanding power use. (This is also probably the reason for the predominance in the past of once-through cooling as opposed to cooling with towers or ponds, which is more expensive.) Because power companies are regulated as to their percentage return on investment, the way for them to make more money is to expand their operation—that is, sell more power. Higher rates discourage growth.

How then shall the needed funds be obtained? In this case, new charges for the individual electric consumer may serve two purposes. A 2 percent charge on the cost of electric consumption would generate over $400 million annually, which would be earmarked for a federal government R&D program in power technology and effects. This would about double the current federal R&D power expenditures. The charge would also have a modest dampening effect on the growth of power demand, an environmentally desirable side effect. A larger charge could accomplish more. The charge need not be imposed at low levels of electrical consumption, but could be applied above some level of utilization. This differential would serve to counteract the present rate structure in which companies offer lower rates (in cents per kilowatt-hour) at higher consumption levels. That structure now encourages electric heating and other large uses of electricity. In fact, as a means of curtailing growth of power demand, regulatory bodies could insist that a single rate apply for all domestic usage no matter what the level of consumption. Finally, the rate could *increase* with increased levels of consumption as a means to actually discourage excess utilization. It is unlikely that any power company would find such proposals pleasing.

The problem of power and its environmental effects is quite new in the sense that our awakening took place only in the last decade. The opportunity still exists for positive action.

REFERENCES

1. *Engineering for the Resolution of Energy-Environment Dilemma*, National Academy of Engineering, Washington, D.C., 1972.
2. F. Culler and W. Harms, "Energy from Breeder Reactors," *Physics Today*, May 1972.
3. *Science*, November 19, 1971, p. 807.
4. *Science*, April 28, 1972, p. 391.
5. *New York Times*, January 5, 1972, p. 11.
6. *New York Times*, October 19, 1971, p. 28.
7. *New York Times*, May 30, 1972, p. 30.
8. "MHD for Central Station Power Generation," prepared for the Executive Office of the President, Office of Science and Technology, June 1969.
9. R. Tybout and G. Lof, "Solar House Heating," *Natural Resources Journal* 10, no. 2 (April 1970).
10. P. Glaser, "Power from the Sun: Its Future," *Science*, November 22, 1968, p. 857.
11. P. Glaser, "Power Without Pollution," *Journal of Microwave Power* 5, no. 4 (December 1970). The entire issue is devoted to solar energy use via satellite.
12. *Advanced Nonthermally Polluting Gas Tur-*

bines in Utility Applications, Water Pollution Control Research Series 16130 DNE, U.S. Environmental Protection Agency, 1971.

13. "The Environmental Effects of Producing Electric Power," *Hearings Before the Joint Committee on Atomic Energy*, 91st Congress, 1st Session, October–February 1969–1970.
14. W. Gough and B. Eastlund, "The Prospects of Fusion Power," *Scientific American*, February 1971, p. 50.
15. J. Barnea, "Geothermal Power," *Scientific American*, January 1972, p. 70.
16. W. Cherry, "The Generation of Pollution-Free Electric Power from Solar Energy," *Journal of Engineering for Power*, April 1972, p. 78.
17. "Electricity Demand Growth and the Energy Crisis," *Science*, November 17, 1972, p. 703.
18. "Advanced Power Cycles," *Hearings Before the Committee on Interior and Insular Affairs*, Serial no. 92-21, U.S. Senate, 92nd Congress, 2nd Session, February 1972.
19. T. Gray and O. Gashus, eds., *Tidal Power*, Proceedings of a Conference, Halifax, Nova Scotia, May 1970, Plenum, N.Y., 1972.
20. *Summary Report of the Cornell Workshop on Energy and the Environment*, Washington, D.C. 1972.
21. "Geothermal Energy Resources and Research," *Hearings Before the Committee on Interior and Insular Affairs*, Serial no. 92-31, U.S. Senate, 92nd Congress, 2nd Session, June 1972.
22. *Solar Energy as a National Energy Resource*, NSF/NASA Solar Energy Panel, December 1972. Available from Dept. of Mechanical Engineering, University of Maryland, College Park, Md.
23. R. Roberts, "Energy Sources and Conversion Techniques," *American Scientist* 61 (January–February 1973): 66.
24. A. Meinel and M. Meinel, "Physics Looks at Solar Energy," *Physics Today*, February 1972, p. 44.

An excellent series of articles concerning new sources of energy appeared in *Science* in 1972–1973. They are listed here in the order of their appearance.

25. "Energy Options: Challenge for the Future," *Science*, September 8, 1972, p. 875.
26. "Geothermal Energy: An Emerging Major Resource," *Science*, September 15, 1972, p. 978.
27. "Solar Energy: The Largest Resource," *Science*, September 22, 1972, p. 1088.
28. "Laser Fusion: A New Approach to Thermonuclear Power," *Science*, September 29, 1972, p. 1180.
29. "Gasification: A Rediscovered Source of Clean Fuel," *Science*, October 6, 1972, p. 44.
30. "Fission: The Pro's and Con's of Nuclear Power," *Science*, October 13, 1972, p. 147.
31. "Magnetic Containment Fusion: What Are the Prospects?" *Science*, October 20, 1972, p. 291.
32. "Magnetohydrodynamic Power: More Efficient Use of Coal," *Science*, October 27, 1972, p. 386.
33. "Fuel from Wastes: A Minor Energy Source," *Science*, November 10, 1972, p. 599.
34. "Photovoltaic Cells: Direct Conversion of Solar Energy," *Science*, November 17, 1972, p. 732.
35. "Hydrogen: Synthetic Fuel of the Future," *Science*, November 24, 1972, p. 849.
36. "New Means of Transmitting Electricity: A Three-Way Race," *Science*, December 1, 1972, p. 968.
37. "Conservation of Energy: The Potential for More Efficient Use," *Science*, December 8, 1972, p. 1079.
38. "Energy Needs: Projected Demands and How to Reduce Them," *Science*, December 15, 1972, p. 1186.

39. "Fuel Cells: Dispersed Generation of Electricity," *Science*, December 22, 1972, p. 1273.
40. "Power Gas and Combined Cycles: Clean Power from Fossil Fuels," *Science*, January 5, 1973, p. 54.
41. "Energy and the Future: Research Priorities and National Policy," *Science*, January 12, 1973, p. 164.

Eutrophication

Many materials considered water pollutants are recognizably objectionable. No one would dispute the undesirability of adding mercury, lead, arsenic, pesticides, or raw sewage to natural waters. Other materials, however, at low concentrations, are neither extremely toxic nor unsightly but nevertheless may lead to objectionable water quality. They have what might be termed a secondary polluting effect. Into this class of pollutants fall nutrients which cause "cultural eutrophication." Examples are phosphorus, nitrogen, and organic matter.

Our discussion considers the natural physical characteristics and biological processes of lakes, describes eutrophication and its problems and control, and investigates the role of phosphate detergents and the controversy over the fate of Lake Erie.

PHYSICAL AND BIOLOGICAL CHARACTERISTICS OF LAKES

Lakes are most often used as a focus for the discussion of eutrophication. Actually rivers, impoundments, estuaries, and other coastal waters can all become eutrophic. *Eutrophic* means "well nourished." The word is from the Greek: *eu* meaning "well" and *trophos* "pertaining to nourishment." Thus, a eutrophic lake is one whose waters contain high levels of nutrients and can support a large number of organisms. An oligotrophic lake, in contrast, is nutritionally relatively poor. This is not to say that oligotrophic lakes are less desirable. On the contrary, their waters tend to be clearer than those of eutrophic lakes, and they are not as subject to nuisance growths of algae. Oligotrophic lakes, although they have smaller populations, are likely to show a greater diversity of species. Eutrophic lakes tend to have larger populations of the species present but only a few species.

Deep lakes often exhibit a temperature differential with depth. During the warm months a warm upper layer of water called the *epilimnion* floats above a cold lower layer called the *hypolimnion.* Between the two is sandwiched a layer in which temperature decreases rapidly with depth; it is called a *thermocline.* The lesser density of the warm water in the epilimnion ensures that it will "float" on top of the denser hypolimnion during the warm months. As winter approaches, however, the epilimnion becomes progressively cooler and denser. When the density of the epilimnion becomes greater than the density of the hypolimnion, the waters of the lake overturn and mix. This event usually happens in the fall and is known as the "fall overturn." [1]

During the warm months, when the layers do not mix, the growth of algae and other aquatic plants is often limited to the epilimnion, where sufficient sunlight penetrates the water to allow photosynthesis. There is little transfer of nutrients from the hypolimnion or from bottom sediments to the epilimnion. The stability resulting from the thermal stratification

[1] In some lakes there is both a fall overturn and a spring overturn. For an additional description of these phenomena see [24].

of the lake prevents an effective barrier to upward circulation of nutrients. However, debris from dead plants and animals, as well as from other waste materials, does sink to the bottom of the lake; there it is decomposed by the lake microorganisms. Desirable fish such as cisco and lake trout, which require cold water, reside in the hypolimnion during the warm months.

In oligotrophic lakes a plentiful supply of oxygen remains in these deeper waters. In eutrophic lakes, on the other hand, oxygen is removed from the hypolimnion by the decay of the quantities of plant and animal debris that have fallen to the lake bottom. The waters of the hypolimnion in eutrophic lakes may thus be devoid of oxygen and unable to support cold-water fish. Eutrophic lakes, then, are characterized by warm-water fish, often described as rough fish; these include carp, crappie, and perch.

Eutrophic bodies of water are also noted for greater numbers of blue-green algae than oligotrophic ones. Blue-green algae, named for their bluish-green chlorophyll, are often more noxious than green or yellow-green algae. Blue-green algae can produce chemicals which cause unpleasant tastes and smells in water and which in some cases are toxic to animals. The fact that these algae are also more resistant to grazing by other aquatic creatures explains in part why they can build up to such an extent in polluted waters.

In eutrophic lakes both blue-green and green algae can occur in enormous masses called blooms. The growth of a bloom is a very rapid and often dramatic phenomenon. A bloom may cause a net production of oxygen in the daytime. The oxygen is a by-product of photosynthesis. At night, though, when the algae respire (use oxygen) and do not photosynthesize, there is a net depletion of oxygen. Huge swings in the oxygen content of the waters are the result. Low levels of oxygen induced by the algae may be detrimental to aquatic life. (See **Organic Water Pollution**.) Table E.3 compares the characteristics of oligotrophic and eutrophic lakes.

TABLE E.3
CHARACTERISTICS OF OLIGOTROPHIC AND EUTROPHIC LAKES

Oligotrophic	Eutrophic
Often deep, with temperature stratification	Often shallow
Low levels of plant nutrients	High levels of plant nutrients
Few individuals but many species	Many individuals of a few species
Oxygen in hypolimnion	Hypolimnion, if present, has little or no oxygen
Fish likely to be cold-water varieties such as cisco, lake trout	Warm-water fish such as carp, crappie, perch, bullhead
Little plant and algal growth	Extensive plant growth; algal blooms common
Examples: Lake Tahoe, Nevada; Finger Lakes, New York	Example: Lake Mendota, Wisconsin

NATURAL VERSUS CULTURAL EUTROPHICATION

Bodies of water may be fertile when they are originally formed, or they may become fertile naturally with age. (This is not to say that they will necessarily become more fertile as they age. The flow of nutrients may be out of a lake's watershed, making the lake more oligotrophic with the passage of time.) When a lake becomes shallow because of the deposition of sediments, it may lose its temperature stratification. Nutrients will then cycle more freely between the lake sediments and the waters, since the barrier of stratification is gone (see Figure E.1). The process of natural eutrophication speeds up when the lake becomes shallow enough for higher rooted plants to occupy most of the lake basin. The degradation of the higher plant remains accelerates the filling in of the lake basin. Such natural eutrophication normally requires from hundreds to thousands of years.

It is possible, however, to produce the state of eutrophication artificially by the addition of large amounts of plant nutrients to the water. Inputs of sewage, industrial wastes, and agricultural runoff may accomplish within a few years a transition which would naturally take thousands of years. It has been estimated that we have accelerated the eutrophication of Lake Erie by 50,000 years. Lake Constance, between Switzerland and Germany, is said to have been aged 10,000 years in the past 20 years by sewage inflows. Change in a lake's trophic state (i.e., in

FIGURE E.1

THE AGING OF A LAKE

(1) Nutrients which settle to the lake bottom are usually trapped beneath the thermocline as long as temperature stratification persists. (2) The natural aging of a lake is accompanied by the filling of the lake basin with sediments. (3) As the lake becomes more and more shallow, it loses its temperature stratification, and free circulation of nutrients between the waters and the sediments becomes possible. (4) Eutrophication becomes very rapid when the lake basin becomes so shallow that a major portion of it is occupied by rooted plants.

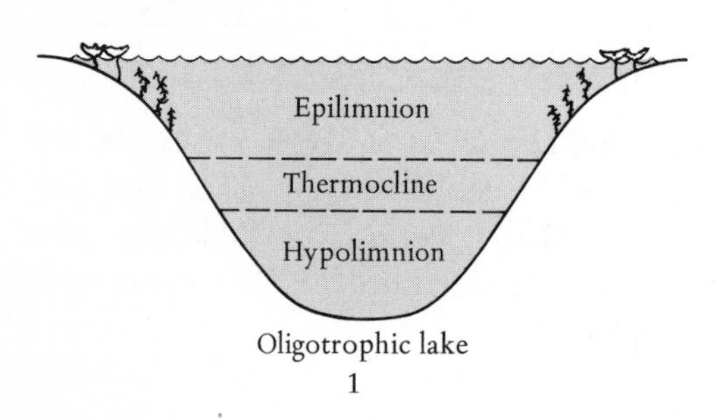

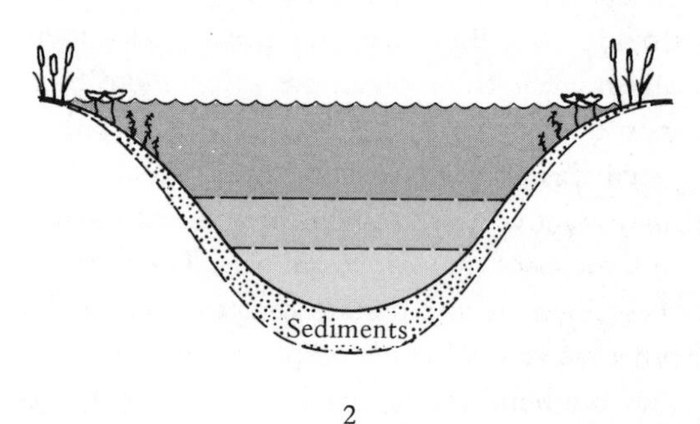

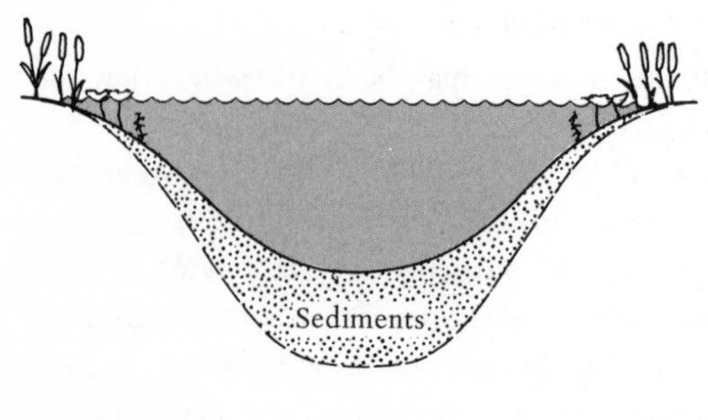

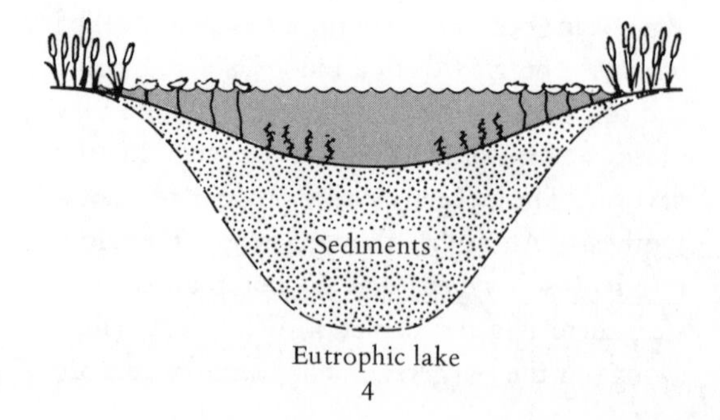

its nutrient content) within the last 50 years is probably due to human activities. This phenomenon of artificial aging has come to be called "cultural eutrophication." The phrase denotes enrichment of a body of water, the subsequent growth of plants and animals, and the accompanying deterioration in water quality.

PROBLEMS CAUSED BY EUTROPHIC WATERS

The use of eutrophic waters for domestic and industrial purposes poses a number of economic problems. Higher water treatment costs will result from the removal of tastes, odors, and colors. Filters[2] become clogged more readily as the water's turbidity increases. Since ammonia and organic materials combine with chlorine, there will be an increase in the consumption of the chlorine used to kill pathogenic bacteria. If the hypolimnion is devoid of oxygen, iron, magnesium, and sulfides may concentrate there. These chemicals may make the water unsuitable for industrial and domestic uses. As a consequence, water from the epilimnion, which is rich in algae, may have to be utilized instead, bringing with it the problems noted above.

If recreational waters become eutrophic, the floating and rooted aquatic plants and algae will make swimming and boating unpleasant. The wind can pile the dead and dying algae along the shore, where they will decay and give off hydrogen sulfide gas. Property values may diminish in response. There is a possible additional economic effect on the fishing industry. Valuable fisheries may collapse as the species formerly caught disappear from the lake.

SOURCES OF NUTRIENTS

Plants require a number of nutrients for growth. Carbon, nitrogen, hydrogen, oxygen, and phosphorus are needed in large quantities. Lesser amounts of trace minerals such as molybdenum, zinc, iron, and magnesium are necessary. Aquatic plants differ somewhat from land plants in their nutritional requirements. For instance, aquatic plants may use nitrogen either in the form of ammonia or as the nitrate, whereas land plants require nitrate.[3] Certain algae prefer ammonia to nitrates. Some species of blue-green algae are even capable of using nitrogen gas from the atmosphere as their source of nitrogen. Phosphorus is used by aquatic plants mainly in the form of phosphate, although some algae can use organic phosphorus compounds.

One of the largest sources of the augmented nutrient input to natural waters is domestic waste water. Sewage may contain 5 to 20 milligrams per liter of phosphorus and 20 to 100 milligrams per liter of nitrogen. The less the degree of treatment of the waste water, the greater the amount of organic carbon compounds which may be added to the receiving water. About half of the phosphorus in domestic waste waters is from detergents; the remainder is from human wastes. It has been estimated that each person contributes 1 to 2 pounds of phosphorus and 7 pounds of nitrogen per year in this source.

Solid waste materials, including sludge from sewage treatment plants and garbage in landfills and dumps, also contain high levels of nitrogen and phosphorus. These nutrients can leach into the groundwater or be carried into natural waters as runoff.

In areas where there are septic tanks rather than central sewage treatment facilities, groundwater can be contaminated by large amounts of nitrates. Phosphates, in contrast, are usually absorbed by the soil unless it is very sandy. If septic tank outlets become plugged, surface

---

[2] See **Water Treatment and Water Pollution Control** for a description of sand filters.

[3] The leguminous land plants do contain bacteria which are capable of converting nitrogen from the air to nitrates.

discharges will carry both phosphates and nitrates into watercourses.

The wastes from certain industries contribute nutrients to water. Pulp and paper industry wastes are high in carbon but low in phosphorus and nitrogen. Phosphate mining wastes, on the other hand, are high in phosphates.

Storm waters from urban areas carry nitrogen and phosphorus at levels of a few milligrams per liter. Sources of these nutrients are not known with certainty but may include lawn fertilizers, pet wastes, leaves, dust, and combustion sources such as the automobile. (See **Air Pollution, photochemical pollutants,** for the origin of nitrate compounds in the air.) It is estimated that runoff in Cincinnati derives an average of 9 pounds of nitrogen from each acre and 0.8 pound of soluble phosphorus from each acre, annually.

Agricultural runoff can also contribute substantial amounts of nitrogen and phosphorus to water, but probably because of improper management practices such as overuse of fertilizers or application of fertilizers before rainfall. Runoff concentrations from agricultural lands can be kept as low as 0.03 pound of nitrogen per acre per year and 0.003 pound of phosphorus per acre per year. These amounts are comparable to natural runoff values.

An example of poor management: If manure is spread on the land in the winter when the ground is frozen, nutrients will not leach into the ground. Instead they will remain until the spring thaw, when they are carried via runoff into natural waters. Consequently, farmers are urged to store manure over the winter and spread it only on thawed ground. Feedlots can present a serious nutrient runoff problem because of the high animal densities involved. A feedlot may have 50 to 100 animals per acre, and each head produces 15 tons of manure annually. In such cases, regular sewage treatment procedures are necessary to prevent serious water pollution. (See **Water Treatment and Water Pollution Control** for a description of sewage treatments.)

Nitrogen, phosphorus, and organic materials as well as trace nutrients are also contributed to water from a number of natural sources. Naturally eutrophic lakes drain watersheds rich in nutrients. For instance, small lakes in the Midwest drain soils naturally rich in nitrogen and phosphorus. Another source of nutrients is lake sediments, which contain nitrogen and phosphorus at levels of a few milligrams per kilogram. These nutrients may be unavailable to lake organisms under certain conditions. Anaerobic conditions (low or no oxygen) contribute to the release to the water of phosphorus from the sediments. Some lake sediments seem to act as phosphate buffers, maintaining a fairly uniform level of phosphorus in the lake waters.

Nitrogen is often carried down in rainfall; concentrations in rain reach up to 1 milligram per liter. If marshes are drained for development, large amounts of nitrogen as nitrates may be released to the water. Wildfowl add organic waste materials to natural waters; their contributions are termed "bombed-in nutrients." The net contribution from wild ducks is probably small because they feed from the water in the first place. Domesticated fowl or deer in parks

TABLE E.4

PHOSPHORUS INPUTS TO LAKE ERIE, 1967

Source	Amount (Short Tons)	Percent
Lake Huron	2,240	7
Municipal wastes	19,090	63
Industrial wastes	2,030	7
Land drainage	6,740	22
Total	30,100	

From Report to the International Joint Commission on the Pollution of Lake Erie, Lake Ontario, and the International Section of the St. Lawrence River, vol. 1, Summary, 1969.

TABLE E.5
SOURCES OF NITROGEN AND PHOSPHORUS IN LAKE MENDOTA

Source	Phosphorus	Nitrogen
Municipal and industrial wastes	36%	10%
Urban runoff	17	6
Rural runoff	42	11
Precipitation	2	20
Groundwater	2	52
Nitrogen fixation		0.4

From G. Fred Lee, "Eutrophication," in Supplement to *Encyclopedia of Chemical Technology*, Wiley, New York, 1970.

and zoos, on the other hand, may make a significant contribution, as may geese, which feed on land.

An example of the relative values of the various phosphorus inputs is given in Table E.4. In the Potomac River basin the percentages are: (1) municipal wastes, 78.0; (2) industrial wastes, 8.6; (3) runoff, 13.4. In contrast, in areas where phosphate is mined, as around Hillsborough Bay, Florida, mining wastes total 43,407 pounds a day, or 94 percent of the total, and municipal wastes come to 2,700 pounds a day, or 6 percent. Table E.5 shows the sources of nitrogen and phosphorus inputs to Lake Mendota in Wisconsin.

THE LIMITING NUTRIENT

In order to grow, plants require their nutrients to be in a balanced ratio. If any one nutrient is missing or in inadequate supply with relation to the others, plant growth will be limited. No additional amounts of the other substances will be able to encourage further growth. There is at the present time a scientific controversy about which of the nutrients required for aquatic plant growth are limiting in natural waters.

In practical terms, if the limiting or key nutrient could be identified, efforts could be focused on decreasing the supply of that nutrient in order to control the noxious plant growth accompanying eutrophication. Suggested candidates for the limiting nutrient are phosphorus, nitrogen, iron, and carbon.

Phosphorus is an attractive candidate for "limiting nutrient." It is normally present in small quantities in unpolluted fresh waters. In the northeastern section of the country the concentration is generally less than 5 micrograms per liter. Oligotrophic lakes often have 1 microgram per liter; sometimes the phosphorus level is so low it is indetectable. Phosphorus is likely to be found in nature as the phosphate compound (a phosphorus atom plus four oxygen atoms). The words *phosphorus* and *phosphate* are frequently used interchangeably. Problems will probably not arise except when the concentration of phosphorus is given. Then the percentage of the element phosphorus will be less than the percentage of the phosphate compound since the latter weighs more. For example, many detergents are 50 percent phosphate compound by weight. This corresponds to about 9 to 10 percent of the element phosphorus by weight. Major sources of phosphorus pollution are generally known (e.g., sewage), and, as will be discussed later, techniques are available to reduce many of them.

In contrast, nitrogen enters natural waters from diverse and relatively uncontrollable sources. Nitrogen is found in large quantities in the air as nitrogen gas. Some blue-green algae, including a number of nuisance species, are capable of using it in this form ("fixing" it). Lightning can produce nitrogen compounds useful to cells; so can volcanoes. From these sources nitrogen compounds are carried into lakes and streams by rainwater. Nitrogen compounds are also easily leached from the soil and find their way into groundwaters. Human and animal wastes are high in nitrogen, and sewage treatment procedures to remove nitrogen are not well advanced.

Carbon is contributed to natural waters by waste materials—from sewage or from the death and decay of plants and animals. Atmospheric carbon dioxide which diffuses into water also provides a ready source of carbon for plant growth. Sewage inputs can be controlled by secondary sewage treatment (see **Water Treatment and Water Pollution Control**).

If phosphorus were found to be the limiting nutrient in natural waters, rather than nitrogen or carbon, efforts could then be directed at controlling phosphorus inputs and reducing total phosphorus in lakes. The definitive experiment pinpointing phosphorus as the limiting nutrient has not yet been done. Experiments point strongly in that direction, but conclusive proof is wanting. A few scientists, along with the soap and detergent industry, challenge the idea entirely.

In 1947 Sawyer [16] did a study of 17 lakes in Wisconsin. He concluded that when phosphorus was present in a concentration above 10 micrograms per liter in these lakes algal blooms could occur. The study has become something of a classic, in that 10 micrograms per liter has achieved status as the magic number above which lakes bloom and below which they do not. Actually the limited nature of the study (17 lakes, all in one geographic area) does not really allow such a sweeping generalization.[4] Nor do Sawyer's data actually support such a definite dividing line. However, a general trend toward algal blooms at higher phosphorus content is apparent. Vollenwieder [26], in a more recent study of Swiss lakes, also found that above 10 micrograms of phosphorus per liter a lake is more likely to show excessive plant growth. Similarly, evidence drawn from the fertilization of fishponds indicates that phosphorus is a limiting nutrient.

Some workers contend that phosphorus has not been shown specifically to be the limiting nutrient in any studies. It is true that many experiments deal with additions of sewage or other nutrient sources to algal cultures and subsequent measurement of algal growth. Although such studies sometimes determine the phosphorus and nitrogen levels in the added nutrients, the concentrations of other factors which may be present, such as organic carbon, vitamins, and trace minerals, are generally not determined. Shapiro and Ribiero [12] have done experiments with Potomac River water to which they added waste effluents whose phosphorus content was determined. Growth of algae was proportional to the addition. Pure phosphate added to river water in similar quantities again spurred growth but only after a time lag. The authors explain this finding by experiments which show that, as increasing quantities of phosphate are added, nitrogen becomes limiting, and the time lag is the delay necessary for the algae to synthesize a system for using atmospheric nitrogen. Only those algae capable of using atmospheric nitrogen (the blue-greens) responded to the pure phosphate.

Experiments designed to show that phosphate is or is not the limiting nutrient can run into several pitfalls. Algae may be able to absorb quantities of phosphorus far in excess of the amount they need to grow. This uptake is a function of phosphorus concentration, more being absorbed as the concentration rises. Some workers believe this phosphorus is released when the algae stop growing. Several results follow. Test algae must be starved for phosphorus before the growth-promoting ability of a particular lake water or sewage effluent is investigated. Otherwise the algae may grow in very low concentrations of phosphorus using

---

[4] Sawyer himself does not extend his work in this manner.

the phosphorus they have stored up. An experiment which did not starve test algae would tell nothing about whether phosphorus is limiting in the water sample.

If the phosphorus concentration is measured in the waters of a lake during a bloom, very low values are often obtained. This finding does not mean that algae are growing in very low phosphorus concentrations but rather that most of the lake phosphorus is at that time in the algae rather than in the water. Thus, timing can be extremely important when waters are sampled. A maximum of 10 percent of the total phosphorus in a lake is probably in solution at any one time. The rest is in the plants and animals or the lake sediments.

Some scientists feel that by comparing the ratio of nitrogen to phosphorus (N/P) in living organisms with the ratio in a body of water, one should be able to specify whether the water is nitrogen or phosphorus limiting. For specific organisms the N/P ratio varies from approximately 6 to 1 to 25 to 1, with an average of about 15 to 1. Thus waters with an N/P ratio greater than 15 to 1 would have an excess of nitrogen and be phosphorus limited. On the other hand, waters with a ratio less than 15 to 1 would have an excess of phosphorus and be nitrogen limited. The latter condition might be expected to encourage the growth of certain blue-green algae which require phosphorus in the water but are capable of extracting nitrogen from the air.

Certain factors, however, may limit the usefulness of this method. As mentioned earlier, algae have been found to concentrate phosphorus in excess of their needs for the building of tissue. Furthermore, measurement of the concentration of phosphorus in a body of water at any particular time is subject to all the difficulties described previously.

Kuentzel [14, 28] makes a good case for the limiting effect of carbon dioxide in some waters. Algae require large amounts of carbon dioxide (a ratio of 1 part phosphorus to 106 parts carbon, according to one author) for photosynthesis and growth. The possible limiting effect of carbon dioxide has generally been ignored, since large amounts are present in the atmosphere. However, this carbon dioxide must diffuse into the water before algae can use it to grow on. Kuentzel calculates that during a bloom the rate of diffusion may be much too slow to supply the necessary amounts. Organic matter decomposed by bacteria can supply the necessary amounts of carbon dioxide to the algae. Some data have been obtained, in fact, showing large bacterial blooms preceding algal blooms. Lee [1] feels that water quality has deteriorated to such an extent by the time blooms are present in lakes that the additional algal growth resulting from increased carbon dioxide supplies is not significant.

Some authors believe it likely that many nutrients become limiting in rapid succession during the growth explosion called a bloom.

Thus, at present, there is not an unequivocal answer to the question of what is the limiting nutrient in natural waters. Work currently in process should give more definitive results. A large group of scientists are working to develop a standard procedure for determining the effect of any given sample on the growth of algae. The procedure would make it possible to analyze lake water samples in terms of their limiting nutrients. Then, waste water samples could be examined in relation to their eutrophication potential. Furthermore, chemicals which are likely to find their way into waters in large amounts (such as the phosphates in detergents) could then be examined before their use was sanctioned. The procedure is called the Provisional Algae Assay Procedure (PAAP). Several years of testing may be required, however, before it becomes reliably predictive.

The most likely situation is that natural

waters in different areas of the country will be found to be individualistic in their nutrient states. Lakes which have limestone beds (Columbia Plateau and the Pacific Northwest) contain a great deal of calcium, which can cause the precipitation of phosphorus as calcium phosphate. These lakes may thus tolerate larger amounts of phosphorus inputs. Lakes fed by steep mountain streams in mountain and wilderness areas may well be nitrogen limited. Stream inflows to Lake Tahoe have been shown to be nitrogen limited, as are some (possibly all) salt waters, particularly in the North Pacific Ocean. All the others, however, are predicted to be phosphorus limited.

CONTROL

Attempts to halt or reverse the process of eutrophication fall into two categories: methods designed to reduce nutrient inputs and methods for lowering nutrient concentrations already present. Although it would be most desirable to prevent excess nutrients from ever reaching natural waters, the current state of the technology does not provide a way to control urban drainage or to decrease the high concentrations of nutrients in groundwater.

In several areas lakes have been saved by diversion of sewage inflows. The best example is probably Lake Washington in Seattle. During the 1950's, 11 different municipalities were dumping sewage with secondary treatment (see **Water Treatment and Water Pollution Control**) into the lake. Blooms began to occur with increasing frequency, and oxygen in the hypolimnion was decreasing. A recognition of the problem led to the formation of the "Metro" federation and the eventual passage of a $120 million revenue bond issue. The money was used to construct facilities to divert the sewage effluent from Lake Washington into Puget Sound. By 1963 one-third of the sewage was being diverted, and by 1968 all of it was diverted. The lake has already begun to recover. In a few more years it is expected that the state of the waters will be comparable to what it was in the 1930's. One wonders, of course, about Puget Sound. Although diversion has been tried in Madison, Wisconsin, for Lake Tahoe in Reno, Nevada, and for Lake Zurich in Switzerland, the solution is obviously not permanent; the problem has only been moved to a new locale.

Some attempts have been made to remove nutrients in lakes by dredging out the nutrient-rich sediments. Unless the concentration of nutrients decreases with depth, however, this procedure will not help. Besides, the dredged material must be disposed of in such a way that its nutrients do not leach back into the water.

Harvesting the plants, algae, and fish in eutrophic lakes has also been proposed to remove nutrients. In general, this is not found to have significant effect except in an aesthetic sense. Lakes whose nutrient inflows have been reduced to very low levels might conceivably benefit. If large quantities of nutrient-poor waters are available, it would be possible to flush nutrients out of a eutrophic lake. This is not felt to be a common situation.

Chemicals have been used to control plant growth in lakes, **Reservoirs,** and ponds. Copper sulfate controls algal growth; 2,4-D and arsenite kill rooted plants. Chemical methods require repeated applications; they are also expensive and are rarely environmentally sound practice (see **Pesticides**).

Lakes are sometimes mixed with compressed air, or by pumping, to break up the stratification. The practice follows from the idea that aquatic nutrients such as phosphate are not released under aerobic (adequate oxygen) conditions. Lee has stated that this is not the case and suggests that destruction of the thermocline may actually increase the flow of nutrients from

Weed harvesting in eutrophic Lake Wingra, Wisconsin. The machine was developed by the agricultural engineering department at the University of Wisconsin.

the sediments into the lake waters. He proposes instead oxygenation of the hypolimnion to allow survival of cold-water fish. On the other hand, withdrawal of the hypolimnion waters in highly eutrophic lakes could help since these anaerobic (insufficient oxygen) waters may contain large amounts of nitrogen and phosphorus, released from the sediments.

There seems to be some promise to the technique of adding aluminum and iron salts directly to lake waters to sediment out phosphorus. Iron and aluminum will combine with phosphate to produce insoluble compounds which will settle to the bottom of the lake. Preliminary results indicate removal can proceed without harm to desirable aquatic life, but further study is necessary.

Zoning restrictions which prevent marsh drainage would reduce nitrogen inputs from this source. Good farm practices will reduce or eliminate phosphorus from agricultural drainage in areas where this is an important source. Methods include strip cropping and terracing to prevent soil erosion, and proper timing of fertilizer application or the proper disposal of animal wastes to prevent runoff. Urban runoff does not seem amenable to control by treatment or diversion at the present time because of the large volume involved.

In areas where eutrophication is due primarily to the input of domestic and industrial wastes, tertiary sewage treatment to remove phosphates and possibly nitrates would appear to be an appropriate procedure. (See **Water Treatment and Water Pollution Control** for a description of tertiary treatment.) There are a number of ways to process municipal and industrial sewage to remove phosphorus. A simple method is precipitation with iron salts or alum. This can be done as part of secondary treatment (as in the "activated sludge" process) or in a separate step. Other methods are more complex. It is likely that individual solutions will vary according to locality, but results reaching levels of 80 to 95 percent phosphorus removal seem quite feasible. Costs appear to be a few cents per thousand gallons. Nitrogen removal, on the other hand, is currently an expensive procedure involving ion exchange, gas stripping at high alkalinity, or bacterial denitrification.

The possibility of control of eutrophication by reduction of phosphorus inputs to natural waters leads inevitably to consideration of one of the major sources of phosphorus in domestic sewage, the phosphates in detergents.

THE IMPACT OF PHOSPHATE DETERGENTS

There is no question that detergents contribute large amounts of phosphate to natural waters. Depending on the area, the detergent contribution to total phosphorus input will vary. For Lake Erie the detergent contribution is 30 to 40 percent while for Lake Mendota, Wisconsin, it may be closer to 20 percent. Americans currently use between 2 and 5 billion pounds of detergent a year. It is generally conceded, even by the detergent industry, that, of the phosphorus in municipal sewage, 50 to 70 percent has been contributed by detergents.

DETERGENT COMPOSITION

A typical heavy-duty laundry detergent has the following formulation [9]:

1. *Active ingredient* (20%). This is also called a surface-active ingredient or surfactant. It acts to allow the surface to become wet and removes dirt.
2. *Phosphate builder* (50%). The major function of this ingredient is to "complex" or tie up the ions responsible for water hardness, mainly calcium and magnesium. If uncomplexed, the calcium and magnesium ions would be free to combine with molecules of the surfactant and prevent the surfactant from removing dirt. Thus, phosphate builders allow a much smaller amount of the active ingredient to be used. Phosphates also supply alkalinity, which seems to help in removal of dirt.
3. *Carboxymethyl cellulose* (*CMC*) (0.5%). This ubiquitous ingredient (it is also used in ice cream and other foods to impart a creamy texture) acts to prevent redisposition of dirt on material after the dirt has been removed.
4. *Sodium silicate* (6%). Functions of sodium silicate include inhibition of corrosion and prevention of caking in solid detergents.
5. *Optical brighteners* (0.3%). These compounds absorb ultraviolet light (invisible to the human eye) and reflect it as visible light. They are the substances which make washes "whiter than white."
6. *Perfume.*
7. *Bleach.* Sodium perborate (oxygen bleach) is often added. Magnesium silicate may be included as a stabilizer to even out the release of oxygen.
8. *Enzymes.* Some detergents contain enzymes to help in the removal of protein stains (see **Detergents**).
9. *Fillers.* Sodium sulfate and sodium carbonate are often added as "inert" ingredients; the sodium carbonate does have water-softening ability.
10. *Minor ingredients.* Several minor ingredients may be included, such as preservatives to slow the oxidation of organic materials in the detergent.

Light-duty detergents, intended for hand dishwashing, are all surfactant with little or no builders. Heavy-duty detergents have contained up to 78 percent phosphate by weight (about 15 percent phosphorus), but typical values are about 50 percent phosphate which is 9 percent phosphorus.

DETERGENT CONTROVERSY

The effect of phosphorus in promoting eutrophication cannot be defined exactly for reasons already mentioned. Thus, some room for doubt exists about whether banning phosphates in detergents will achieve any results. The aim of such a ban would be to prevent further eutrophication of our waters and to reverse partially the present conditions.

One effect of Congressional hearings on the subject of detergents has been to give the controversy the definite flavor of an adversary procedure. Thus, environmental groups, most ecologists, and the chairman of the subcommittee, Representative H. S. Reuss, are acting to prosecute the case for removal of phosphates from detergents, while the detergent industry and a smaller group of public officials and scientists defend the presence of phosphates in detergents. It is most interesting to examine the case for and against the removal of phosphates from detergents by recounting the arguments presented by the two sides.

Those who oppose the removal of phosphate from detergents argue that since phosphorus can be removed by sewage treatment we need not worry about the contribution of detergent phosphates. In addition, they express concern that the removal of phosphates from detergents would allow people to relax, in the belief that the problem of eutrophication was solved. Impetus for the building of sorely needed sewage treatment plants, they assert, would then be lost.

Those who favor removing phosphates note that only about 65 percent of the population at present has sewage treatment.[5] The rest have septic systems or no system at all. In many cities there are combined storm sewers and waste water sewers. During periods of rapid flow, the water from such sewers cannot be treated by sewage treatment plants. Thus, much sewage never goes through a treatment plant. It was estimated at the beginning of the decade that 213 million pounds of detergent phosphorus from the sewered population and 157 million pounds from the unsewered population enter the waters each year.

The detergent industry contends that phosphates are strongly bound to soil particles and that tile fields in septic systems will remove the compounds. In actual practice, though, the systems leak and are not that effective. Septic systems in sandy areas may not absorb phosphates at all.

Of the sewered population, about 70 percent have only primary treatment, which removes solid materials and little more. A further 20 percent have secondary treatment by the activated sludge process, which removes soluble organic matter. A number of communities utilize the "trickling filter" process to remove soluble organic matter. (See **Water Treatment and Water Pollution Control** for a description of these processes.) Probably less than 1 percent of the people in the United States are served by treatment plants removing phosphates, although the number of such plants is growing.

The New York State Health Department approved a policy of nutrient removal for metropolitan systems in 1969. It is hoped that this will be extended to include the Finger Lakes–Ontario watershed by 1975.

As noted, methods used for phosphate removal are of several types. A simple and relatively cheap procedure uses alum or ferric chloride to precipitate out the phosphate. This can be done in a separate step before primary treatment or in the activated sludge process. If the removal is done in a separate step before

[5] One estimate of the cost for needed treatment facilities for the entire country is $10 billion.

primary treatment, it will require capital expenditures for pretreatment facilities. If the removal is to be done in the activated sludge step and there is currently no secondary treatment, capital outlays will be needed for the activated sludge facilities as well. Annual operating expenses for phosphate removal using the simple procedure might total about $370 million. A number of other procedures are available. Cost varies according to the procedure. The method chosen will probably be different in various areas of the country. It is generally felt that if detergent phosphates were absent from sewage, the efficiency of treatment procedures would be increased. Substantial savings, possibly reaching one-half to one-third of the total annual operating costs, could be effected in most cases.

To summarize, not all detergent wastes find their way into sewage treatment plants, and most of the population is not currently served by plants able to provide phosphate removal.

Advocates for leaving detergents in phosphates have stated that phosphates are familiar substances and their effect on the health and welfare of the public is well documented. If phosphates are removed, they continue, other substances will be substituted and the public will be subjected to unknown hazards. Carl Klein, Assistant Secretary of the Interior for Water Quality and Research, points out that a substitute for phosphates, used in the same quantity, could result in the dumping of as much as 2 billion pounds of possibly hazardous material into the waters per year.

The possible effect of phosphate substitutes is a serious problem. The experience with nitrilotriacetate (NTA) is illustrative. For a number of years detergent companies and scientists have been looking for chemicals which might prove to be better or less expensive water softeners than phosphates. One such compound developed was NTA. Growing concern over the effects of detergent phosphates coupled with advances in methods of production for NTA stimulated great interest in the late 1960's in its possibilities as a phosphate substitute. Several detergents were marketed in Sweden using NTA as a builder, and a few companies in the United States began to use NTA to substitute for a portion of the phosphate in their products. Preliminary testing uncovered no hazards to human or animal life, and NTA seemed unlikely to act as a significant nutrient source for nuisance algae. Some concern was voiced about the apparent failure of systems to degrade NTA if oxygen was in short supply. Some people also felt that NTA had not been adequately tested from a health standpoint.

In December, 1970, the United States government temporarily banned the use of NTA in detergents on the basis of tests performed by the National Institute of Environmental Health Sciences. The tests showed that NTA, when administered in high doses to test animals in combination with **Cadmium** or methyl-**Mercury,** caused a significant increase in birth defects in the animals. The compound therefore appears to be teratogenic in combination with certain heavy metals. Whether this would be a hazard at the levels NTA and heavy metals are found in water is not clear. At present one can only say that NTA may be a potentially hazardous chemical which could be quite unsuitable for addition to natural waters in large quantities. Any substitute for detergent phosphates must be very carefully tested for such hazards.

It has been suggested that since all areas of the country do not have comparably hard waters, detergent manufacturers could vary the phosphate content of their detergents, using less phosphate for soft water areas. Detergents now appear to be formulated for the hardest water that may be encountered. Yet only 7 percent of the country has waters classed as very hard. The detergent industry responds that water hardness is not uniform within geographical areas; that pockets of hard water exist in many soft water areas. Industry spokesmen feel that regional formulation of detergents would be discriminatory to housewives in these areas. At least one company has attempted to vary the formulation of its machine dishwashing detergents. In 1970, Economics Laboratory, Inc., reported that it was distributing products containing different phosphate concentrations; three formulations of Finish and two of Electrasol were provided.

It is further argued by the industry that clothes contribute a significant amount of hardness to water. A load of heavily soiled clothes could raise water in a soft water area to a medium or even hard level. One solution would entail labeling of detergents as to their phosphate content to help the consumer choose the amount of phosphate necessary for her water hardness and laundry soil. New York State has passed legislation requiring such labeling, and the Federal Trade Commission is considering requiring it on a national level.

Dishwasher detergents present a special problem. No method or substitute has been found to replace high-phosphate products. Commercial dishwashers do run on a caustic solution which manufacturers feel would not be suitable for home use. The suggestion has been made that phosphates could be left in dishwasher detergents, which constitute only a small fraction of the total detergent use. The addition of a dye could prevent use of dishwasher detergents on clothing but not affect dishes. The current price of machine dishwashing detergents would seem to remove these products from the clothes washing market in most cases.

CAN LAKE ERIE RECOVER?

The battle over detergent phosphates rages particularly around their effect on Lake Erie. One of the smaller and shallower of the Great Lakes, Erie is also surrounded by areas of high population density. These factors have combined to cause a rapid eutrophication of the lake in the past 30 years. Nutrient enrichment and altered oxygen levels have in part changed the types of organisms which inhabit the lake. Commercial fisheries report no change in the pounds of fish caught, but the species have changed from whitefish, lake herring, and blue pike to less valuable kinds. In fact, the blue pike is now on the list of **Rare and Endangered Animals.** Floating rubbish and localized bacterial and pesticide chemical pollution are also problems in Lake Erie. One of the purposes for which the International Joint Commission on the Pollution of Lake Erie (etc.) was formed was to determine whether a reversal of the eutrophication of the lake is possible. The commission concluded that reversal is feasible and it recommended removal of phosphate from detergents as a major part of the reclamation program. The arguments for and against removing detergent phosphates bring out many points that should be applicable to lakes less well studied than Erie. The phosphate level in Lake Erie has been steadily increasing. The data in Table E.6 indicate the extent of the increase.

Those who argue for the retention of phosphates in detergents point out first that the present concentration of phosphorus is well above the level Sawyer determined to cause nuisance algal growths in Wisconsin lakes (10 parts per billion). They further calculate that the concentration of the inflow to the lake

TABLE E.6

ACCUMULATION OF PHOSPHORUS IN LAKE ERIE, 1942–1968

Year	Parts per Billion (or $\mu g/l$)
1942	14
1958	33
1959	36
1967–1968	40

contains 132 parts per billion phosphorus.[6] Accepting 70 percent as the proportion of the phosphates in municipal sewage stemming from detergents, the soap and detergent industry states that removing phosphates from detergents would reduce the influent concentration by 58.5 parts per billion. Further removal of all the rest of the phosphate in sewage by tertiary treatment would reduce the concentration by another 30 parts per billion. Thus, 132 minus 58.5 (detergent phosphorus) minus 30 (remaining phosphorus in sewage) equals 40 parts per billion or 40 micrograms per liter, *well above the so-called critical phosphorus level of 10 micrograms per liter* (italics from the soap and detergent people). The influent would thus have a concentration easily able to maintain the lake at or above its current 40-parts-per-billion level. They conclude then that removal of the phosphates from detergents will not help to prevent eutrophication of Lake Erie.

Critics of this argument do not agree with the premise that the annual phosphorus input to the lake has such direct control over the equilibrium concentrations in the lake. If the premise were true, the rise in concentration between 1942 and 1968 should have been much greater, considering the high concentration in the input. Furthermore, if the premise is correct, then decreasing the input from 132 to 73.5 parts per billion by detergent removal should decrease the lake concentration. Since the lake concentration is about 30 percent of the influent concentration, an influent concentration of 73.5 parts per billion ought to result in a lake concentration of about 20 parts per billion. Removal of phosphorus in sewage by tertiary treatment would, by the same reasoning, lower the lake concentration to about 12 parts per billion, a figure lower than the 1942 level. Those favoring detergent controls point out, moreover, that remaining inputs would be due to land drainage, which is controllable by improved farming practices.

The blooms which cover Lake Erie are increasing in area each year. Although there may be a phosphorus level at which blooms start to occur, evidence suggests that the size of the bloom may still be a function of increasing phosphorus concentration. Certain laboratory experiments show that, in addition to being limited in growth below 9 to 10 parts per billion phosphorus, blooms achieve optimal growth at levels of 90 to 180 parts per billion phosphorus.

The Great Lakes Water Quality Agreement, signed by the United States and Canada on April 15, 1972, is expected to reduce the phosphorus input to Lake Erie from 32,000 tons in 1972 to 16,000 tons in 1976—mainly by sewage treatment to remove phosphates. Canada has, furthermore, legislated a decrease in the allowable amounts of phosphate in detergents. Some conservationists feel that if the United States did the same, the annual phosphorus load on Lake Erie could be decreased to 11,000 pounds.

There is another and variable source of phosphates in lakes. As was mentioned earlier, a large portion of the phosphorus input to a lake may be captured in the bottom sediments. For example, the total phosphorus concentration of Lake Minnetonka remains constant despite an annual addition of as much phosphorus as is already present in the lake. Fertilization of lakes with superphosphates also suggests capture in the sediments, since results do not last more than one season. Removal of detergent phos-

[6] The calculation presented by the Soap and Detergent Association, using International Joint Commission data, involves dividing the annual input of phosphates from all sources by the annual inflow of water from all sources. Thus:

$$\frac{\text{phosphorus}}{\text{water}} = \frac{(30{,}100 \text{ tons of phosphorus/year})(2{,}000 \text{ lbs/ton})}{(1.25 \times 10^{12} \text{ lbs/day})(365 \text{ days/year})}$$

$$= 132 \text{ lbs phosphorus}/10^{9} \text{ lbs of water}$$

$$= 132 \text{ parts per billion phosphorus.}$$

phates from inflows may therefore have a much greater effect on the lake concentration than the soap and detergent industry is willing to concede. Seizing on the point that phosphorus is present in bottom muds, however, some scientists state that lake bottoms are a reservoir of phosphorus. If water concentrations are decreased, this reservoir could act to restore former levels.

On the other hand, there is evidence that the phosphorus tied up in bottom sediments is not necessarily readily available. A layer of iron salts in the top of the mud layer helps prevent the dissolving of phosphate salts, although under low-oxygen conditions this layer can break down and allow phosphate release. The potential release of phosphate from sediments means that, once eutrophication has progressed to the point of creating low-oxygen conditions, it may be accelerated. Erie is approaching but has not yet reached this state. To compound the issue, experiments with radioactive phosphorus have shown that once buried under a quarter-inch to a half-inch of mud, the phosphate is essentially no longer available because low-phosphorus muds cover the old higher-phosphorus muds. Thus, it is possible that the older deposits can be discounted. Since differing experimental results have been obtained for different bodies of water, it is treacherous at present to predict the situation for any particular body of water.

Estimates vary for the time necessary for recovery of Lake Erie, following the removal of detergent phosphates and installation of tertiary treatment. The time required for phosphate-free new bottom sediments to be deposited could be 2 to 50 years; flushing of the lake waters might take 6 to 20 years.[7] Arguments for retaining phosphate detergents may cite the fact that phosphorus is recycled between living and dead organisms; however, these commonly ignore the phenomenon of precipitation and sedimentation of phosphorus. Consequently, if lakes which are now eutrophic on account of their phosphorus content receive fewer or no inputs of phosphorus over an extended period, their state of eutrophication may be reversible.

[7] Although the time required for one complete volume of lake water to flow in is much less than this, mixing of inflows and water already present means that removal of nutrients in the lake will take longer than just the time required to fill and empty the lake once.

ALTERNATIVES

The subject of alternatives must be approached from two directions. First, what possible substitutions can the detergent industry make for phosphates? Second, what can an individual do to avoid using phosphate-containing detergents?

The detergent industry is experimenting with products containing various substitutions:

1. *Polyelectrolytes such as carboxymethyl cellulose.* Some controversy exists about whether these are biodegradable. They appear to be better at preventing soil redisposition than phosphates and have previously been included in detergents for this purpose. At present they cost two to three times as much as phosphates (based on sodium tripolyphosphate), and their use would bring a 13 to 21 percent cost increase to the consumer for dry detergents. No increase would be expected for the substitution of cmc in liquid detergents, which currently contain a more expensive phosphate builder, tetrapotassium pyrophosphate. One scientist contends that consumers can actually afford to pay more for phosphate-free detergents since they will be saving money on phosphate removal or other water pollution abatement procedures.
2. *Sodium silicates.* These compounds cost less than phosphates now in use.
3. *Carbonates.* Washing soda, or sodium carbonate, has been used for a long time as a water softener. Magnesium and calcium carbonates are precipitated out as small solid particles which sink to the bottom of the wash. This

process may not be as desirable as the formation of complexes which remain in solution and disappear with the spent wash water. Also, the effect of the discharge of carbonates in quantities equal to present phosphate levels is not known. Some scientists feel the high alkalinity and the carbonate ions might alter the solubility of phosphates in water.

4. *Borates.* These compounds have also been in use for some time. Definite concern has been voiced about possible borate pollution since boron may be utilized as a plant nutrient and is toxic to both plants and animals at *high* concentrations.

The Environmental Protection Agency is funding research on phosphate-free laundry detergents. One project, for instance, involves the use of surfactants which do not combine with the calcium and magnesium ions in hard water. Thus, they do not require the addition of a builder, such as phosphate, to soften the wash water.

The market for nonphosphate detergents seems to be burgeoning, if the number of different products is any indication. For purposes of illustration, here is a list of the ingredients of two nonphosphate detergents sold in New York state in 1971:

20 Mule Team Detergent (solid)
- sodium carbonate monohydrate, sodium polysilicate, sodium gluconate — 86.1%
- wetting and emulsifying agents — 12.0%
- soil suspenders, brighteners, perfume — 1.9%

Miracle White (solid)
- sodium linear alkylate sulfonate
- sodium silicate
- sodium carboxymethyl cellulose
- fluorescent whitening agents
- sodium sulfate
- perfume
- anticaking agents
- sodium carbonate
- sodium borate
- sodium perborate

The housewife choosing a nonpolluting detergent would do well to be somewhat wary, however. (*How* is not clear unless ingredients are required by law to be given on labels.) Some of the new detergents are highly alkaline and may cause the buildup of inorganic ash in clothes. The ash may lead to skin irritation and also may harbor bacteria. The Food and Drug Administration recently required one chemical company to relabel its no-phosphate detergents with a warning to consumers. The products contained sodium metasilicate, which FDA tests showed to be "toxic, corrosive to intact skin and the cause of severe eye irritation."

There is, besides, no assurance that these new formulations will not contribute to eutrophication in some as yet undetermined manner. As mentioned above, carbonates and borates may not be any more desirable than phosphates as additions to natural waters, in the quantities in which they can be found in clothes washing products.

The consumer has several other alternatives open. If one lives in a soft water area, one can use soap products like Ivory or Duz soap for all except very soiled or greasy clothes. Even in hard water areas, ion-exchange water softeners will produce water soft enough for soap to clean well. Washing soda can also be added to soften water to allow the use of soap.

Despite the claim of a few detergent manufacturers that there is not enough tallow to produce soap for the whole country, it appears that the soap industry could meet the challenge. Approximately 3 billion pounds of tallow are used annually for cattle feed and export while 2.5 billion pounds is the estimate of the amount needed for the manufacture of enough soap to replace detergents. Vegetable fats might provide another source for soaps.

Finally, a method tested in Sweden might prove helpful. There, a small amount of detergent combined with a larger amount of soap in the washer gave results at least as good as detergent alone. This method requires high temperatures and so might not be suitable for some synthetic fabrics.

THE CHANGING SCENE

Several state legislatures and local governments have assessed the evidence available and decided that detergent phosphates must be removed or limited. For instance, Indiana outlaws detergents containing more than 12 percent phosphorus as of January 1972 and more than 3 percent phosphorus as of January 1973. Maine prohibited the sale of detergents containing more than 8.7 percent phosphorus after June 1972 and New York after January 1972.

The detergent industry has in some cases begun to lower the phosphate content in detergents. The response is partly due to these legislative restrictions, as well as consumer pressure, but it is also the result of competition from nonphosphate detergents. Tables E.7 and E.8 compare the phosphorus content of a few detergents sold in a supermarket in New York state in August 1971 and August 1972. The amounts given in both tables appeared on the detergent boxes. The brands listed in Table E.7 were chosen only from the group that contained phosphorus. Some brands, including 11 soap formulations and so-called nonpolluting detergents, claimed to contain no phosphorus at all.

The industry is much concerned about the number of regional restrictions which are being imposed and hopes to have a uniform federal law passed specifying the maximum phosphate level allowable in detergents. It apparently would like to see the level set at 8.7 percent

TABLE E.7

PERCENTAGE OF PHOSPHORUS IN CLOTHES WASHING COMPOUNDS

		Phosphorus	
Brand	Phosphate (August 1971)	*August 1971*	*August 1972*
All, regular	51.0	9.7	8.7
Axion	46.0	8.7	8.7
Biz	43.4	8.2	8.7
Bold	46.0	8.2	8.7
Burst	59.3	11.2	8.7
Cheer	46.0	8.7	8.7
Cold Power	46.0	8.7	8.7
Cold Water All, liquid	25.0	4.8	4.8
Dash	77.2	14.6	8.7
Duz detergent	46.0	8.7	8.7
Oxydol	46.0	8.7	8.7
Rinso	46.0	8.7	8.7
Salvo	78.1	14.8	6.1
Tide	46.0	8.7	8.7
Wisk, liquid	18.0	3.5	3.5

TABLE E.8

PERCENTAGE OF PHOSPHORUS IN DISH WASHING COMPOUNDS

Brand	August 1971	August 1972
Automatic dishwasher detergents		
All	12.9	7.5
Calgonite	12.7	8.7
Cascade	12.9	8.7
Electrasol	5.6	7.1
Finish	7.3	10.9
Hand dishwashing liquids		
Gentle Fels	0	0
Ivory	0	0
Joy	0	0
Thrill	2.3	2.3

phosphorus, still close to 50 percent phosphate builder. This level would not be expected to be satisfactory to scientists and conservationists who feel detergent phosphates make a large and unnecessary contribution to eutrophication.

REFERENCES

1. G. Fred Lee, "Eutrophication," in Supplement to the *Encyclopedia of Chemical Technology*, Wiley, New York, 1970.
2. G. Fred Lee, "An Approach to the Assessment of the Role of Phosphorus in Eutrophication," Paper presented before the Division of Water, Air and Waste Chemistry, American Chemical Society, Los Angeles, Calif., April 1970.
3. G. Fred Lee, "Factors Affecting the Transfer of Materials Between Water and Sediments," *Literature Review No. 1*, Eutrophication Information Program, Water Resources Center, University of Wisconsin, Madison, July 1970.
4. *Eutrophication in Coastal Waters: Nitrogen as a Controlling Factor*, U.S. Environmental Protection Agency, Water Pollution Control Research Series, Report EP 1.16.16010 EHC, December 1971.
5. "Great Lakes Water Treaty Signed," *Science*, April 1972, p. 390.
6. H. J. Stueck, *New York Times*, July 17, 1972, p. 15M.
7. T. E. Brenner, "Biodegradable Detergents and Water Pollution," in *Advances in Environmental Sciences*, vol. 1, ed. J. N. Pitts and R. L. Metcalf, Wiley, New York, 1961, pp. 147–96.
8. "Phosphates in Detergents and the Eutrophication of America's Waters," *Hearings Before a Subcommittee of the House Committee on Government Operations*, 91st Congress, 1st Session, December 15–16, 1969.
9. *Phosphates in Detergents and the Eutrophication of America's Waters*, Twenty-third Report by the Committee on Government Operations, Washington, D.C., 1970.
10. *Toxological and Environmental Implications of the Use of NTA as a Detergent Builder*, Staff Report prepared for the Senate Committee on Public Works, Washington, D.C., December 1970.
11. Report to the International Joint Commission on the Pollution of Lake Erie, Lake Ontario and the International Section of the St. Lawrence River, vol. 1, Summary, 1969.
12. J. A. Shapiro, "Statement on Phosphorus," *Journal of the Water Pollution Control Federation* 42 (1970): 772.
13. K. M. Mackenthun and W. M. Ingram, *Biological and Associated Problems in Freshwater Environments*, U.S. Dept. of the Interior, Federal Water Pollution Control Administration, 1967, pp. 103–42.
14. R. F. Legge and D. Dingledein, "We Hung Phosphates Without a Fair Trial," *Canadian Research and Development*, March–April 1970, p. 19.
15. K. M. Mackenthun et al., "Nutrients and Algae in Lake Sebasticook, Maine," *Journal of the Water Pollution Control Federation* 40, no. 2 (1968): R72.
16. C. N. Sawyer, "Some New Aspects of Phosphates in Relation to Lake Fertilization," *Sewage and Industrial Wastes* 2, no. 6 (1952): 768.
17. Summary Report, Conference on Pressing Problems Related to Basins, Geneseo, N.Y., June 4–5, 1970.
18. *Science Journal*, January 1971, pp. 12–13.
19. *Chemical and Engineering News*, January 4, 1971, pp. 15–16.
20. *Chemical and Engineering News*, September 14, 1970, pp. 50–54.
21. Marjorie Hunter, *New York Times*, October 28, 1971, p. 28C.
22. G. K. Ashforth and B. Calvin, "Safety

Evaluation of Substitutes for Phosphates in Detergents," Paper presented before the International Association on Water Pollution Research, Conference on Phosphorus in Fresh Water and the Marine Environment, University College, London, April 11–13, 1972.

23. Paul S. Welch, *Limnology*, 2nd ed., McGraw-Hill, New York, 1952.
24. Franz Ruttner, *Fundamentals of Limnology*, trans. D. G. Frey and F. E. J. Frey, University of Toronto Press, Toronto, 1963.
25. J. W. G. Lund, "Eutrophication," *Proceedings of the Royal Society of London* 180 (1972): 371.
26. R. A. Vollenweider, "Scientific Fundamentals of the Eutrophication of Lakes and Flowing Waters, with Particular Reference to Nitrogen and Phosphorus as Factors in Eutrophication," DAS/CSI/68.27, Organization for Economic Cooperation and Development, Directorate for Scientific Affairs, Paris, 1968.
27. G. E. Hutchinson, *A Treatise on Limnology*, vol. 1 (1957), *Geography, Physics and Chemistry*, vol. 2 (1967), *Introduction to Lake Biology and the Limnoplankton*, Wiley, New York.
28. L. E. Kuentzel, "Bacteria, Carbon Dioxide and Algal Blooms," *Journal of the Water Pollution Control Federation* 41 (1969): 1737.
29. C. N. Sawyer, "Basic Concepts of Eutrophication," in *Readings in Conservation Ecology*, ed. G. Cox, Appleton-Century-Crofts, New York, 1969, p. 462.

F

Fossil Fuel Power Plants

See also Air Pollution; Electric Power; Fuel Resources and Energy Conservation; Nuclear Power Plants; Reservoirs; Thermal Pollution.

Fossil fuel power plants include power plants fired by coal, oil, and natural gas. The emphasis here is on the environmental effects of these plants, which contribute to both **Air Pollution** and **Thermal Pollution.** Fossil fuels are utilized to boil high-purity water for steam, which is used to turn the turbine for power generation (see Figure F.1). Spent steam from the turbine is condensed and cooled in a heat exchanger, and the water is recycled to the boiler. The cool water for the heat exchanger is drawn from and returned to the ocean, a lake, or a river. In consequence of its use, the cooling water is returned to the body of water at an elevated temperature. Alternatively, the heated water may be cycled through a cooling tower or cooling pond prior to discharge or reuse. Figure F.1 shows the principle of the power plant, which except for the fuel very much resembles the boiling water nuclear reactor.

THERMAL IMPACT

The water returned to the lake or river at an elevated temperature may increase the average temperature in a portion of the body of water, bringing about a number of biological and physical effects (see **Thermal Pollution**). The temperature of the returned cooling water depends on the thermal efficiency of the plant. Thermal efficiency is the fraction of the heat

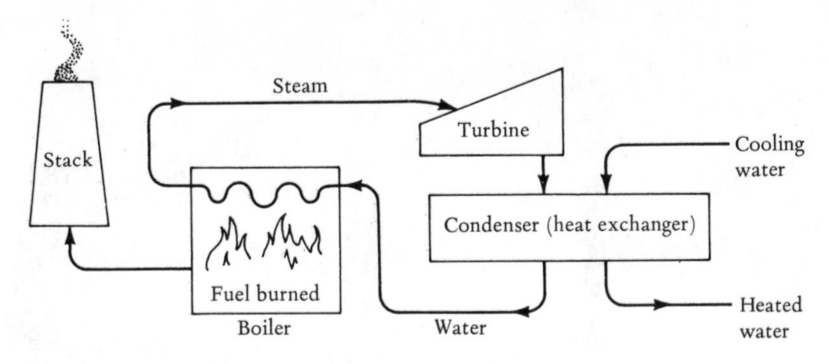

FIGURE F.1
OPERATIONS AT A FOSSIL FUEL POWER PLANT

energy generated which is converted to electrical energy. If a conventional steam plant reaches 40 percent efficiency (which is about an upper limit), 40 percent of the heat produced is being converted to electric power and 60 percent of the heat is wasted to the environment. Some of the heat appears in the hot combustion gases vented through the smokestack. The remainder appears in the cooling water dumped back in the river. New fossil fuel plants do reach 40 percent efficiency by superheating steam; older versions are less efficient.

Today's nuclear plants, in contrast, have thermal efficiencies in the range of 30 to 32 percent. Taking 40 percent for the best conventional fossil plant and 32 percent for the best nuclear power plant,[1] we find that the rate of waste heat production per unit of electric power is about 40 percent greater for the latter. In addition, the entire impact of wasted heat from the nuclear plant is directed toward the water because virtually no heat is vented through the nuclear plant's stacks. The conventional plant may lose 25 percent of the *waste* heat up the stack. We can therefore conclude that **Nuclear Power Plants** put nearly 90 percent more waste heat into the water per unit of electrical energy than conventional fossil plants do.

[1] Superheating of steam in nuclear power plants may be some time away, although two prototypes exist.

AIR POLLUTION IMPACT

The air pollution from a conventional power plant depends on the specific fuel being utilized and on the control devices used to diminish emissions. The sources of each of a number of pollutants are traced in the following discussion, and the contributions of the several types of conventional power plants are noted.

SULFUR OXIDES

Coal combustion for power generation accounted for about 40 percent of the 28 million tons of sulfur oxides emitted in 1966. Low-sulfur coals, when available, are more costly but are commonly tied up by long-term steel industry contracts. Oil-fired boilers in power plants produced much less sulfur oxide than coal (about 5 percent of the 1966 total). Natural gas contributes very little in the way of sulfur oxide emissions. In general, gas is more costly for power production than oil, and oil is more costly than coal. Three processes for cleaning stack gases of sulfur oxides have evolved to the stage of preliminary use, but their application is

Coal-fired power plant in Madison, Wisconsin.

not widespread. Sulfur oxides have been firmly linked to respiratory ailments in the population. Other effects and possible controls are discussed in detail in **Sulfur Oxides.**

NITROGEN OXIDES

Of the nearly 21 million tons of nitrogen oxides produced in the United States in 1968, about 90 percent arose from the combustion of fossil fuels. Motor vehicle operation contributed 35 percent of the total, but fossil fuel power plants supplied about 20 percent of the total. Techniques for controlling the emission of nitrogen oxides at power plants are not well advanced. Nitrogen dioxide plays an important role in photochemical air pollution, acting to stimulate atmospheric reactions to produce such pollutants as ozone, aldehydes, and PAN compounds. Additional information on the effects and the controls of nitrogen oxide emissions may be found in **Photochemical Air Pollution.**

PARTICULATE MATTER

The combustion of fossil fuels for power generation accounted for slightly over 25 percent of the total particle emissions in the United States in 1966, and most of this quantity stemmed from coal-fired plants. Combustion at industrial sources made up the dominant portion of the remaining 75 percent. As with sulfur oxides, particle levels have been linked to health effects, in this case to stomach cancer and respiratory illnesses. Numerous techniques are available for control, but the electrostatic precipitator appears the prime choice for removing fly ash from the gas stream of coal-fired power plants. Greater detail on effects and control may be found in **Particulate Matter.**

CARBON DIOXIDE

Power generation is relatively innocent with regard to the emission of hydrocarbons and **Carbon Monoxide**, but **Carbon Dioxide** is another matter. In the mid-1960's, speculation abounded about the buildup of carbon dioxide in the earth's atmosphere. A "greenhouse effect" warming the earth was envisioned. Nuclear power advocates appear fond of citing this concern and pointing out how nuclear power can save the day. However, they do not usually cite the potential for a counteracting cooling trend for the earth due to particulates in the atmosphere. Particles are evidently reflecting solar radiation from the earth to a certain extent. Do these effects cancel? We don't know, but it is doubtful that either issue is of current importance to the debate over the preferability of nuclear versus fossil fuel power, given the more serious and immediate issues raised above.

RADIOACTIVE EMISSIONS

When comparing nuclear power and fossil fuel power, the advocates of nuclear power do point to an emission from *coal*-fired and *oil*-fired power plants which until the last decade had gone unnoticed. There are radioactive elements in the stack gases of these plants; the elements are radioactive daughters of natural components of coal and oil. Appearing in the fly ash in the stack gases of coal-fired and oil-fired plants are radium-226, radium-228, thorium-228, and thorium-232. Although these substances appear in the fly ash from oil-fired plants, the emissions are nearly three orders of magnitude less than those from coal plants. This difference is due to the small amount of ash emanating from oil-fired plants and the fact that the parent radioelements are present in oil at much lower concentrations.

As to the radioactivity of emissions from coal plants, it is useful to compare them to the releases from nuclear plants. Speaking roughly, the coal plants' gaseous release of radioactivity seems to exceed that from the nuclear pressurized water reactor (PWR) by several orders of magnitude. It is less than the release from the nuclear boiling water reactor (BWR) by several orders of magnitude. Since the BWR's release is generally less than a percentage point of the maximum allowable, the health significance of such emissions is considered to be very small. But the realization that radiation is emitted by our longtime smoky friend does bring some perspective to the contest between nuclear and conventional power.

REFERENCE

1. "The Environmental Effects of Producing Electric Power," *Hearings Before the Joint Committee on Atomic Energy,* 91st Congress, 1st Session, October–February 1969–1970.

Fuel Resources and Energy Conservation

It has been alleged that the United States is facing an energy crisis. The word *crisis* has been used to imply a shortage of fuel resources. In fact, what the nation is experiencing is a phenomenon better termed "an energy policy conflict." Is there a current shortage of fuel *resources* either in the United States or in the world? No. Fuel resources are vast. But in the last several years there have been recurring shortages of deliverable supplies of certain fuels in the United States. Many distribution companies are not taking new customers for natural gas. Residual oil, a petroleum product used as a fuel in electric generating stations, is mostly imported. Number 2 fuel oil used in home heating has been in insufficient supply in New England and the upper Midwest. Gasoline rationing has occurred in one section of the country and is threatened in others. Nevertheless untapped reserves of the fossil fuels will be adequate far into the future. Coal, in particular, is abundant.

What are the forces at play which have caused the current shortfall in fuel supplies? First, the demand for **Electric Power** has been increasing dramatically, and the generation of electrical energy from the heat energy of fuels is at this time inherently inefficient. When a fossil fuel is burned at the power plant, no more than 40 percent of the heat energy produced is converted to electrical energy. As the utilization of electric energy grows, wastefulness grows. Second, within the framework of electric power growth, state plans to control air pollution have led to a preference on the part of electric utilities for certain fuels. Because of the need to decrease **Sulfur Oxides** from power plant emissions to meet the recently established **Air Quality Standards,** utilities have been turning increasingly away from coal to low-sulfur residual oil and to natural gas, a virtually sulfur-free fuel. The shift is explained by several facts. First, although naturally occurring low-sulfur coals exist, they are frequently inaccessible from the standpoint of transport costs. Low-sulfur coals can be manufactured, but the process is not well advanced and it appears that only about half of the sulfur can be removed. Second, the processes which cleanse stack gases of sulfur oxides are still largely in the demonstration stage, and few utilities are willing to risk installing expensive equipment when its success is uncertain. The potential application of other options is even more remote.

Thus, in the short run, electric utilities have three fundamental fuel options when firing their steam boilers. The nuclear option (see **Nuclear Power Plants**) has been chosen by many, but licensing delays have an inhibiting effect. The other two options are low-sulfur residual oil and natural gas. Although the production of residual oil by petroleum refiners has steadily declined in the United States, states on the eastern seaboard have been able to import residual oil practically without limit for use in electric utilities. Unfortunately, of the residual oils from the two principal suppliers, Venezuela and the Middle Eastern nations, the oil from the Middle East is lower in sulfur. It is also thought to be the less "secure" supply.

The third option is natural gas, historically a low-priced fuel and a logical choice for utility application because its combustion produces virtually no **Particulate Matter** or sulfur oxides. However, it has recently been unavailable to new customers, whether residential or industrial. Its lack of availability for new uses stems from its price, perhaps regulated at too low a level, and a burgeoning demand.

The three fossil fuels—coal, oil, and natural gas—form together with hydropower and nuclear power the energy base of industrial societies. In 1971 petroleum, used for power generation and heating as well as for transport, provided about 44 percent of the nation's energy. Coal, once the motive source of railroads and the source of heat for homes, now is principally used in electric power generation and in the iron and steel industry. Coal furnished 18 percent of our energy. Gas, used widely to heat residences and industries and to generate power, accounted for 33 percent of the nation's energy budget. Hydropower accounted for 4 percent and nuclear about 1 percent.

NATURAL GAS

RESOURCES

The judgment of the viability of the oil- and gas-producing industries has historically been based on the ratio of proven reserves (R) to annual production (P) of the fuel (R/P). Table F.1 provides historical data on production, consumption, and proved reserves of natural gas through 1970. In addition, production and consumption are projected through 1985, and the imbalance is indicated. Demand, of course, is a function of price. These figures appear to be

TABLE F.1

SUPPLIES OF NATURAL GAS IN THE CONTIGUOUS UNITED STATES, 1966–1985 (IN TRILLIONS OF CUBIC FEET AT 14.73 PSIA)

	1966	1967	1968	1969	1970	1975	1980	1985
Total production	17.5	18.4	19.3	20.6	21.8	24.7	20.4	18.5
Consumption	17.9	18.9	19.9	21.3	22.6	29.8	34.5	39.8
	Imported from Canada					*Deficit*		
Difference	0.4	0.5	0.6	0.7	0.8	5.1	14.1	21.3
Production for the interstate market	11.1	11.8	12.6	13.4	14.1			
Reserve additions	19.2	21.1	12.0	8.3	11.1			
Year-end reserves	286.4	289.3	282.1	269.9	259.6			
Reserves to production ratio	16.4	15.8	14.6	13.1	11.9			

From testimony of J. Nassikas, in "Fuel and Energy Resources, 1972," *Hearings Before the Committee on Interior and Insular Affairs*, Parts 1 and 2, U.S. House of Representatives, 92nd Congress, 2nd Session, 1972; "Natural Gas Regulation and the Trans-Alaska Pipeline," *Hearings Before the Joint Economic Committee*, U.S. Congress, 92nd Congress, 1972.

projections of the demand which would occur if price changes did not alter the competitive position of natural gas.

Reserve additions have been falling, as has the R/P ratio. Furthermore, the producers have not been commiting new gas to the interstate market. There have been new commitments within states, however. This is because the Federal Power Commission (FPC) sets the price to be paid for gas in interstate commerce but does not regulate the price of gas produced and consumed in the same state. In the free market situation which results, the gas finds a higher price than it does in the regulated interstate market. Hence, industries in producing states (e.g., Texas, Louisiana, Oklahoma) may siphon off new supplies because of higher internal prices. The fraction of natural gas consumed in intrastate sales was about one-third in 1970. The intrastate market has consumed nearly 30 percent of the increase in production since 1966. The deficit indicated in 1975 and beyond is far too large to be supplied by Canada.

Against this background of growing demand and declining reserve additions one must examine estimates of the potential as well as proved reserves of natural gas (see Table F.2). It has been alleged that gas-producing companies hold greater proved reserves than the statistics have indicated. The reasoning behind the allegation is that the producers are aware that large price increases would result if the FPC no longer set the price for gas sold interstate. Such decontrol may not be far off. In the interim, it is argued, instead of selling more gas at the lower price, producers are not committing new supplies. The present and former chairmen of the FPC think the allegation is untrue, and a study of reserves by the FPC buttresses the contention of falling reserves. As shown in Table F.2, figures for the proved reserves have been supplied by the American Gas Association (AGA), an industry organization. Access to individual company estimates to verify data has not been available; the company figures have been regarded as proprietary.

TABLE F.2

PROVED AND POTENTIAL NATURAL GAS RESERVES IN THE UNITED STATES, 1970 (IN TRILLIONS OF CUBIC FEET)

		Potential Categories[b]			
	Proved Reserves[a]	*Probable Reserves*	*Possible Reserves*	*Speculative Reserves*	Total of Potential Reserves
Onshore		179	227	207	613
Offshore		39	99	100	238
Total for the lower 48 states	259	218	326	307	851
Alaska	31	39	61	227	327
Total	290	257	387	534	1,178

From testimony of W. Rogers and J. Nassikas, in "Fuel and Energy Resources, 1972," *Hearings Before the Committee on Interior and Insular Affairs*, Parts 1 and 2, U.S. House of Representatives, 92nd Congress, 2nd Session, 1972.

[a] From the American Gas Association.
[b] From the Potential Gas Committee.

PRICE REGULATION

The rationale for regulating the price of gas while no limit is set on coal or oil prices derives from the nature of the supply mechanism of natural gas. Natural gas usually arrives at a city through the pipeline of a single supplier. Investment in expensive gas-fired equipment such as furnaces, stoves, and water heaters effectively makes gas customers the captive of the supplier. To prevent exploitation of the consumer, the Federal Power Commission has set "just and reasonable" rates at the wellhead for gas moving in interstate commerce. (Presumably competition at the local level by suppliers of oil acts to protect the consumers' interests and makes regulation unnecessary.) The rates were designed to provide to a producer a fair return based on the average costs of production in his location. The return aimed for by the FPC was on the order of 12 percent, which is fairly standard for corporations in the United States.

The response of the producers to price regulation since the FPC began setting prices at the wellhead has been consistent. Beginning in 1955, the industry warned of impending shortages, claiming that the prices set would not give producers sufficient incentive to find new gas. Interestingly, the shortage did not arrive until the late 1960's. The onset of the shortage corresponds with a growing demand for natural gas as an environmentally clean fuel. This explains in part the depletion of existing proved reserves but does not offer a satisfactory explanation of the rapid decline in reserve additions.

To explain why reserve additions have declined, industry spokesmen point to a long-term decrease in the number of wells drilled because of a lack of economic incentive. In 1970 and 1971 the FPC allowed increases in the wellhead price of gas to promote increased supplies and reserve additions. For example, the increase above the previously allowed price for new gas in the southern Louisiana area was 6 cents per thousand cubic feet, raising the wellhead price to 26 cents per thousand cubic feet.

Would nationwide decontrol of gas prices lead to sufficient exploratory activity to meet demands? This question has two aspects. If prices rise, demands will fall. Hence, the price rise could have two effects—increase in supply and a decrease in projected demand. Should the price of interstate natural gas be *fully* decontrolled? How high would prices rise? What economic impacts would this have on the consumer? Are there other mechanisms to stimulate exploratory activity?

The Federal Power Commission in its Rule 441 *proposes* an optional rate-making procedure to apply to producers' sales of *new gas* to pipeline companies. New gas is gas over and above the amount currently allotted to such sales. Under the rule, the price agreed upon by producer and pipeline would be reported to the FPC, which would make sure it is in the public interest. In theory, the FPC may disapprove. In fact, the Commission's insistence on a rate lower than the going price in the state of production (the intrastate rate) would likely induce the producer to sell his gas within the state. Thus, if the goal is to stimulate interstate flows, the Commission will find it difficult to insist on rates lower than the unregulated price within the state. It is this *proposed* rule which producers are alleged to be waiting for.

In 1973, the chairman of the FPC indicated that he favored deregulation for new gas and gradual decontrol for flowing gas. Shortly thereafter the FPC began proceedings to establish a uniform national ceiling price for natural gas, apparently willing to leave behind the old method of area rate-making.

REMEDIES FOR THE SHORTAGE

Political and Economic Alternatives

Numerous plans to alleviate the shortage, in addition to the decontrol advocated by the gas industry, are proposed. Prominent among them is some form of end-use control. It is argued that residential uses should have preference over industrial uses because it would be uneconomical for the homeowner to install equipment to control the particles and sulfur oxides produced in the combustion of home heating oil. The utility or industry would then buy a fuel either more costly than gas or more expensive from the standpoint of emissions control.

A method proposed by Charles Frazier[2] to achieve end-use control is a tax on the use of natural gas. The tax could be selectively applied to large industrial users such as utilities, instead of to residential uses. Such a tax would have the desired effect of influencing end use, and it might help to diminish the use of natural gas in electric generation in the producer states. Suggested uses of the revenue are exploration for new gas and research and development in technologies for producing "artificial" gas.

One of the most widely advocated methods to encourage gas exploration and production is simply to deregulate, to allow the free market to determine the price of natural gas. With the higher prices thus achieved, advocates argue, sufficient development will follow. The argument is countered by the observation that oil is unregulated, protected in the past by a quota system and now by a fee system, and yet is reported in short supply. Between the opposing fronts are those who argue for experimental decontrol.

Lee White, a former Chairman of the Federal Power Commission, has suggested the formation of a public corporation, a national, government-owned energy resources corporation, to take up the slack in exploration and development where profits are insufficient to attract the private sector.[3] The corporation's function would be "to explore for and develop petroleum resources on the publicly owned lands." Frazier

---

[2] Testimony before Congress [3].

[3] Testimony before Congress [3].

proposed that the taxes he suggested might be channeled to such a corporation to finance its activities. In Britain, a public corporation, the Gas Council, controls all transmission of gas. It even engages in exploration in the North Sea and is associated with Amoco in one venture. Another suggestion of White is "field pricing," which would allow an individual producer to come forward with his costs and show that to earn his desired return he will need a higher price.

Another proposal is to prevent large volume uses within the state of production and to channel more gas to the interstate market. The proposal would necessitate extending the jurisdiction of the FPC to intrastate sales. Such additional powers for the FPC might require an act of Congress. If this were the *only* mechanism employed, it would likely be ineffective, for it provides neither economic incentive to the producer nor stimulus for greater exploration.

Since about 25 percent of our natural gas is discovered in the search for oil, gas finds could increase under a policy designed to encourage domestic petroleum exploration. This could be achieved by modifying current taxation policy as it affects oil companies. Royalties paid to states in the United States are treated as business deductions. Presently royalties paid to other nations for the extraction of oil are considered taxes paid by the corporation, and companies are given a *tax credit* for them, producing greater revenues. If royalties paid both here and abroad were treated alike, as deductions, some of the stimulus for foreign oil ventures would be removed.

This large array of political and economic alternatives to influence the supply of natural gas is matched on the technological side with an equally long list.

Technological Alternatives

About 31 trillion cubic feet of gas are estimated to be associated with (actually dissolved in) the oil find at Prudhoe Bay, Alaska. Another 327 trillion cubic feet of potential gas reserves are estimated to exist there. Additional quantities are likely in nearby portions of Canada. Apparently the least expensive method to bring this gas to the areas of need in the United States (the Midwest and the Northeast) is by pipeline overland *through* Canada. The politics of the ultimate decision on the pipeline to bring *oil* from Alaska may influence whether the gas pipeline will be built. In turn, the need for an overland gas pipeline ought to influence the route chosen for the oil pipeline.

Nuclear explosive stimulation of natural gas deposits is the object of investigations by the U.S. Atomic Energy Commission (AEC). Low permeability reservoirs of natural gas are found in the Rocky Mountain area; an estimate by the Bureau of Mines places the reserves there at 317 trillion cubic feet. Two tests have been conducted thus far. A 26-kiloton device was detonated in Project Gasbuggy in 1967. About two years later Project Rulison utilized a 40-kiloton device. The gas made available by this technique contains low levels of radioactivity, principally in the form of tritium. Given the application of natural gas to residential heating, large populations could be exposed to these low levels, if such gas were utilized. The factor of public acceptance is significant.

Natural gas also may be imported, but to bring it by tanker, it is most economical to liquefy it first. At the port of arrival, the liquefied natural gas (LNG) is regasified for entrance to the pipeline. In the short run, this alternative approved by the FPC will be filling the excess demand on the East Coast. Deliveries on short-term contracts have already been made. Approval by the FPC of long-term agreements by Distrigas and Columbia LNG corporations with Algeria were pending in May 1973.

It is also possible to manufacture pipeline-quality gas from oil. Given that LNG does not fall under the oil import quota system, it is debatable whether oil imported for gas manufacture will be subjected to the quota. While such manufacture is new, synthetic gas has been made from naphtha in England and can be made from other liquid hydrocarbon feedstocks.

Another process with the potential to provide high Btu (high heat content) gas suitable for pipelines is coal gasification. The use of coal for producing a gaseous fuel is not new. In the 1920's, a by-product of coking called "illuminating gas" or "water gas" was in use in the United States. The gas, unfortunately, contained **Carbon Monoxide** and is no longer considered suitable. New processes for coal gasification produce a sulfur-free fuel whose methane content approaches that of natural gas. In 1971, El Paso Natural Gas announced plans for a coal gasification plant in northwestern New Mexico that would deliver 250 million cubic feet per day of high Btu gas. Another venture of similar and perhaps greater capacity is projected in New Mexico by a trio of firms including a pipeline company, a mining company, and a utility.

A number of other processes are under development. The Department of the Interior and the American Gas Association are cooperating in pilot plant studies of promising processes. Plans call for an effort costing $30 million per year for four years. The projects mentioned above, which use the Lurgi process, could be on stream by 1975. Other processes may not be ready for commercial application until 1980. Given the massive coal resources of the nation, coal gasification is regarded by many as an important way to supplement energy resources. The prices of gas under these various technological alternatives are indicated in Table F.3.

Typical prices in 1972 charged by pipelines to distributors were 40 cents per 1,000 cubic feet in Chicago and 50 cents in New York City. The

TABLE F.3

OUTLOOK FOR NATURAL GAS PRICES (PER 1,000 CUBIC FEET)

Source of Gas	Currently Estimated Price Range	Earliest Estimated Period of Availability
Alaska via pipeline	80¢–$1.00	1980
Coal		
Lurgi process	85¢–$1.10	1975
Other processes	50¢–90¢	1980
Naphtha and other hydrocarbons	$1.10–$1.80	1975
Liquefied natural gas		
Short-term	$1.10–$2.20	Available now
Long-term	65¢–85¢	Available now

From "Natural Gas Policy Issues," *Hearings Before the Committee on Interior and Insular Affairs*, Parts 1 and 2, U.S. Senate, 92nd Congress, 2nd Session, 1972.

prices of these alternatives may reach three times the price of domestic natural gas. The FPC approved an LNG import plan in 1972 in which prices would be at least double the current rates for domestic gas. This does not, however, mean that the price of gas to the consumer will double. Because the gas will be mixed with flowing gas, the price of the mixture will only increase in proportion to the fraction of LNG included. Since the contribution of LNG to the total will be small, the price rise will not be quite so precipitous. This method of determining price is called "rolled-in" pricing. It is argued by some that *new* industrial customers should pay for gas at the high price of the LNG so that individual consumers are not unfairly burdened and so that demand growth does not continue accelerating. In the face of the exceedingly expensive importation of gas and the Commission's approval

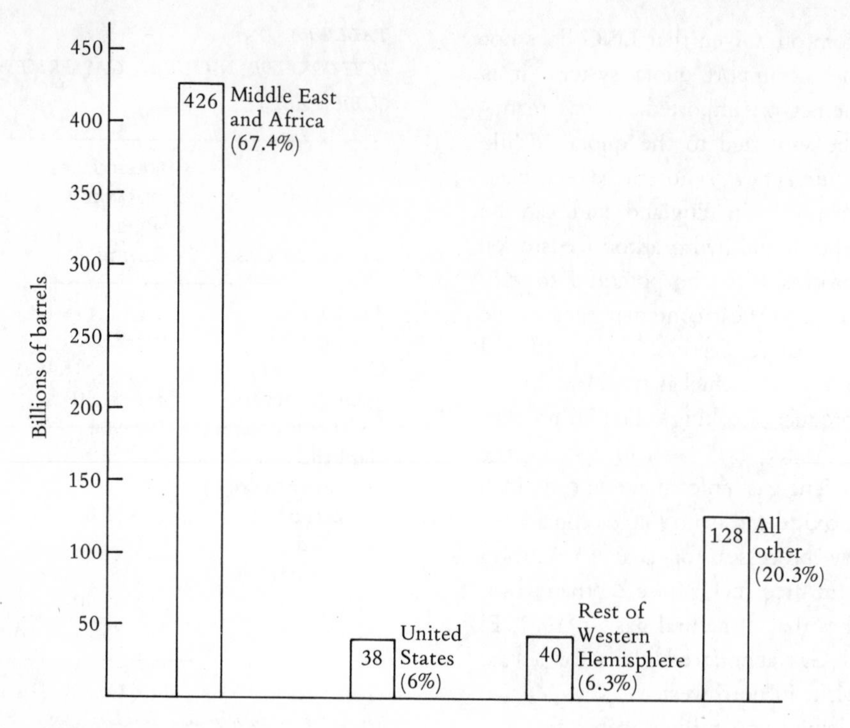

FIGURE F.2

PROVED WORLD PETROLEUM RESERVES, JANUARY 1, 1972

Data from testimony of F. Ikard, "Natural Gas Regulation and the Trans-Alaska Pipeline," *Hearings Before the Joint Economic Committee,* U.S. Congress, 92nd Congress, 1972; quoted sources are American Petroleum Institute (U.S. data) and "World Wide Oil" issues of *Oil and Gas Journal.*

thereof, it is understandable that the FPC is considering price increases for new gas commitments by domestic producers.

OIL

The shortages in the United States of gasoline and heating oil reported in 1971, 1972, and 1973 represent a remarkable confluence of national security politics, air pollution control, the pricing of natural gas, and the profit motive. Two features characterize a shortage—demand and supply. On each side events conspired to lead to an imbalance, and the factors which contributed to the imbalance have not yet been alleviated.

U.S. AND WORLD RESOURCES

Proved world petroleum reserves have been rising. Figure F.2 indicates the division of the world's 632 billion barrels of proved reserves of petroleum at the beginning of 1972. The proved oil reserves in the United States of 38 billion barrels (including 9.6 billion barrels from Prudhoe Bay, Alaska) are complemented by a large store of potential reserves, perhaps 417 billion barrels, believed to exist but not yet discovered.[4] From 160 to 190 billion barrels of this total may be in the Outer Continental Shelf.[5] In addition,

[4] R. Morton, Secretary of the Interior [3].

[5] F. Ikard, President of the American Petroleum Institute [3].

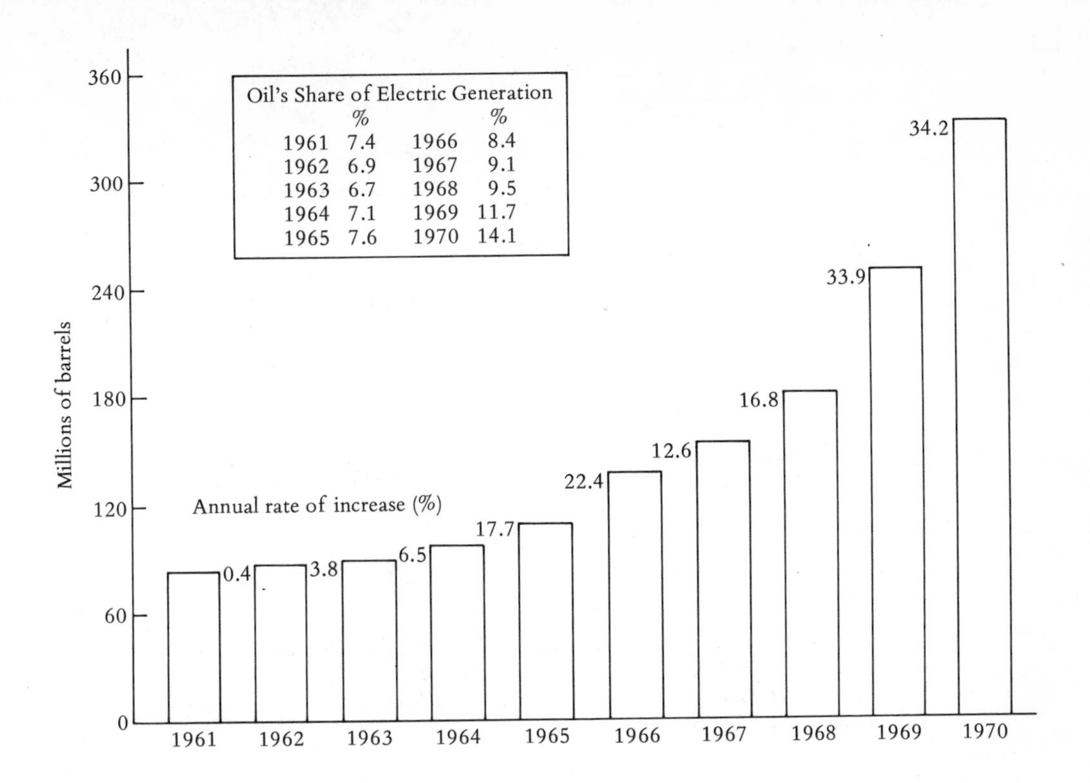

FIGURE F.3

CONSUMPTION OF FUEL OIL BY STEAM ELECTRIC PLANTS, 1961–1970

From National Oil Fuel Institute, Inc., New York, 1971.

oil in shale rock, not currently available with proved economic technology, may amount to 500 or 600 billion barrels. The shale oil may not be usable in the near future, however, because of the problem of handling the vast quantities of waste rock in an environmentally acceptable way. Canada has similarly hidden resources; the tar sands of Athabasca are thought to harbor 300 billion barrels of oil. A commercial venture has already begun there. Neither the United States nor the rest of the world is running out of oil.

U.S. DEMAND

During the past two decades, the reliance of the United States on world oil producers has grown. In 1950 the United States used 2.38 billion barrels of oil, in 1960 3.54 billion barrels, and in 1970 5.37 billion barrels.[6] Estimates of use from 1971 through 1975 are (in billions of barrels): 1971, 5.54; 1972, 5.84; 1973, 6.31; 1974, 6.68; 1975, 7.19.[7] In 1971 the United States consumed one-third of all the oil used in the world. The burgeoning demand for oil is a function of many variables. One is the supply of gas, which has been a preferred fuel for commercial and residential heating. Homebuilders in states which have prohibited new gas customers are turning to oil, and even to electricity, for heat.

Another variable exerting its influence on oil

[6] Ibid.

[7] *Oil and Gas Journal*, August 28, 1972.

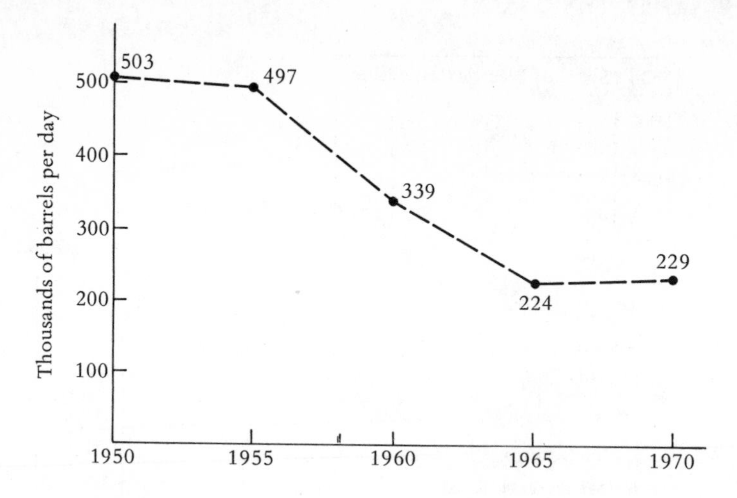

FIGURE F.4

PRODUCTION OF RESIDUAL FUEL OIL IN EAST AND GULF COAST STATES, 1950–1970

From testimony of J. Lichtblau, "Cost and Adequacy of Fuel Oil," *Hearings Before the Subcommittee on Small Business of the Committee on Banking, Housing and Interstate Commerce,* U.S. Senate, 92nd Congress, 1st Session, 1971.

demand is the control of air pollution. The implementation plans adopted by states to achieve federal **Air Quality Standards** have called for significant reductions in the emission of **Sulfur Oxides.** Since coal-burning power plants have accounted for about 40 percent of the annual sulfur oxide emissions in the United States, they are a principal target for control.

Utilities may follow several avenues to reduce these emissions. One option, almost the only possibility for utilities on the East Coast, is to convert existing plants to oil and build new capacity to use oil. The impact on oil demand is significant (see Figure F.3). The petroleum fraction usually used in electric power generation is known as residual oil (No. 6 oil). It is not in short supply. Figure F.4 shows that residual oil has not been a favorite of American refiners in recent years, having declined to 5 percent of the refinery yields (on average) in the United States in 1969. Its competitor for the electric power market has been coal. The abundance and relatively low price of coal decreased the demand for residual and its price. The result was that the oil companies largely ignored residual, emphasizing their higher priced fractions such as gasoline and distillate oil. Refineries in the Caribbean, however, have specialized in this fraction, and since 1966 residual oil has been freely available for importation to the East Coast from the world market.[8]

The fuel oil that has been reported in short supply is distillate oil (No. 2 oil), a more refined fraction, used principally for home heating. The sulfur content of No. 2 fuel oil is quite low, on the order of 0.25 percent. In contrast, residual oil has an average sulfur content in the range of 1.6 to 1.8 percent. The pressure for utilities to reduce their sulfur oxide emissions has led them to use residual oil

[8] As will be explained shortly, oil imports have been subject to a quota. Residual oil imported to the East Coast was an exception.

blended with distillate oil. A recent poll conducted in the Northeast indicated that from 5 to 10 percent of No. 2 sales were going into blending. Furthermore, No. 2 is being consumed by some utilities to drive turbines during periods of peak electric demand. This function used to be served by natural gas. Where residual oil was used for heating in industrial and institutional situations, conversions to No. 2 are taking place.

It is easy to draw the wrong conclusion, to link efforts to control air pollution with an oil shortage. Air pollution control requirements have increased the demand for oil, but supplies potentially available from the world market have been sufficient to meet this demand. Until 1973, however, a program of protection for the domestic oil industry, justified by the government on the grounds of national security, blocked U.S. access to the world market. The program was so constructed that needed supplies required approval for entry by a Washington bureaucracy. Supplies could not be directly imported except under special circumstances. Distillate oil (No. 2), one of the fuels reported in short supply, was a victim of the awkward import control apparatus. The quantities which could be imported and the source of the imports of this fuel were limited. The amounts sought in New England to prevent a potential shortage during the heating season of 1972–1973 were minor; yet before supplies could be obtained, a proclamation by the President was required.

THE QUOTA SYSTEM FOR IMPORTATION

A quota system was begun by Presidential Proclamation 3279 in 1959 with the intent of protecting the national security. According to the former director of the Office of Emergency Preparedness, the "national security, in this context, relates directly to the indispensability of petroleum to our economy, our way of life, and our strength as a nation." The concept is not limited, he stated, "to current national defense requirements or to future mobilization needs" [9]. By limiting imports of crude oil and derivatives, the quota system was to enhance the profitability of the domestic oil industry. The quota was thus to encourage exploration and development of the oil resources of the United States, and the nation's self-sufficiency in oil was to be assured.

The aims of the quota system, which is gradually being eliminated, are achieved by the distribution of "quota tickets" (actually import licenses) to companies operating refineries. The allocation is a fraction of the total refinery input but favors small refiners. The tickets may also be traded for domestic crude or for a product from some other domestic company; the company receiving the ticket on trade acquires the right to import foreign oil. Such a trade would likely be utilized by an inland refiner who did not wish to incur the transport costs of bringing foreign oil to his refinery.

Let us examine the impact of the quota as measured against its stated function. To do so, we must presume that the "vigorous, healthy petroleum industry" sought by President Eisenhower is one which is capable of meeting the nation's needs, and not just the needs of corporations with large annual profits. In 1972, in the face of growing demand, only one refinery was under construction (in the Midwest) in the contiguous United States; no new refineries were scheduled for 1973, but some expansions probably were occurring. With a lead time for refineries of three years, it appeared that badly needed capacity was not being added. New refineries were especially needed in New England, but the quota system would have limited such refineries' access to foreign markets, making the territory less desirable than location near domestic sources of petroleum.

There have been shortages of home heating oil and could be additional ones. It is not entirely clear how the potential home heating oil shortage in New England in the fall of 1972 came about. Refinery switchover from emphasis on gasoline to emphasis on heating oil occurred late even though inventories entering the heating season were low. The failure to build heating oil stocks during the summer has been attributed to the control of prices during that period. In May 1973, gasoline was being rationed in one portion of the United States, and advertisements were warning of a spreading scarcity.

Oil imports have been rising since 1950 (see Table F.4), and the outlook is for even greater imports. The pattern could lead to an increasing proportion of the imports being derived from the Middle East. In 1972, according to estimates of the National Petroleum Council, the United States consumed 15.9 million barrels per day of oil, and imports accounted for about 27 percent of that demand. Oil from the Eastern Hemisphere accounted for about one-third of the imports. By 1985, according to the Office of Oil and Gas, demand will reach 26.5 million barrels per day and imports may provide 57 percent of that quantity, if Alaskan oil does not reach the states. Eastern Hemisphere oil is seen by the Office of Oil and Gas as providing 35 percent of the total U.S. consumption by 1985. The remainder of the imports come from Canada and Latin America. The projection indicates an actual decrease in domestic production from 1972. If Alaskan oil entered the picture, U.S. reliance on oil from the Eastern Hemisphere could decrease to 23 percent of the nation's annual consumption. "Eastern Hemisphere" is not synonymous with "Middle East." Nigeria is entering an era of growing production from rich resources, and the Indonesia-Malaysia region may become a large producer.

Where does the cause of the shortages lie? Are the recent shortages the result of insufficient imports? Or were they the result of allowing too much oil to be imported, thus depressing the domestic oil industry? Oil companies claim that higher prices are needed to stimulate their expansion. They also argue that the Alaska pipeline is needed and that the Outer Continental Shelf could provide greater abundance if it were made available for leasing more quickly.

The quota system has had an effect on the consumer in addition to making his sources more secure. Oil on the world market has been less expensive than domestic crude. The companies who obtained this crude petroleum and refined it to products could sell at prices near those prevailing for refined domestic oil. The differential went to the oil companies.

If the less expensive world oil had been allowed free access to the U.S. market, the price of oil would have fallen to the world level. This, of course, is what the quota system sought to prevent, for it would have depressed profits and hence the viability of the domestic

TABLE F.4

BILLIONS OF BARRELS OF PETROLEUM IMPORTED TO THE UNITED STATES, 1950–1970

Source	1950	1955	1960	1965	1970	Percent in 1970
Western Hemisphere	265	356	534	718	1,042	83.6
Middle East and Africa	43	102	122	156	113	9.1
All others		12	31	26	91	7.3
Total	308	470	687	900	1,246	100.0

From U.S. Bureau of Mines, *Minerals Yearbooks*, various years.

oil companies. It is worth examining, nevertheless, what the consumer has paid for the security afforded by the quota system. Speaking roughly, the annual excess cost borne by business, industry, and individuals would be calculated by multiplying the difference between domestic price and world price by the annual consumption—that is, the dollar difference between the price under the quota and the price under a "free trade" situation.

The Cabinet Task Force on Oil Import Control and others have made estimates of the cost of oil security.[9] The Task Force calculated that in 1969 about $5 billion more was paid for petroleum products in the United States because of the quota system. Because the producer nations have combined to raise prices, the future estimates of consumer costs are no longer applicable. It is correct though to assume that at constant world prices the cost will rise in time because demand (or consumption) is growing.

REMEDIES FOR THE SHORTAGES

With Presidential Proclamation 4210 issued in April 1973, the system of controlling imports entered an era of change. The new system, administered by the Secretary of the Interior, requires payment of fees to the U.S. government by importers of crude, unfinished, and finished oils, after which payment the stated quantity of petroleum may be imported freely. The fees for crude oil increase from 10½ cents per barrel in 1973 to 21 cents per barrel in 1975. The fees for gasoline started at about 1¼ cents per gallon in 1973 and increase to 1½ cents per gallon by 1975. Other unfinished oils and finished products will have fees increasing from about a third of a cent per gallon in 1973 to 1½ cents per gallon in 1975.

The quota system, or rather the system of allocating tickets which entitle a company to import oil, remains but is being eliminated in steps. The tickets allocated to refiners via quota in 1973 are reduced to 90 percent in 1974, 80 percent in 1975, 65 percent in 1976, etc., until in 1980 no imports will be allocated under the system. Imports under the quota will not require payment of the fees which are being established.

The new fee payments enter the Treasury of the United States and are not designated for any specific purposes. The Cabinet Task Force had recommended that such funds be allocated to the acquisition and storage of a "strategic petroleum reserve" designed to temper the impact of a supply interruption. Such a reserve could have a deterrent effect, serving notice that an interruption would have to be long and costly to those who would attempt it.

To encourage the building of new refining capacity, the payment of the import fees may be waived for a period of five years for additions to current refining capacity. The new system allows the importation of finished oils such as gasoline and home heating oil. This new system means that independent marketers of petroleum products (such as heating oil suppliers) will have another source of product besides domestic refiners.

The fees do not discriminate by source; imports from Canada and other places in the Western Hemisphere receive no preferential treatment. The Cabinet Task Force had proposed that oil other than that from the Western Hemisphere be discouraged by a tariff, bringing the delivered price to a par with domestic oil. To Latin American countries a lower tariff would have been granted, while for Canada no tariff barrier would exist. Under these conditions, the Task Force predicted minimal reliance on Middle Eastern oil.

The new system has a number of virtues. Appeals to a bureaucracy for quota alterations

---

[9] See [10] for a comparison and for details of the calculation.

will become unnecessary. Artificial shortages, which the nation has been experiencing, will be less likely to occur. Under the quota system, the difference between domestic and world prices on a barrel of oil accrued to the refiner who received the import ticket. Under the fee system, some revenues will go to the government.

The impact of future shortages might also be softened by remedying the natural gas shortage. Many methods to achieve this have been mentioned. It bears repeating, however, that the lack of gas for home heating and for power generation has put a stress on oil which could account for a portion of the shortage. If new gas were imported or produced from imported petroleum, supply interruptions could hit this sector as well.

The balance of payments issue has been raised repeatedly in arguments against increasing oil imports. Oil is not the only good we import, however, and if we are worried about balance of payments, other goods—such as foreign automobiles and electronic equipment—could be also restricted, perhaps by means more reasonable than a quota. Furthermore, the annual trade deficit incurred in importation is not the product of barrels per year times price paid but a lesser figure. First, American products may be purchased directly or indirectly with revenues from oil.[10] Further, exporting companies purchase American oil-field equipment. Also, many of the firms from which we purchase oil are American firms, and their profits, if they bring them back into the United States, are a dollar inflow. These counterflows offer considerable compensation according to Task Force calculations. If the tariff system were modified to hold down imports from Eastern Hemisphere sources, America's trading partners in oil would be principally Canada and Venezuela, two nations who historically have spent money on United States goods, providing a diminished impact on the balance of payments.

The oil companies see the energy crisis being solved by "simple" measures. The building of the Alaska pipeline is one; faster leasing of offshore areas is another; incentives for refinery construction is a third. It is true that allowing production from areas off the East Coast, if they are found to contain oil, could work to stimulate some of the needed refinery capacity. Whether the contribution of an Alaska pipeline would obviate imports, however, is moot.

The Alaska pipeline as presently advocated by the Department of the Interior would extend from Prudhoe Bay south across Alaska to the port of Valdez. At Valdez, oil would be loaded on tankers for shipment to the West Coast. Predictions of the oil demands of the West Coast states differ; oil companies expect relatively larger demands than others do. Some predict that Alaskan oil could make the region an area with a large surplus. Given that the East Coast has been seen as the region with the greatest need for importation in the early 1980's, it may be that the proposed overland-tanker supply route would take the oil to the wrong place.

Economist Charles Cicchetti of Resources for the Future argues for an overland pipeline route through the Mackenzie River valley to central Canada. From there the oil could go by pipeline either to the Midwest or to the western United States. Since natural gas from Alaska may ultimately be transported by pipeline to the same destination, the two pipelines could share the same right-of-way. If the overland-tanker supply route were established, according to Cicchetti, some of the surplus oil probably could be exported to Japan. The possibilities for oil companies to reap excess profits are discussed in Cicchetti's *Alaskan Oil* [11]. These

[10] The calculation is complex. See [10].

possibilities may have been limited, however, by the recent changes announced in the oil import system. A summary of environmental impacts of the pipeline is included in Cicchetti's book.

One other way to increase domestic production is to "bring the oil companies home." Obviously this is an oversimplified notion, but a stimulus does tend to send the oil companies overseas. Royalties which the companies pay the producing nations for the oil they extract receive special tax treatment in the United States. The usual tax owed the United States government is first computed; then royalty payments are deducted directly from the taxes owed. This advantageous method of computation is used only for royalties paid to foreign countries. In the United States, royalties paid to the states are not deducted directly from taxes paid the federal government. Instead, the royalties are considered business deductions, expenses of doing business; these and other items are deducted from gross income to determine the taxable income. The tax is then calculated on the taxable income. Here is a simplified example:

Domestic Oil Production

Gross income	$10,000
Royalties	−3,000
Other deductions	−4,000
Taxable income	3,000
Tax at 50%	1,500

Overseas Oil Production

Gross	$10,000
Other deductions	−4,000
Taxable income	6,000
Tax at 50%	3,000
Royalties	−3,000
Tax	0

Two points remain to be made on the subject of oil. The first is that we have two alternatives or a combination of two alternatives if our energy demands continue to rise. Either we enhance domestic production by various measures, perhaps by increasing production from the offshore areas, or we import oil from overseas. Each choice has environmental risks. Offshore production risks blowouts. Importation chances increased tanker accidents as the tanker fleets grow (see **Oil Pollution**). Importation also brings political and economic dilemmas. In the short run, fuel shortages do not appear to be inevitable. In the long run, we should be looking to new sources of energy and to energy conservation.

The second point applies to both oil and gas; the goal of self-sufficiency through exploitation of domestic resources *cannot* be a long-term strategy if oil and gas resources are to continue to play their present role. Our resources of these fuels are finite and will run out. A blind strategy of self-sufficiency today will lead to dependence tomorrow. If research on coal gasification, coal liquefaction (oil from coal), solar power, fusion power, MHD (magnetohydrodynamics), and the like, does not proceed, we will be choosing the worst of paths, and a national energy crisis may occur "for real" in the future.

COAL

Of all U.S. fossil fuel resources, coal is the most plentiful. Coal resources are so enormous that perhaps one-third of the world's coal is thought to be within the continental United States. The coal deposits are concentrated principally in four regions (see Figure F.5). The Department of the Interior estimates that 390 billion tons of coal can be recovered under "current technological and economic conditions." At the 1970 rate of production, about 603 million tons, the reserves are sufficient for centuries.

The abundance of coal makes it far too easy to say, "This is the way to go." At least two major considerations dampen our enthusiasm for coal. The need to control air pollution

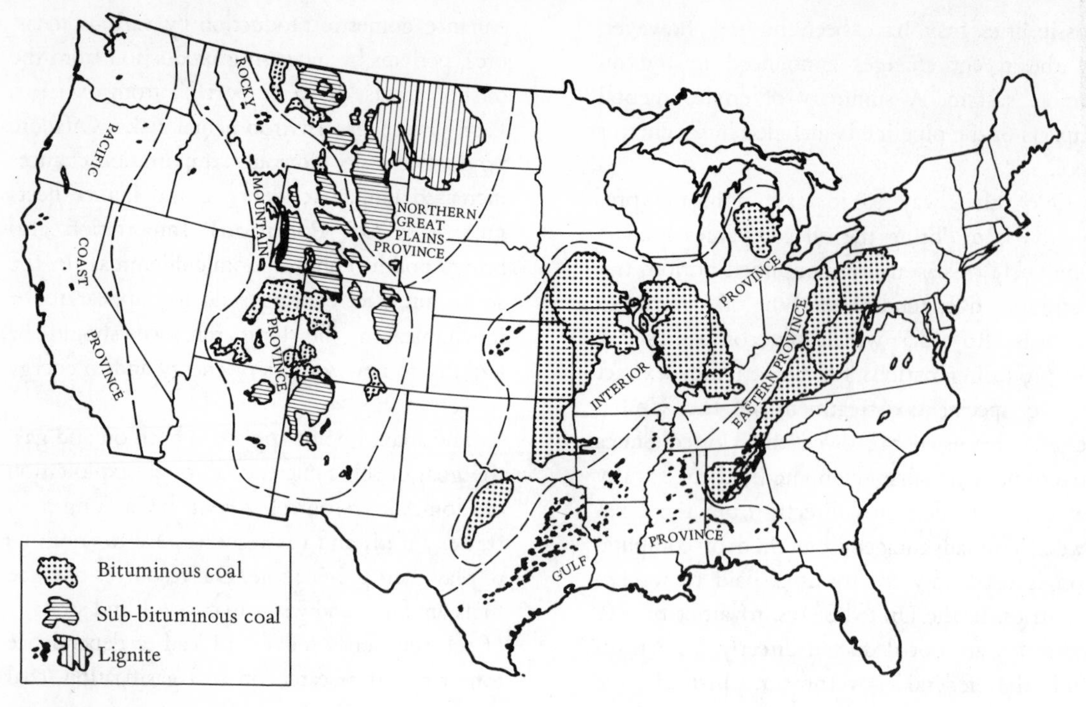

FIGURE F.5

BITUMINOUS AND SUB-BITUMINOUS COAL AND LIGNITE FIELDS IN THE CONTIGUOUS UNITED STATES

From "Strippable Reserves of Bituminous and Lignite Coals in the United States," Bureau of Mines Information Circular 8531, U.S. Dept. of the Interior, 1971.

makes much of the resource unacceptable according to recently announced emission standards for **Sulfur Oxides.** The destruction and the creation of derelict land by **Strip Mining** cause us to view this fuel's potential still more cautiously.

Coal is utilized in electric generating stations and in other industries to produce steam; it is used extensively in the iron and steel industry; and it is exported in quantity. In 1970, 339 million tons of coal were consumed by the electric utilities. (Approximately 1 pound of coal is consumed in burning a 100-watt bulb for 10 hours.) About two-thirds of the quantity sold to utilities was coal which had been strip-mined, according to estimates of the National Coal Association, an industry organization. About 95 million tons of the total production went to plants which carbonized the coal to coke; most of the coke, in turn, was consumed in the blast furnace production of pig iron. Another 71 million tons were exported in 1970, principally to Canada, Japan, and Europe. The export of coal is an indication of the abundance of the nation's reserve. Most of the exported coal is also destined for coking.

SULFUR CONTENT

The average sulfur content of U.S. coal used for generating steam is about 2.7 percent, which is equivalent to 2.1 pounds of sulfur per million Btu of heat energy, on the average. There is, however, a wide range in sulfur content and heat content in U.S. coal. The amount of sulfur may vary from 0.5 percent to 6 percent, heat from 15,000 Btu per pound for high-grade bituminous coal down to 7,000 Btu per pound for lignite with a high moisture content. The U.S. Environmental Protection Agency has established standards for sulfur oxide emissions from new facilities which generate steam by burning coal. The sulfur emissions are limited to 0.6 pound of sulfur per million Btu of heat from coal, which corresponds to about 0.8 percent sulfur by weight. Such coals are in limited supply for utility use. Only about 18 percent of the current supply of steam coal meets this requirement, and the fraction of the supply which qualifies for such use is not expected to grow in the decades ahead.

Metallurgical coke manufacture is one of the largest consumers of low-sulfur coal (having less than 1 percent sulfur). Because of the steel companies' need for stable supplies, a great deal of the low-sulfur coal is locked out of the utilities' market by long-term contracts or by actual ownership of supplies. About three-fourths of the low-sulfur coal produced has been going to coke manufacture and export. In addition, its production is limited principally to Kentucky, Tennessee, Virginia, and West Virginia. Even so, low-sulfur coal constituted over a third of production in 1964, but very little of it reached the electric utilities.

Western coal has been touted for its low sulfur content. Unfortunately, although much of the western coal does have a low sulfur content, it also has a lower heat content than eastern coal does. That means that more coal must be burned to produce the same amount of heat. Since the New Source Performance Standard is set on the basis of pounds of sulfur per million Btu, western coal may not be as advantageous for use as it first seems.

The effect of a sharply limited supply of low-sulfur coal in the East has brought about a dramatic shift in the fuel used to generate electric power. As already mentioned, residual oil has made large inroads into coal utilization on the East Coast. The National Coal Association cites a study by the Pennsylvania Railroad of coal transport to East Coast users from 1968 to 1972. The study indicates that by the end of the period, the annual rate of coal consumption on the East Coast had declined by a total of 33 million tons, the loss being attributed principally to conversions to oil.

Coal may be cleansed of some of its sulfur by a washing process. Sulfur is present in coal in two forms—organic sulfur and pyritic sulfur, in about equal amounts. Whereas organic sulfur is associated with the carbon in the coal itself, pyritic sulfur is present in the form of copper and iron sulfides known as *pyrites,* which are responsible for **Acid Mine Drainage.** Washing removes only a portion of the pyritic sulfur; crushing and grinding may be used to expose pyrites in veins or crystals, after which the washing process is more effective. Economical methods to remove organic sulfur have thus far not been found.

Sulfur oxides may be removed from stack gases by a number of processes, but the technology cannot be said to be in the production stage (see **Sulfur Oxides**). Research and development sponsored by the EPA were being pursued through 1972, but budget cuts threatened federal support for the projects.

There is another way to continue to use coal and avoid the emission of sulfur oxides. Coal gasification, accomplished by passing water and oxygen through intensely hot coal, is approaching commercial feasibility. Pipeline-quality gas from coal, which nearly duplicates the heat and methane content of natural gas, may fill home, utility, and commercial needs. Both low-Btu and high-Btu gas may be produced by the processes envisioned. In the production of high-Btu gas, oxygen is used to combust coal, and steam is injected to react with the burning coal. The reaction produces a number of compounds: carbon dioxide, carbon monoxide, hydrogen, methane, ammonia, and hydrogen sulfide. A second stage "cleans" the gas, leaving methane, carbon monoxide, and hydrogen. Hydrogen and carbon monoxide are converted to methane in a final stage, yielding a gas of high heat content suitable for long-distance transport through pipelines. If air, rather than oxygen, were fed to the first reaction, the gases produced would be diluted by nitrogen from the air, resulting in a low heat content fuel. Only the gas-cleaning stage would follow in this case, and the product, low in heat value, would be used in power plants. It would not be economical to transport fuel with such a low heat content over long distances.

The Environmental Protection Agency has sponsored research on an electric generating process which would combine coal gasification producing low-Btu gas, a gas turbine plant, and a steam turbine plant. The process, referred to as "advanced power cycle," is thought to have potential not only for controlling sulfur oxide emissions but for a high thermal efficiency, approaching 50 percent (see **Thermal Pollution**). Sulfur is removed in the gasification process and hence is not combusted. Moreover, because the burning of the coal gas produces less nitrogen oxide (see **Photochemical Air Pollution**) than coal combustion does, these pollutants will be diminished as well. Further, since gas, rather than coal, is burned, emissions of **Particulate Matter** will be nearly eliminated.

But simple answers are not easily obtained. Coal gasification seems to be about 65 percent efficient; only about 65 percent of the heat energy of the coal is captured in the gas produced. Speaking only of gasification and not of "advanced power cycle," which is more distant, we can predict large demands for coal if such gas entered the market, a prospect which now seems likely. Because of the inefficiency of the gasification process, demands for coal could call forth increased strip mining because stripping is presently the least costly method of coal extraction. There has been discussion of "in-situ" or underground gasification, but research on such a method is apparently not proceeding.

STRIP MINING

Sulfur content is not the only factor which makes coal a less than desirable fuel. The Bureau of Mines estimated that about 44 percent of the coal in 1970 was derived from **Strip Mining.** In 1941, only about 11 percent of the yearly production derived from surface mining. New machinery and the cost of making mines safer have combined to shift the industries' emphasis from deep mines. The lower cost of strip-mined coal has meant larger profits for operators. The lower cost is commonly attributed to the greater productivity of a surface "miner," but in fact it results in part from the frequent failures of operators to reclaim land which has been strip-mined.

Unreclaimed land may lie barren because pyritic substances exposed by the mining can turn the soil acid. Water running off strip-mined land may also be highly acidic, and **Acid Mine Drainage** may destroy aquatic life. High walls left when coal is cut from a hillside may isolate land from use by people or animals. Strip mining on slopes may leave behind loose rubble which can wash down in rain and choke streams. Rubble on slopes becomes increasingly unstable as it is wet by rain. The threat of landslides thus is added to the list of ills wrought by strip mining.

Whereas the Appalachian region up through Ohio has in the past been the scene of most strip mining, in the future greater activity may take place in the northern Great Plains and

Rocky Mountain states, where vast reserves of sub-bituminous coal and lignite underlay the land. The Northern Great Plains and Rocky Mountain coal provinces (see Figure F.5) have an estimated 26.6 billion tons of strippable coal. In contrast the Interior and Eastern provinces respectively have only about 8.6 billion and 5.2 billion tons of strippable reserves. Much of the western coal land has already been leased by the coal and oil companies. There is concern that the land is more fragile in this region than in the Appalachians and Midwest, for rainfall is low and the return of plant species to land which has been regraded may be hindered.

Proposals to deal with strip mining range from outright banning, a proposed federal law, to the posting of modest reclamation bonds. In the latter case, if the land is not reclaimed to a satisfactory condition, as judged perhaps by the state, the operator forfeits his bond. Increasingly, it is becoming apparent that the early cost estimates of $200 to $600 per acre for reclamation are inadequate for restoring the land. Nor do such costs check the growth of strip mining. It appears that the reclamation of land in which high walls have been created is more difficult than is restoring level land. This recognition has prompted pressure for laws which prohibit stripping on steep slopes.

Turning coal production back to deep mining would take time and money. The National Coal Association estimates that 132 additional underground mines, each of 2 million tons annual capacity, would be required to match the 1971 output from strip mining. About 80,000 additional underground miners would be required to accomplish the shift, which could cost from $3 billion to $5 billion. This demand for new miners, however, is the other side of an argument used by the coal industry to support strip mining. The industry has warned that if strip mining were halted, about 25,000 men would be out of work.

URANIUM

The Atomic Energy Commission's Division of Industrial Participation publishes estimates of the anticipated nuclear power generation requirements for uranium (in the form of uranium oxide, U_3O_8) as well as the current status of uranium reserves. Table F.5 indicates that there are ample estimated reserves to meet demand to 1980. At the present time AEC policy is that a safe reserve, to ensure continued operation of plants, is equal to 8 years of projected demands. On this basis 600,000 tons of U_3O_8 would be needed to cover the safe reserve requirement as well as U.S. demand through 1980. New discoveries will have to be made if we are to meet both of these needs entirely from domestic production.

TABLE F.5

URANIUM REQUIREMENTS AND ESTIMATED RESERVES THROUGH 1980 (IN TONS OF U_3O_8)

	United States	Foreign[a]	Total
Reserves[b]	340,000[c]	589,000	929,000
Requirements			
1970	7,500	6,000	
1975	17,000	17,000	
1980	34,000	38,000	
Cumulative total (1970–1980)	208,000	208,000	
Cumulative grand total (1970–1980)			416,000

From *The Nuclear Industry, 1970*, U.S. Atomic Energy Commission, 1970.

[a] Free world.

[b] These figures represent reasonably assured uranium reserves priced at less than $10 per ton. There are also believed to be another 100,000 tons worldwide in less assured reserves and about 150,000 tons available at $10 to $15 per ton.

[c] U.S. reserves were also estimated at 204,000 tons, representing uranium available at $8 per ton.

Exploration and drilling for uranium by industry in 1970 was expected to be down about 20 percent from 1969, but substantial additions to reserves were still expected. It is also believed that additional reserves exist and will be found by future explorations. Whether the new reserves found will cover both demand and reserve needs has been questioned. In any case there are other factors to consider. The amount of uranium required in the future will depend on the success of new schemes for generating nuclear power. The breeder reactor, for instance, produces more radioactive fuel than was originally put in, while at the same time generating electricity. Successful development of breeder reactors would thus greatly increase the life of uranium reserves. The liquid metal fast breeder reactor will probably not be ready for large-scale commercial electric generation until the 1990's. After the year 2000, fusion power may have been developed into a usable technology. The fuel in this case would not be uranium but a mixture of the hydrogen isotopes deuterium and tritium. Deuterium is found in almost unlimited supply in the oceans, and tritium can be made from lithium, which is also abundant.

ENERGY CONSERVATION

There is an approach to help solve the problem of meeting the nation's energy needs other than increasing supply. Demand might be decreased. Techniques to achieve a decrease fall under the heading of energy conservation. Although the expression "conservation of energy" can have unpleasant connotations, numerous opportunities for the saving of energy would not deprive individuals of their creature-comforts. The opportunities for conservation have come about because energy has been a relatively cheap commodity and hence has been used freely. In fact, relative to the prices of other commodities, the price of most types of energy has declined since the 1940's. Such a circumstance has provided no incentives for either residential or industrial consumers to attempt to conserve energy.

A change has begun to occur, however, with the appearance or prediction of oil and gas shortages. In 1971 the price of electricity increased for the first time since 1946, and gas companies in some areas were unable to accept new customers because of shortages of natural gas. Further, awareness of the pollution generated by fuel production and energy consumption is growing. Legislation designed to protect the environment is expected to increase the cost of production of various types of energy, particularly electricity. High-sulfur coals, America's largest fossil fuel resource, will in many states be unacceptable for use in electric power generating stations because of new emission standards. Thus, the stage has been set for intelligent consumers and cost-conscious industrial decision-makers to look for and accept methods of energy conservation.

In the United States energy use can be divided approximately into the following categories: (1) space heating, 20 percent; (2) transportation, 25 percent; (3) industry, 35 to 40 percent. The remaining energy consumed is devoted to a variety of uses, none of which amounts to more than a small percent of the total. It should be mentioned, however, that air conditioning is in this group, and its share of the total is increasing two and a half times as rapidly as the sum of all uses. Efforts at conservation can be expected to achieve the greatest results in the areas where the largest amounts of energy are being used: space heating and cooling, transportation, and industrial applications.

IN SPACE HEATING AND COOLING

Buildings in most areas could be insulated more thoroughly than they are at present. The thickness of the insulation applied to walls and ceilings could be increased, and floors could be insulated with foil. In some cases, weather stripping needs to be added around doors and windows and storm windows should be installed. Such techniques alone might, according to some estimates, result in overall savings of 7 percent of the country's total energy budget. In fact, this figure might be too low, for energy savings would be realized on air conditioning as well as on heating.

Better insulation, far from being onerous to the average homeowner or commercial building owner, should bring welcome monetary savings. A well-insulated building is less costly to heat and cool. In 1971, the Federal Housing Authority issued revised insulation guidelines for new construction designed to conserve energy. A homeowner living in New York in a house insulated according to the new guidelines might save a net of $30 to $75 per year more than someone in a home meeting only earlier requirements. Additional insulation might save him $4 to $80 per year more. The addition of an attic fan can often save energy and money by reducing cooling loads on air-conditioning systems.

A number of architectural features can be designed into a building to conserve heat or coolness. For example, fewer windows can be used in commercial buildings, and these can be of reflective glass or the type that open to allow the use of natural rather than artificial ventilation. Roofs may have large overhangs to provide shade and reduce air-conditioner loads. Buildings might be positioned to use more of the sun's warmth in the winter. Devices such as evaporative coolers can be used in arid zones to precool air passing through an air conditioner. Water-cooled lighting fixtures can reduce cooling loads and transfer heat to other areas. Department of Defense Manual 4270.1M specifies a large number of these techniques that might be incorporated into new buildings to reduce energy consumption from space heating or cooling.

The equipment chosen for heating and cooling has a large effect on the amount of energy consumed. For instance, heating a home electrically requires about three times as much fuel as heating it with natural gas or oil, since 60 to 70 percent of the energy in the fuels used to generate electricity is wasted as heat at the power plant. A further 10 percent of the electricity generated is lost in transmission. In a gas or oil furnace about 75 percent of the energy in the fuel may be used directly to heat the building. Hot water is also heated more efficiently by gas- or oil-fired burners than by electricity. Poorly maintained furnaces or heaters or ones running at low capacity may fall to 50 percent efficiency. Although direct fuel heating will result in significant savings to the owner, the higher initial cost of the equipment often obscures these future savings. Contractors are usually reluctant to put in oil-fired water heaters because of their cost, and consumers have not been aware enough to demand them.

Air conditioners vary widely in the efficiency with which they use electricity in cooling. Among the approximately 1,400 models of room air conditioners sold, the efficiency ranges from 4.7 to 12.2 Btu of cooling per watt-hour of electricity. Legislation requiring the efficiency to be stated in a prominent place on the air conditioner could help consumers choose the model which would use less electricity. They could thus save money over the long run despite a possibly higher initial cost.

IN TRANSPORTATION

A number of opportunities for the conservation of energy can be identified in the transportation sector. However, they in some cases involve a change in life style. Tables F.6 and F.7 show the relative efficiencies of various methods of trans-

TABLE F.6
EFFICIENCY OF FREIGHT TRANSPORT

Method	Btu Utilized per Ton-Mile
Pipeline	450
Railroad	670
Waterway	680
Truck	2,800
Airplane	42,000

From data in E. Hirst and J. Moyers, "Efficiency of Energy Use in the United States," *Science*, March 30, 1973.

TABLE F.7
EFFICIENCY OF PASSENGER TRANSPORT

Method[a]	Btu Utilized per Passenger-Mile
Intercity bus	1,600
Intercity railroad	2,900
Intercity automobile	3,400
Urban mass transit	3,800
Urban automobile	8,100
Airplane	8,400

From data in E. Hirst and J. Moyers, "Efficiency of Energy Use in the United States," *Science*, March 30, 1973.

[a] It is assumed that buses are 45 percent filled; railroads, 35 percent; automobiles in intercity travel, 48 percent; planes, 50 percent. Urban mass transit is assumed to operate at 20 percent of capacity; automobiles in urban travel, at 28 percent of capacity.

portation. It is obvious that freight transport by truck or plane uses much more energy than railroad and barge transport do. However, truck transport is more flexible in terms of pickup and delivery points, and plane transport is the most rapid. For these reasons the proportion of freight trucked or sent by plane has been increasing. Further, commuters for a variety of reasons have developed preferences for their own automobiles or for plane travel over railroad or bus travel.

To change these patterns of freight and passenger travel, various governmental incentives appear to be necessary. The government could help to finance mass transit and improve the traffic flow of buses by using priority bus lanes. To decrease the consumption of gasoline a special sales tax on automobile horsepower or weight could be levied. Wider use of small cars could make a significant difference, since a 5,000-pound car uses 100 percent more gas than a 2,500-pound car. Even a 500-pound decrease in weight from 3,500 pounds to 3,000 pounds should increase gas mileage about 16 percent.

The government at present indirectly subsidizes both truck and plane transport by constructing roads, using the highway trust fund, and by subsidizing airport construction. Diversion of a portion of the highway trust fund to mass transit could stimulate growth of this energy-efficient transport method. (See **The Automobile and Pollution** for further justification of mass transit.) Over the long term, urban cluster development involving walkways and bicycle paths could help to decrease urban automobile use.

Emission control devices for automobiles and trucks, designed to reduce air pollution, will increase fuel consumption, possibly by as much as 15 percent. However, power and weight options already use even more energy than this. Air conditioning adds 9 percent to fuel use, while automatic transmission adds another 5 to 6 percent. In addition, heavier cars are extremely inefficient in their use of fuels.

IN INDUSTRY

Six industries comprise a group using over half of the energy consumed in industrial applications. They are (1) primary metal producers, (2) chemical manufacturers, (3) petroleum and coal products manufacturers, (4) paper manufacturers, (5) the food industry, (6) and the stone, glass, and clay industry. Some experts estimate that the industrial use of energy might be reduced by as much as 30 percent, primarily in response to energy cost increases. Since industries pay special attention to ways of decreasing costs of production, it is expected that measures designed to force decreases in energy use by price increases would have their most rapid effects in the industrial sector. Savings might be effected by techniques such as the replacement of inefficient, old equipment, better maintenance of boilers and heat exchangers, and demands for energy-conscious designs.

The electric utility industry has achieved over the years an increase in the thermal efficiency of electric generation. The most modern fossil fuel plant is now about 40 percent efficient, but further development has slowed considerably. New experimental modes of electric generation such as MHD (see **Electric Power**) and "advanced power cycle" have the potential for much higher fuel efficiencies. Near-term savings might be achieved by more rapid replacement of outdated equipment and by efforts to cut down peak demands for electricity. The latter causes the use of inefficient peaking generators.

IN OTHER AREAS

Because the pilot lights in gas stoves burn continuously, they use a significant amount of energy. Electronic ignitors could be substituted to save fuel. Outdoor gas lights similarly use excessive amounts of fuel. Self-cleaning ovens, oddly enough, may use less electricity than their non-self-cleaning counterparts because of the extra insulation which is necessary because of the self-cleaning feature. On the other hand, another time-saving kitchen appliance, the frost-free freezer or refrigerator, can use up to twice the electricity of the manual defrost type, for heat is employed to prevent frost buildup. It has been suggested that a tax be levied on electric appliances such as water heaters and stoves to decrease their use or force a switch to nonelectric models.

Fluorescent lights consume only one-quarter of the electricity used by incandescent lights. Commercial buildings and schools may currently use more energy for lighting than is necessary. It is possible that areas could be lit according to how much light is needed for the activity to be carried out rather than to a uniform level. A possible savings of 4 percent of the total use of electricity might result from such measures. The World Trade Center in New York City uses more electricity than the entire city of Schenectady, New York, which has a population of 100,000. This excessive use of electricity could have been reduced, perhaps by as much as one-half, by the use of fossil fuels for heating, central absorption air conditioners operating on heat rather than electricity, reflective glass, and reduced ventilation and lighting.

REDUCTION OF ELECTRIC POWER USE

In 1971 consumption of energy to produce electric power for industrial, residential, and commercial uses amounted to about 25 percent of all the energy consumed in the United States. The two factors which have the greatest effect on electric demand are price and population growth. As prices rise demand should decrease. Thus it is possible that electricity use could be decreased by legislative means designed to increase the cost of electricity. For example, instead of giving high-volume customers a discount by decreasing the price per kilowatt-hour with increasing usage, "flat" pricing structures could be used. Surcharges could be levied during peak months, or prices could increase with increasing usage. Charging the electric power industry for the thermal pollution it causes would help reflect the true cost of generating electricity by increasing the cost per kilowatt-hour.

DECREASING DEMAND FOR OTHER FUELS

Although studies comparable to those on the demand for electric power have not been published, rising gasoline prices may be expected to decrease demand for gasoline in a similar fashion. Withdrawal of depletion allowances, capital gains structures, and tax deductions from the energy-producing industries would help reflect the true cost of producing energy, while at the same time decreasing demand with increased prices. Significant response to factors which increase prices is not expected to be seen until about 1980 because of things such as equipment changeovers, which take time.

RECYCLING AND ENERGY "SYSTEMS"

Recycling certain materials appears to afford a definite energy savings over the continual use of "virgin" materials. For instance, recycled copper can be produced for 600 to 1600 kilowatt-hours of electricity per ton; if produced from copper ore, the same ton would require 13,000 to 25,000 kilowatt-hours. It takes only 5,000 Btu of energy to produce a pound of recycled aluminum; it requires 62,000 Btu to produce a pound from ores. This calculation, however, presupposes starting with relatively pure aluminum scrap. Some articles—aluminum engine blocks, for instance—could be easily obtained for this use. In other articles the aluminum would be mixed with other materials and recovery would be less efficient.

Glass returnable bottles for soft drinks were found in one study to use three times less energy than nonreturnable bottles. Recycling the throwaway bottles to make new bottles was, in this study, not found to save any energy. Assuming that cans are throwaway, steel and aluminum cans used three times as much energy as returnable glass bottles, while the use of all aluminum cans required four times as much energy.

Paper can be recycled to save both trees and energy. The suggestion has been made that the government specify recycled paper in all its contracts to encourage paper recycling. Where distribution problems can be solved, organic wastes used in agriculture provide an energy savings over the use of chemical fertilizers (as well as helping to solve solid waste disposal problems).

This leads to the consideration of energy systems designed to utilize energy in as efficient a manner as possible. Industrial-residential complexes can be designed so that waste heat from electric power generation can be used for space heating and cooling, water heating, and certain industrial processes which use low temperature steam. Up to 85 percent of the energy available from fuels could be utilized in such systems. In addition, municipal wastes could be burned to generate electric power. With glass and metals removed, municipal wastes were shown in one case to have a heating value one-third that of high value bituminous coal. The steam produced by burning refuse would be of a relatively low temperature, but refuse could be mixed with coal to overcome this problem. Some scientists believe it will eventually be practical to convert organic wastes into oil or gas to be used in power generation.

IN THE FUTURE

Increasing energy prices can be expected to spur the development of devices for more efficient energy use, which would act to conserve energy supplies. Examples of such devices include microwave ovens and stoves, ultrasonic dishwashers and washing machines, electrochemiluminescent lighting, and thermoelectric refrigerators and air conditioners.

Technological improvements in the production of electricity are also possible. In addition to MHD, which could increase the fraction of energy available from coal to 50 or 65 percent, fuel cells generating electricity from chemical energy are possible. Breeder reactors can increase the life of uranium reserves. There are also methods of producing power which do not use up limited fossil fuel or uranium reserves. These include fusion, geothermal, and solar power. Geothermal power, from the use of the heat buried in the earth, may be practical in certain areas of the country; solar power, from the trapping of part of the enormous amount of energy radiated to the earth by the sun, appears to be practical for space heating and hot water heating.

How much of a reduction in energy requirements can we effect by employing energy conservation methods? According to a staff study by the Office of Emergency Preparedness, these methods could decrease the projected energy demand for 1980 (equivalent to 45.4 million barrels of oil per day) by the equivalent of 7.3 million barrels per day, in other words by 16 percent. The savings could amount to two-thirds of the projected oil imports for 1980. Methods designed to conserve energy are important not only as a means for averting an energy crisis. As pointed out by J. C. Swidler, former Chairman of the Federal Power Commission [2]:

> Energy conservation will do more than extend the life of our hydrocarbon resources. It will also dampen the environmental impact that follows unavoidably from supplying and using energy—the air pollution that comes from burning fuels, the thermal pollution of surface waters that comes from generating electricity, the impairment to land that comes from mining for coal or drilling for oil.

REFERENCES

1. Presidential Proclamation 4210, *Federal Register* 38, no. 75 (April 19, 1973): 9645.
2. "Fuel and Energy Resources, 1972," *Hearings Before the Committee on Interior and Insular Affairs*, Parts 1 and 2, U.S. House of Representatives, 92nd Congress, 2nd Session, 1972.
3. "Natural Gas Regulation and the Trans-Alaska Pipeline," *Hearings Before the Joint Economic Committee*, U.S. Congress, 92nd Congress, 1972.
4. "Natural Gas Policy Issues," *Hearings Before the Committee on Interior and Insular Affairs*, Parts 1 and 2, U.S. Senate, 92nd Congress, 2nd Session, 1972.
5. "Advanced Power Cycles," *Hearings Before the Committee on Interior and Insular Affairs*, U.S. Senate, 92nd Congress, 2nd Session, 1972.
6. "Review of the Developments in Coal Gasification," *Hearings Before the Subcommittee on Minerals, Materials and Fuels and the Full Committee on Interior and Insular Affairs*, U.S. Senate, 92nd Congress, 2nd Session, 1971.
7. "Underground Uses of Nuclear Energy," *Hearings Before the Subcommittee on Air and Water Pollution of the Committee on Public Works*, Parts 1 and 2, U.S. Senate, 91st Congress, 2nd Session, 1970.
8. "Cost and Adequacy of Fuel Oil," *Hearings Before the Subcommittee on Small Business of the Committee on Banking, Housing and Interstate Commerce*, U.S. Senate, 92nd Congress, 1st Session, 1971.
9. "Adequacy of Home Heating Oil Supplies," *Hearings Before the Subcommittee on Small Business of the Committee on Banking, Housing and Interstate Commerce*, U.S. Senate, 92nd Congress, 2nd Session, 1972.
10. *The Oil Import Question*, Cabinet Task Force on Oil Import Control, 1970.
11. C. Cicchetti, *Alaska Oil: Alternative Routes and Markets*, Johns Hopkins Press, Baltimore, 1972.
12. "Toward a Rational Policy for Oil and Gas Imports," Paper prepared for the Committee on Interior and Insular Affairs, U.S. Senate, 1973.
13. M. Adelman, *The World Petroleum Market*, Johns Hopkins Press, Baltimore, 1972.

14. K. Brown, *Regulation of the Natural Gas Producing Industry*, Johns Hopkins Press, Baltimore, 1972.
15. M. Adelman, "Is the Oil Shortage Real?" *Foreign Policy*, no. 9, Winter 1972–1973, p. 69.
16. "Oil Shale," *Hearings Before the Subcommittee on Minerals, Materials and Fuels of the Committee on Interior and Insular Affairs*, U.S. Senate, 92nd Congress, 1st Session, 1971.
17. "Oil Shale Development," *Hearings Before the Subcommittee on Minerals, Materials and Fuels of the Committee on Interior and Insular Affairs*, U.S. Senate, 91st Congress, 2nd Session, 1970.
18. "Surface Mining," *Hearings Before the Subcommittee on Minerals, Materials and Fuels of the Committee on Interior and Insular Affairs*, Parts 1, 2, and 3, U.S. Senate, 92nd Congress, 1st Session, 1971, 1972.
19. H. Risser, "Power and the Environment: A Potential Crisis in Energy Supply," *Environmental Geology Notes*, no. 40, Illinois State Geological Survey, Urbana, Ill., 1970.
20. *Strippable Reserves of Bituminous Coal and Lignite in the United States*, Bureau of Mines Information Circular 8351, U.S. Department of the Interior, 1971.
21. *The Nuclear Industry, 1970*, U.S. Atomic Energy Commission, 1970.
22. *Conservation of Energy*, Committee on Interior and Insular Affairs, U.S. Senate, 92nd Congress, 2nd Session, 1972.
23. *Energy Demand Studies: An Analysis and Appraisal*, Committee on Interior and Insular Affairs, U.S. House of Representatives, 92nd Congress, 2nd Session, 1972.
24. J. C. Moyers, *The Value of Thermal Insulation in Residential Construction: Economics and the Conservation of Energy*, Oak Ridge National Laboratory Report ORNL-NSF-EP-9, Oak Ridge, Tenn., December 1971.
25. D. Chapman et al., "Electricity Demand Growth and the Energy Crisis," *Science*, November 17, 1972.
26. A. L. Hammond, "Conservation of Energy: The Potential for More Efficient Use," *Science*, December 8, 1972.
27. A. L. Hammond, "Energy Needs: Projected Demands and How to Reduce Them," *Science*, December 15, 1972.
28. G. A. Lincoln, "Energy Conservation," *Science*, April 13, 1973.
29. M. Altman et al., "The Energy Resources and Electric Power Situation in the United States," *Energy Conversion* 12 (1972): 53.
30. G. Lof and R. Tybout, "Cost of House Heating with Solar Energy," *Solar Energy* 4 (1973): 253.
31. H. Garg, "Design and Performance of a Large-Size Solar Water Heater," *Solar Energy* 14 (1973): 303.
32. *The Potential for Energy Conservation: A Staff Study*, Executive Office of the President, Office of Emergency Preparedness, October 1972.

H

Herbicides

Herbicides are chemicals used to kill plants. Sometimes they are utilized selectively to kill one kind of plant; for instance, dalapon can be used as a grass killer. They may also be employed nonselectively, as when herbicides like dinoseb are applied along the sides of highways and along railroad right-of-ways to kill overgrowths. The ability of a compound to be used selectively may depend on the amount used as well as on the specific herbicide. Dalapon is a selective grass killer at 3 to 6 pounds per acre but a soil sterilant at 10 to 50 pounds per acre. Herbicides have found extensive use in modern mechanized agriculture as weed control agents. Human help for weeding is said to be too expensive or even unavailable in some areas. Certain herbicides cause the leaves to drop off plants. These are called defoliants and have found use in such diverse activities as warfare and the harvesting of cotton.

At least 100 different herbicides or combinations are available. In the United States in 1969, 85 million acres were treated with herbicides, a 60 percent increase since 1959. The sales of herbicides in this country have risen 271 percent since 1963, according to U.S. Department of Agriculture surveys, and predictions are that herbicide usage will be twice that of other pesticide usages by 1975.

Herbicides affect plants in three different ways. First, they may kill all parts of the plant with which they come in contact. These are the contact herbicides and are often used for clearing right-of-ways. Second, they may be absorbed through the roots or foliage and transported to

At left, a cornfield treated with herbicides to control weeds. At right, an untreated cornfield.

all parts of the plant. These are systemic herbicides. Finally, some herbicides are soil sterilants; they kill seeds and roots in the soil. Soil sterilants may be effective anywhere from a few days to a few years. Pre-emergence crabgrass killers are in the third group. At the proper dosage they kill the overwintering seeds of crabgrass before they sprout but do not affect the roots of more desirable grasses.

Herbicides have been found to change the chemical content of certain plant species. The herbicide 2,4-D increases the protein content of wheat but decreases the amount of protein in beans. It has also been reported to change the potassium nitrate content of sugar beets from 0.22 percent (dry weight) to a toxic level of 4.5 percent. In addition, 2,4-D may increase the sugar content of ragwort, a weed which is toxic to cattle. Some plants appear to be developing resistance to 2,4-D. In one experiment, another herbicide, 2,4,5-T, increased the level of poisonous hydrogen cyanide in Sudan grass by 69 percent.

Because there is such a great physiological difference between people and plants (seaweed-man excepted), herbicides may be expected to exhibit low toxicity to man. This has usually proved to be the case in terms of immediate effects. However, some disturbing evidence has come to light on the possible carcinogenic, teratogenic,[1] and mutagenic[2] effects of herbicides (see **Pesticides**). Since these effects take a long time to manifest themselves (generations for mutagenic effects), herbicides cannot be considered innocuous compounds.

Particular controversy has surrounded the use of the phenoxy herbicide 2,4,5-T, or Agent Orange, which was used by the United States

[1] Teratogenic substances cause malformations in fetuses during gestation.

[2] Mutagenic substances are detrimental to the hereditary material contained in cells.

for a time as a defoliant in Vietnam. Laboratory studies in 1969 indicated that abnormal fetal development resulted when 2,4,5-T was administered to pregnant rats and mice. Subsequent studies showed that the 2,4,5-T used had been contaminated with dioxin, the most toxic chlorine-containing compound known, a compound which itself may be teratogenic. Thus early studies on the teratogenic effects of 2,4,5-T were inconclusive. It is possible that both 2,4,5-T and dioxin are teratogenic, but further study is needed.

Almost 3 million acres were sprayed with 2,4,5-T in Vietnam in 1967–1968. However, it is felt that the concentrations used were probably not sufficient to cause human birth defects if one calculates possible doses from food, water, and direct skin contamination. The possibility exists nevertheless that doses might have been large enough in some instances. In June 1970 the Department of Defense ordered cessation of 2,4,5-T defoliating programs in Vietnam.

In May 1970 registration was suspended[3] for the inclusion of 2,4,5-T in products used in lakes and ponds and in liquid products for use around homes and recreational areas. Registration was canceled for 2,4,5-T in products intended to be applied to food crops, and in granular formulations for use in homes or recreational areas. In June 1970 the Department of the Interior forbade use of 2,4,5-T on lands under its control.

Residues of 2,4,5-T have been found in United States waters (28 out of 235 samples were positive in a study carried out from 1966 to 1968). Levels were 0.01 to 0.07 parts per billion and are not considered hazardous. The herbicide remains for about three weeks after application as a residue on forage crops and persists for three to four months in soil. The Panel on Herbicides of the President's Science Advisory Committee in its 1971 report noted that residues of 2,4,5-T are found in the tissues of animals grazed on treated pasture and range lands. The panel therefore recommended surveillance programs for the presence of 2,4,5-T in meat and milk.

Chemically, the herbicides are a diverse group including both inorganic compounds, such as sodium chlorate and calcium arsenate, which "burn" plants and are dangerous to man on contact, and organic compounds (See Table H.1).

In addition to chemicals that kill plants outright, some compounds are known to induce plants to behave (in the broadest sense) in a more desirable manner. These are generally plant hormones:

1. *Auxins.* Very small amounts of these compounds are used to control the setting, thinning, and early dropping of fruit and to promote rooting of cuttings. Two auxins are β-indolebutyric acid and silvex.
2. *Gibberellins.* Gibberellins vary in effect depending on the plant species and dosage. Some beneficial effects include early flowering, early germination of seeds usually slow to germinate, and growth at lower than normal temperatures. Gibberellic acid is the most commonly used chemical in this group.
3. *Cytokinins.* These hormones may increase the possible storage life of green leafy vegetables. Zeatin is an example as is N^6-benzyladenine.
4. *Growth retardants.* If further growth is undesirable, as when potatoes or onions sprout, growth retardants are very useful. In addition, ornamental plants and flowers can be controlled in size. Commonly used compounds are maleic hydrazide, Cycocel, Alar, and Phosphon.

[3] For the definiton of this term see **Pesticides, standards, laws, and monitoring programs.**

TABLE H.1

HERBICIDAL CHEMICALS FROM ORGANIC COMPOUNDS

Chemical Group	Example	Comments
Metallic compounds	Phenylmercuric acetate, disodium methane arsenate	Often used on right-of-ways; very persistent
Carboxylic aromatics		
Phenoxy compounds	2,4,5-T; 2,4-D	Most widely used herbicides
Phenylacetic compounds	Fenac	
Benzoic acid compounds	Dicamba	
Phthalic acid compounds	Dacthal	Pre-emergence herbicides
Aliphatic acid derivatives	Dalapon	
Phenol derivatives	Dinoseb	Contact herbicides; toxic to man
Heterocyclic nitrogen compounds	Atrazine, Simizine	
Aliphatic nitrogen derivatives		
Urea derivatives	Monuron	Structural similarity to the plant fertilizer urea
Carbamates	CDEC	Moderately toxic
Miscellaneous amides	Cypromid	Pre-emergence herbicides
Nitroaniline derivatives	Trifluralin	Pre-emergence weed control agents
Benzonitrile derivatives	Ioxynil	Pre-emergence broadleaf weed control agents

REFERENCES

1. Report of the Secretary's Commission on Pesticides and Their Relationship to Environmental Health, U.S. Dept. of Health, Education and Welfare, December 1969.
2. N. N. Melnikov, "Chemistry of Pesticides," *Residue Reviews* 36 (1971).
3. D. Pimentel, *Ecological Effects of Pesticides on Non-Target Organisms*, Executive Office of the President, Office of Science and Technology, June 1971.
4. "Reasons Underlying the Registration Decisions Concerning Products Containing DDT, 2,4,5-T, Aldrin and Dieldrin," U.S. Environmental Protection Agency, March 18, 1971.
5. Philip H. Abelson, "Pollution by Organic Chemicals," *Science*, October 30, 1970.
6. Report on 2,4,5-T, A Report of the Panel on Herbicides of the President's Science Advisory Committee, Executive Office of the President, Office of Science and Technology, March 1971.

Hydrocarbons

See Air Pollution, contaminants—photochemical pollution; Photochemical Air Pollution, hydrocarbons.

Hydroelectric Power Plants

See Reservoirs.

I

Insecticides

See Pesticides.

Inversion

See Air Pollution, meteorological aspects.

L

Lead

Gasolines formulated with lead additives for antiknock properties have seriously polluted urban air with lead particles. Lead also occurs in foods and in water. In deteriorating housing, paint chips containing lead are eaten by young children with disastrous consequences. Thus, lead exposures surround urban man. One way to control exposures is to remove lead from gasoline. Although lead is gradually being eliminated from gasoline, surprisingly, the circumstances which require its removal are in part fortuitous.

LEAD POISONING

The toxic effects of lead may be classified by severity. The most severe effect is encephalopathy, or damage to the brain. The disease is rarely seen in adults; children and infants are the principal victims. Two forms of encephalopathy are distinguished. *Acute* encephalopathy is characterized by the sudden onset of seizures and coma, although sluggish behavior, apathy, and mild colic may predate the arrival of these violent symptoms. Of the young children who survive such an attack, a quarter or more are afflicted with some permanent nervous system effects. A frequent result is mental impairment. *Chronic* encephalopathy involves a gradual loss of mental ability, dexterity, and speech; behavior problems may accompany the process.

A less severe condition is acute abdominal colic, which may develop over a period of several weeks. In an adult, the disease may begin with tiredness and headache; these symptoms are followed by constipation, stomach cramps, vomiting, and weight loss. In a child, the disease is, at first, characterized by apathetic and aggressive behavior; vomiting and constipation follow. The least severe form of the disease is anemia. The adult has a sickly complexion and headache; he tires easily and is irritable. The child's disease is seen in loss of appetite, in behavior problems, and in a diminished interest in play.

The symptoms of lead poisoning are accompanied by elevated levels of lead in the bloodstream and tissues. Although most of the lead in the body is found in the skeleton, it is the blood which is tested, both because it reflects well the disease state and because it is readily accessible for sampling.

Although the blood level at which symptoms may appear is variable, lead poisoning (plumbism) is generally not noted in adults below concentrations of 80 micrograms of lead per 100 grams of whole blood. Children, however, may begin to exhibit symptoms of lead poisoning at concentrations as low as 50 micrograms of lead per 100 grams of blood. Table L.1 indicates some average blood lead concentrations for nonsmokers.[1] The data provide a baseline for comparison with levels found in occupationally exposed individuals; they may also be compared to levels found in children in the inner city. Although data are from Philadelphia, the values are typical.

Female population groups consistently have lower blood levels than their male counterparts; the difference is unexplained. The upper limit of normal levels is 40 micrograms per 100 grams of whole blood, according to Dr. Jane Lin-Fu, an expert in the public health aspects of lead poisoning [10]. A blood level above this value may then be taken as evidence of undue absorption of lead.

[1] The effect of smoking on lead intake and blood levels is discussed subsequently.

TABLE L.1

AVERAGE CONCENTRATIONS OF BLOOD LEAD IN PHILADELPHIA ADULTS (IN MICROGRAMS OF LEAD PER 100 GRAMS OF WHOLE BLOOD)

Group	Males	Females
Residents of suburban Philadelphia	11	11
Commuters to Philadelphia	13	12
Residents living and working in downtown Philadelphia	22	18

From "Survey of Lead in the Atmosphere of Three Urban Communities," U.S. Dept. of Health, Education and Welfare, Public Health Service, 1965.

A study in Chicago of children under six years old produced an average blood level of 23.5 micrograms of lead per 100 grams of whole blood. The month of June saw the average concentration among the children reach 36 micrograms per 100 grams of whole blood. The children were obviously exposed to lead in the air and were not thought to have any other significant exposure besides lead in the diet. In her discussion of undue absorption of lead, Dr. Lin-Fu posed a question which has not been answered: "Does slight but sustained elevation of blood lead level cause subtle though appreciable impairment of brain functions such as mild retardation and learning defects in young children?"

INGESTED LEAD

Although there is no evidence that lead is an essential trace element, lead residuals in food are natural. The levels are augmented, however, by lead arsenate, a pesticide which is applied to such crops as apples (see **Pesticides**). In addition, when lead particles are present in the air, they may be deposited on the surface of crops. Where the edible portion of a plant is protected (e.g., the carrot), the lead content on that fruit or vegetable will be small. Crops such as lettuce, on the other hand, have the potential for higher concentrations. Since the lead contamination is principally on the surface of such plants, washing will have a beneficial effect.

An average value for the lead content of food is 0.2 part per million or 0.2 microgram of lead per gram of food. The daily dietary intake of an adult is estimated as 300 micrograms; that of a child one to three years old is estimated as 130 micrograms. Of course, the quantity ingested is variable from one individual to the next in accordance with differing dietary habits. An accepted estimate of the portion of the lead intake which is absorbed through the gastrointestinal tract is 10 percent. Thus, an adult would be expected to absorb from the diet 30 micrograms of lead daily and a child about 13 micrograms.

The lead in the glaze on pottery may be a further source of ingested lead. Acidic beverages such as fruit juices and soft drinks may leach lead from improperly glazed pottery articles. The glaze on stoneware is not dissolved by such beverages, but the pottery produced by the hobbyist is generally fired at a lower temperature than stoneware. The product of the hobbyist should therefore not be used to hold such beverages. The release of lead to solution by an improper glaze caused one individual to consume 3,200 micrograms of lead daily; the lead was dissolved in a carbonated beverage which he drank from a homemade mug. Some imported pottery items may also release lead from their glazes. Unfortunately, there is no way to determine visually whether a pottery item is safe. Cases of lead poisoning have been attributed to a release of lead from pottery glaze.

"Moonshine" whiskey may also furnish lead to the diet. Automobile radiators which are used as condensers for the illegal distillation of whiskey are responsible for the addition of lead to the final product. In one investigation of moonshine whiskeys, 26 percent of the samples had lead in concentrations of 1,000 to 10,000 micrograms per liter, and 3 percent had lead levels above 10,000 micrograms per liter. In contrast, of 20 brands of legally produced whiskeys tested by EPA, all samples had less than 400 micrograms of lead per liter, but eight brands had between 200 and 400 micrograms per liter.[2]

Many lead-induced fatalities and numerous cases of lead poisoning have been diagnosed among children who have eaten peeling chips of lead-based paints. Beginning at about one year of age, about 50 percent of children develop the habit of ingesting nonfood items such as paper, string, dirt, cigarette butts, paint chips, etc. This is called "pica," and it is a common although unexplained form of behavior among young children from all economic and social classes. The incidence of pica and use of lead-based paints on interior walls has caused an epidemic of lead poisoning of enormous proportions among children.

The results of mass screening programs give reason to believe that 1 to 2 percent of the children living in deteriorating prewar residences may be afflicted with some form of lead poisoning. A number of studies indicate that more than 20 percent of the children under six living in old urban neighborhoods have blood lead levels greater than 40 micrograms per 100 grams of whole blood. Recall that values greater than 40 micrograms may be taken as evidence of undue absorption of lead. As Dr. Lin-Fu notes, "It is altogether possible that many who do not progress to . . . encephalopathy are never diagnosed and never treated, and eventually appear in school with learning disabilities, . . . and other behavior problems."

Prior to World War II, lead was widely used

[2] It may be of interest to compare these numbers to the **Drinking Water Standards** for lead.

as a pigment in both interior and exterior paints. The paint from that era is today peeling from the walls of the dilapidated and deteriorating housing stock which remains. The lead in such paint may constitute between 5 and 40 percent of the final dried solids; thus, a 1-gram paint chip derived from paint applied in this era might have 50,000 micrograms of lead or more. Interior paints and paints used on children's toys and furniture have been limited since 1955 to 1 percent lead in the final dried solids of fresh paint by adherence to a *voluntary* standard of the American Standards Institute. Even so, 10,000 micrograms per gram is a significant amount. Since about 1940, the use of lead pigments has been declining[3] while the use of titanium dioxide, a less expensive white pigment, has increased.

In 1971 the Food and Drug Administration proposed warning labels on all paints containing more than 0.5 percent lead in the dried paint film [12]. The FDA also published a petition from private parties to ban from interstate commerce paints which contain more than traces of lead [12]; the petitioners scored warning labels as inadequate, pointing out that "the parents of the eighties and nineties will have no opportunity to read the warning labels of the seventies."

The public response to the petition prompted the FDA in March 1972 to abandon the labeling proposed and to issue proposed regulations which would ban leaded paint in two stages [19]. The rules would affect all paints and surface coverings for use in and around homes (i.e., both interior and exterior paints) as well as paints and coverings used on toys. In the first stage, paint with more than 0.5 percent lead in the dried film would be banned from interstate commerce after December 31,

Peeling lead-based paint, a serious environmental threat to the urban child. Mental retardation and death may result from the ingestion of paint chips.

1972. The FDA considered responses to the rule, and in August 1972 affirmed the regulation [20]. The second stage would see paints banned from interstate commerce after December 31, 1973, if they contained more than 0.06 percent lead in the dried film. As of January 1973 the second regulation had not been finalized.

Education, the boarding up of exposed surfaces, the removing of paint, and mass screening of children in high-risk areas are possible points for a comprehensive prevention program. Another avenue of potential control is to diminish lead in urban air by the removal of lead from gasolines. Whether lead from the air contributes to the problem of lead poisoning in children is the subject of debate. The point of view taken here and explained more fully later is that airborne lead from automotive exhaust is

[3] Anyone who thinks we are solving this problem would do well to ponder the current use of **Mercury** as a fungicide in latex paints.

a potentially dangerous source of body lead which heightens background levels in urban children and makes the possibility of achieving toxic levels more likely.

One additional source of lead must be mentioned. A survey of midwestern cities revealed that street dust may contain between 1,600 and 2,400 micrograms of lead per gram of dust. The ingestion of street dust then could be a factor in the high incidence of lead poisoning of children in urban areas. Again, lead particles from the air may settle and contribute to this source.

IN WATER

The sources of dissolved lead in natural waters are industrial and mining effluents and the solution of lead from rock where it occurs as the oxide or sulfide (galena). Lead commonly occurs in natural waters in concentrations up to 40 micrograms per liter; higher concentrations are possible but unlikely because of the insolubility of lead carbonate and lead sulfide. Lead also may occur in the drinking water drawn from lead piping systems. Lead piping is still allowed by plumbing[4] codes in portions of the United States.

The limit for lead according to the Public Health Service's **Drinking Water Standards** is 50 micrograms per liter, and such a concentration is grounds for rejection of a water supply for public use. It was only 1962, however, when this limit was set. Prior to that time, 100 micrograms per liter was a permissible concentration. In a 1969 survey of drinking water, 2 percent of the samples exceeded the lead standard set by the Public Health Service. In irrigation waters, the Committee on Water Quality Criteria of the Federal Water Pollution Control Administration has set a maximum allowable concentration for lead at 5 *milli*grams per liter. A report [7] of the National Academy of Sciences (NAS) assumed for discussion an average concentration of lead in drinking water of 20 micrograms per liter. This report is the source of much of the data presented in this discussion.

[4] Lead is represented by the chemical symbol Pb, which stands for the Latin name for lead, *plumbum. Plumbum* is the root for the word *plumbing.*

IN THE AIR

The amount of lead in the air in the Northern Hemisphere has been steadily increasing since the Industrial Revolution. Lead concentrations in the Greenland ice sheet reflect atmospheric levels over a long period of time; these concentrations have risen by a factor of 400 since 800 B.C. Urban areas now tolerate an average annual lead burden in the air about 20 to 40 times greater than rural areas.

SOURCES

Today, lead enters the atmosphere primarily from the combustion of gasoline which contains the additives lead tetraethyl or lead tetramethyl. There is no possibility that lead from the weathering of soil could account for the concentrations observed over urban areas. This fact is well established by studies on the isotopic concentrations in various lead samples. "Isotopes" of lead are lead-204, lead-206, and lead-207. In lead from a given source the relative amounts of the isotopes are essentially constant; the isotopic ratios then may be used like a fingerprint to identify the origins of lead. The isotopic concentrations of lead in urban atmospheres do not correspond to the concentrations of lead mined in the United States; in fact, they correspond with those of lead mined in Canada and Mexico. Moreover, the isotopic concentrations found in the lead in air are virtually identical with those in gasoline. Finally, atmospheric lead concentrations are highly correlated not only with gasoline sales but also with daily traffic volume.

The contribution of gasoline combustion to atmospheric lead amounts to about 98 percent of the total catalogued lead emissions. The remainder derives mainly from lead smelting, the manufacture of alkyl leads, and the combustion of coal. Isotopic studies indicate that the lead in coal seldom resembles the lead in gasoline. The NAS report cited earlier pointed out other possible sources of lead in the atmosphere, such as the burning of lead-painted surfaces and the wearing off of lead-painted dividing lines on highways. No estimates were made of the relative contributions of these other sources.

The major source of atmospheric lead.

LEAD ADDITIVES IN GASOLINE

Alkyl lead (lead tetraethyl or lead tetramethyl) is added to gasoline to improve its antiknock quality. It is, in effect, a substitute for a more costly refining process. Additional refining can produce a gasoline whose performance is such that it does not require the additive. Standard of Indiana has for many years (since 1915) produced an unleaded gasoline which does not need lead additives to achieve adequate antiknock performance.

The total amount of the two additives in gasolines was limited in 1972 to a maximum of 4.23 grams of lead per gallon of gasoline. This amount is equivalent to a maximum of 4 milliliters of tetraethyl lead (TEL) per gallon, a limit established only in 1965. The maximum allowable concentration had formerly been 3 milliliters of TEL per gallon. In 1958, the manufacturers of TEL asked the Surgeon General of the United States for his advice on the public health effects of an increase in the limiting concentration from 3 to 4 milliliters per gallon. A committee appointed by the Surgeon General suggested a joint investigation by the Public Health Service, TEL manufacturers, and the petroleum industry, among others, to study the issue. The report [2] provided by that investigation was published in 1965, the year the Surgeon General accepted the increase to 4 milliliters of TEL per gallon. Not all fuels utilize lead in the same concentrations; the average concentration in gasoline was 2.3 grams per gallon in 1963. By 1970, the average concentration had risen to 2.59 grams per gallon. In general, premium grades have had higher lead concentrations than regular grades of gasoline.

In the combustion of gasoline the lead in the alkyl compounds is oxidized to lead oxide. Scavengers, such as ethylene dibromide, ethylene dichloride, and phosphorus compounds, react with the oxide to yield lead bromochloride and other lead salts. About 70 to 80 percent of the lead in gasoline is eventually expelled in the exhaust stream in fine particles; the remainder is retained in the oil, the oil filter, and the exhaust system. The amount exhausted to the atmosphere will vary with engine age, design, and condition and with the exhaust apparatus. An estimated 262,000 tons of lead were consumed in 1968 for use in antiknock compounds for gasoline. Of the lead used in additives about 181,000 tons were emitted to the atmosphere on combustion of the gasoline.

The fine particles emitted in the exhaust stream range in size from less than a micron up to 5 microns and larger. The 5-micron particles (and larger ones) settle rapidly. If they are inhaled, they are generally trapped by mucous secretions and discharged or swallowed. Particles of 1 micron or less may penetrate the respiratory tree all the way to the alveoli, where absorption into the bloodstream is possible. Intermediate-size particles are unlikely to penetrate to the alveoli. Unfortunately, 90 percent of the emitted particles are generally less than 0.5 micron in size. In a typical sample of air tested for lead, over 75 percent of the mass of lead is accounted for by particles of less than 0.5 micron.

CONCENTRATIONS OF ATMOSPHERIC LEAD

The study prepared for the Surgeon General has been for a number of years the principal source of data on lead in the atmosphere; EPA, however, is now starting to publish new data. The original study, conducted in 1961 and 1962, sampled the air for lead in three cities; downtown-industrial areas and suburban areas were included. Unless otherwise noted, all concentrations reported here are the average annual concentrations. Cincinnati showed suburban concentrations at about 1 microgram per cubic meter; downtown and industrial concentrations were about 2 micrograms per cubic meter. Philadelphia also exhibited an average concentration of about 1 in the suburbs, but its downtown-industrial concentrations reached about 3 micrograms per cubic meter. The concentration of lead in the air of suburban Los Angeles was at 2, and downtown-industrial concentrations rose to 3 micrograms per cubic meter. The fall and winter concentrations were highest in all locations, and during these seasons average monthly concentrations reached 3.1 in Cincinnati, 4.4 in Philadelphia, and 6.4 in Los Angeles (see Table L.2.).

The highest daily concentrations occurred during the morning rush hours, although afternoon hours showed smaller peaks corresponding to the return-from-work trips. One reason for the observation of higher concentrations in morning rush hours is that inversion conditions are most frequent at that time. An inversion is characterized by a layer of stagnant air over an area (see **Air Pollution, meteorological aspects,** for further discussion of inversions). Pollutants are trapped in this stable layer and are not readily dispersed until the inversion has broken up. Individual measurements in traffic showed lead reaching 25 micrograms per cubic meter in Los Angeles and 14 micrograms per cubic meter in Cincinnati.

TABLE L.2

ATMOSPHERIC LEAD IN THREE CITIES, 1961–1962 (IN MICROGRAMS PER CUBIC METER)

	Annual Average Concentration		Maximum of the Monthly Average Concentrations
City	*Suburban*	*Downtown*	
Cincinnati	1	2	3.1
Philadelphia	1	3	4.4
Los Angeles	2	3	6.4

From "Survey of Lead in the Atmosphere of Three Urban Communities," U.S. Dept. of Health, Education and Welfare, Public Health Service, 1965.

In order to set the urban-suburban levels of atmospheric lead in perspective, one should consider remote readings. An average concentration for samples in remote areas taken by the National Air Surveillance Network was 0.022 microgram per cubic meter. The National Academy of Sciences report indicated that 0.05 microgram per cubic meter might be a typical concentration at a very rural station.

The 1965 report to the Surgeon General was unable to discern a trend in the average yearly concentrations in the cities studied. New data suggest, however, that trends in lead concentrations do exist. Since it is clear that rural and suburban settings experience lower lead concentrations, one may project that sections which were becoming urbanized or industrialized or which were drawing increased traffic were also experiencing higher lead concentrations. This result is, in fact, reported for Mission Valley in southern California, where lead concentrations were rising at a rate of 15 percent per year from 1967 through 1970; the same study noted an average annual rise of 5 percent in the downtown area of San Diego during the period from 1957 to 1966.

A preliminary report in 1971 by the Environmental Protection Agency gives further credence to the argument that lead levels are increasing. The report was unfortunately released too late to be included in the considerations of the NAS panel. Sampling stations were reestablished in 1968 at the same sites used in the 1961–1962 study for the Surgeon General. At all of the 21 stations but two, the monthly average of lead concentrations was higher. At Los Angeles, increases in the monthly average ranged from 33 to 64 percent at eight stations. In Philadelphia, the monthly average at only one station was unchanged, and the values at six other stations rose from 2 to 36 percent. In Cincinnati, the concentration at one station fell by several percentage points; at three other stations the concentration rose by 13 to 33 percent.

The monthly average values at the sites in Los Angeles varied from 2.9 to 7.6 micrograms per cubic meter in December, when atmospheric conditions appeared to be the worst. In Philadelphia the levels ranged from 1.5 to 5.1 micrograms per cubic meter in September, a month when inversions are common. The interim report emphasized that the data could not be used to characterize the lead level of a city; they were specific for the particular set of sites chosen. Data for sites in New York City, Houston, and Chicago taken in 1970–1971 were released in October 1971 by EPA. Monthly average values in New York City and Chicago were in the range of 1 to 3 micrograms per cubic meter depending on the month and site location. Monthly values in Houston varied from about 0.5 to 3.0 micrograms per cubic meter.

BLOOD LEAD CONCENTRATIONS

For a given rate of lead intake by ingestion or inhalation, the body mechanisms respond by removing the lead via urine, feces, and sweat. The greater the rate of absorption, the greater is the rate of removal, but concentrations may build up until the levels of absorption and removal are equal. Such a level is referred to as an "equilibrium" concentration, and such concentrations are generally observed in any sample of the population. The first set of data is drawn from the study submitted to the Surgeon General in 1965. The individuals in the sample populations referred to here are all adults whose contact with lead would primarily be through air pollution and food.

Lead measurements are most commonly made on blood because it is readily accessible for study and because it tends to mirror long-term lead exposure. Recall that a lead level of 80 micrograms per 100 grams of whole blood is the threshold level above which symptoms of lead poisoning may be observed. The Cincinnati study focused on individuals exposed to excessive air pollution in their daily work. The average concentrations of lead for traffic officers, parking attendants, and garage mechanics ranged from 30 to 38 micrograms per 100 grams of whole blood (see Table L.3).

If no one in these samples exceeded the highest noted average concentration, the issue of atmospheric lead might be less urgent. In fact, however, contributing to the average are individuals whose blood levels are above the mean and individuals whose blood levels are below the mean. Of the 152 garage mechanics, 25 had lead concentrations of 50 to 59 micrograms per 100 grams of whole blood and 4 had concentrations exceeding 60 micrograms per 100 grams of whole blood. In all studies the average lead concentration in the blood of smokers exceeded that of nonsmokers.

In the Los Angeles study, policemen headed the list at an average of 21 micrograms of lead per 100 grams of whole blood, and commuters exhibited no particular increase in lead levels with driving time. The Philadelphia study contrasted the lead concentrations in individuals who lived and worked downtown with the levels in commuters and with the levels in people who remained primarily in the suburbs. From downtown to commuter to suburb, the concentrations decreased from 22 to 13 to 11 micrograms of lead per 100 grams of whole blood. These appear to be the data upon which the Surgeon General's acceptance of higher lead levels in gasoline was based.

TABLE L.3

AVERAGE CONCENTRATIONS OF BLOOD LEAD IN INDIVIDUALS EXPOSED TO HIGH ATMOSPHERIC LEAD LEVELS, CINCINNATI

Occupational Group	Concentration (μg lead/100 g whole blood)	Number in Sample
Traffic police (1956)	31	17
Traffic police (1961–1962)	30	40
Parking lot attendants (1956)	34	48
Garage mechanics (1956)	38	152

From "Survey of Lead in the Atmosphere in Three Urban Communities," U.S. Dept. of Health, Education and Welfare, Public Health Service, 1965.

There is one further source of airborne lead to which adults are exposed. The pesticide lead arsenate was formerly used in growing tobacco. This compound left lead residues in the soil which contribute a significant concentration of lead to cigarette tobacco. One study of five brands of cigarettes showed an average of 13 micrograms of lead in a cigarette. The NAS report indicates that of this quantity, 1 to 2 micrograms appear in the smoke from a cigarette. Multiplying 1.5 micrograms by 20 for the pack-a-day man yields a 30-microgram quantity potentially inhaled daily by the serious smoker. If 2.5 micrograms per cubic meter is taken as the concentration of lead in the urban atmosphere and an estimated 23 cubic meters are inhaled daily by the average man at "light work," one could expect that 58 micrograms might be inhaled daily by an average individual who was not occupationally exposed. The potential 30-microgram supplement from cigarette smoke could increase the daily intake significantly.

A more recent study built on the earlier data has related blood lead levels to atmospheric exposures in a quantitative way. The work by Goldsmith and Hexter was regarded by the Environmental Protection Agency as one of the key elements in its arguments for control of

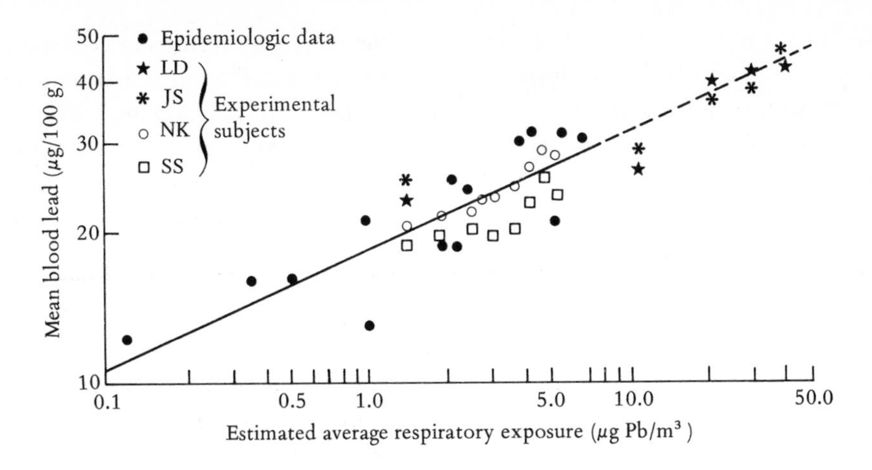

FIGURE L.1

MEAN BLOOD CONCENTRATION FROM EPIDEMIOLOGIC AND EXPERIMENTAL RESPIRATORY EXPOSURES

The regression is from epidemiologic data only.

From J. R. Goldsmith and A. C. Hexter, "Respiratory Exposure to Lead: Epidemiological and Experimental Dose-Response Relationships," *Science* 158 (October 6, 1967): 133; copyright © 1967 by the American Association for the Advancement of Science.

lead in gasoline [16]. The investigators took published data on blood lead levels and estimated average lead exposures of the various population groups. The average exposures took into account not only concentrations in the working environment but also the lower atmospheric levels during nonworking hours. Their graph as redrawn for the NAS report is shown in Figure L.1. The plot shows a direct relation between mean blood lead levels and the average atmospheric lead levels to which the groups were exposed.[5] The NAS report (pp. 62, 63) takes care to point out that the curve may not be reliably used to project blood levels at atmospheric concentrations below 2 micrograms per cubic meter.

[5] Actually the plot is of the logarithms of the data rather than the data themselves.

COMPARISON OF LEAD INPUTS

Lead is naturally present in food and water, but it may be classed as an unnatural component of the air at the concentration levels common to urban atmospheres. It is important to compare the relative contributions of these sources to the lead absorption of adults and children. The assumptions made in the comparison are based on data suggested in the NAS report; these numerical assumptions are indicated in Table L.4.

If the atmospheric lead concentration is 2.5 micrograms per cubic meter, the relative contributions from the air and from food and drink may be derived. From Table L.5 one can see that children receive greater inputs than adults per kilogram of body weight both from food and drink and from the air. The child's *total* input per kilogram of body weight is about 2.5

TABLE L.4
ASSUMPTIONS USED TO CALCULATE LEAD ABSORPTION BY ADULTS AND CHILDREN

Subject	Weight (kg)	Daily Volume Inhaled (m^3)	Daily Lead in Food and Drink (μg)
Standard adult	70	23 (light work)	300
Standard one-year-old	10	6	130

Fraction of ingested lead absorbed by gastrointestinal tract = .10.
Fraction of inhaled lead absorbed by lungs = .37.

times greater than that of the adult. Even if pica were not a factor, then, one would expect to see higher concentrations of lead in the blood of children.[6] Notice that the child receives a greater input per kilogram of body weight from food and drink than the adult receives from both the air and diet.

Children may suffer lead poisoning with concentrations above 50 micrograms of lead per 100 grams of whole blood, in contrast to adults, in whom lead poisoning symptoms are not seen below 80 micrograms per 100 grams of whole blood. That is, the toxic threshold is lower for children. Children apparently are not only more prone to higher blood levels but less distant from the toxic threshold. At the level of atmospheric lead used in these calculations, lead from the air makes up about one-third of the total intake of the child. Lead from food is evidently uncontrollable; it is not clear how to reduce this input. On the other hand, lead in the air is controllable; the removal of lead from gasoline will drastically reduce atmospheric lead exposures and heightened background blood levels. Given the formidable danger posed by pica and the eating of lead paint chips, exposure to lead from the air is a menacing addition to the environment of the urban child.

[6] This would be the case if there were not vast differences in lead metabolism by child and adult. It is true that lead is a bone-seeking element and that lead will be incorporated in the bones of growing children.

REMOVING LEAD FROM GASOLINE

Lead-free and low-lead gasolines are now appearing on the market. In part, this change represents a public-spirited response to the problem of lead air pollution, but it is also a response to the Clean Air Act of 1970. The act does not specifically legislate against leaded fuels. It does, however, call for 1975 cars to produce 90 percent less exhaust hydrocarbons than 1970 models. Auto manufacturers expect to achieve these standards by installing a catalytic converter in the automobile's exhaust system. Hydrocarbons and carbon monoxide will be converted to water and carbon dioxide by the device (see **The Automobile and Pollution**).

Unfortunately (or should we say fortunately?) lead will foul the catalyst and make the device ineffective. Since it appears that the automotive industry is moving toward the catalytic converter, the oil industry has been responding by moving to low-lead and lead-free fuels—in anticipation of being ordered to do so by law. In fairness, though, the notion of "cleaning up the environment" may account in part for the response. The slow sales of the slightly higher-priced lead-free fuels support the explanation that there is an element of public-spirited cooperation involved. The availability of these fuels also predates the proposed action of the Environmental Protection Agency.

In 1972 and again in 1973, EPA *proposed* schedules for the diminishment (not ultimate removal) of lead in gasoline, the later proposal

TABLE L.5
MICROGRAMS OF LEAD ABSORBED PER KILOGRAM OF BODY WEIGHT OF ADULTS AND CHILDREN

Source	Adult	Child
Air	23 × 2.5 × 0.37/70 = 0.30/1.0	6 × 2.5 × 0.37/10 = 0.55/1.0
Food and drink	300 × 0.10/70 = 0.43/1.0	130 × 0.10/10 = 1.3/1.0
Total	0.73/1.0	1.85/1.0

superseding the earlier one [18 and 21]. The schedule outlined in 1973, however, has not been finalized as of this writing. The 1973 proposal would require the average lead content of leaded gasoline grades offered for sale not to exceed 2.0 grams of lead per gallon after January 1, 1975; 1.7 grams of lead per gallon after January 1, 1976; 1.5 grams of lead per gallon after January 1, 1977; 1.25 grams of lead per gallon after January 1, 1978.

In addition to the schedule for lead reduction in gasoline, EPA issued *final* regulations [21] requiring the general availability of lead-free gasoline after July 1, 1974. Under the rules, retailers above a certain size must make available by that date at least one grade of lead-free gasoline of not less than 91 Research Octane Number.[7] About 70 percent of the gas stations in the United States are expected to be involved. Furthermore, it will be unlawful to fill cars designated for unleaded fuel (i.e., with the converter) with leaded gasoline. Conspicuous labels which state "Unleaded Gasoline Only" will be on such cars. Design of the pump nozzle and filler inlet of the car will prevent accidental usage of the wrong fuel.

In issuing the proposed regulations, the administrator cited the necessity of having available a lead-free gasoline to avoid impairing catalyst performance in the catalytic converter. He also noted the health risks associated with airborne lead, averring that the scheduled reduction would decrease atmospheric lead concentrations. One surmises that the continued availability of leaded fuel is in part to provide for older-model cars. Given that the law will also prevent the sale of leaded gas to cars with converters, it may be that, in fact, very little leaded fuel will actually be sold as the years go on.

Most 1971 and later models can use lead-free gasolines. Whether or not earlier models run effectively on lead-free gasoline ought to be determined by trying it. The spark plugs, the muffler, and the tail pipe all may last longer on unleaded gas.

The end of lead in gasoline is indeed strange. Itself a health hazard, lead is being removed partly in order to pave the way for the introduction of new technology to remove another health hazard. At least it is being removed.

[7] With exceptions for higher altitudes, where octane requirements are less.

REFERENCES

1. R. Engels, *Environmental Lead and the Public Health*, U.S. Environmental Protection Agency, 1971.
2. "Survey of Lead in the Atmosphere of Three Urban Communities," U.S. Dept. of Health, Education and Welfare, Public Health Service, 1965.
3. *The Automobile and Air Pollution*, part 2, U.S. Dept. of Commerce, 1967.
4. Report of the Committee on Water Qual-

ity Criteria, U.S. Dept. of the Interior, Federal Water Pollution Control Administration, 1968.

5. John Kopp and Robert Kroner, *Trace Metals in Waters of the United States*, U.S. Dept. of the Interior, Federal Water Pollution Control Administration, 1967.
6. *Public Health Service Drinking Water Standards*, U.S. Dept. of Health, Education and Welfare, Public Health Service, 1962.
7. *Airborne Lead in Perspective*, Committee on Biological Effects of Atmospheric Pollutants, National Academy of Sciences, Washington, D.C., 1972. This reference is the most complete of the works listed and cites others of these articles in its bibliography, but see also reference 1.
8. J. Chisolm, "Lead Poisoning," *Scientific American*, February 1971.
9. T. Chow and J. Earl, "Lead Aerosols in the Atmosphere: Increasing Concentrations," *Science*, August 7, 1970, p. 577.
10. J. Lin-Fu, "Undue Absorption of Lead Among Children: A New Look at an Old Problem," *New England Journal of Medicine*, March 30, 1972, p. 702.
11. *Science*, November 19, 1971, p. 800.
12. *Federal Register* 36, no. 211 (November 2, 1971): 20985–87.
13. T. Chow and J. Earl, "Lead Isotopes in North American Coals," *Science*, May 5, 1972, p. 510.
14. L. Tepper, *Seven-City Study of Air and Population Lead Levels*, Interim Report 1 to the Environmental Protection Agency, May 1971.
15. L. Tepper, *Seven-City Study of Air and Population Lead Levels*, Interim Report 2 to the Environmental Protection Agency, October 1971.
16. *Federal Register* 37, no. 115 (June 14, 1972): 11786.
17. J. Goldsmith and A. Hexter, "Respiratory Exposure to Lead: Epidemiological and Experimental Dose-Response Relationships," *Science*, October 6, 1967, p. 132.
18. *Federal Register* 37, no. 36 (February 23, 1972): 3882.
19. *Federal Register* 37, no. 49 (March 11, 1972): 5229.
20. *Federal Register* 37, no. 155 (August 10, 1972): 16078.
21. *Federal Register* 38, no. 6 (January 10, 1973): 1257.

M

Mercury

Consider a poison which causes symptoms mimicking emotional and psychic disorders: irritability, fearfulness, depression, headache and fatigue, inability to accept criticism or to concentrate or to make decisions, amnesia, drowsiness or possibly insomnia, digestive troubles, and generally exaggerated emotional responses. These are symptoms of chronic mercury poisoning.

Mercury is a familiar substance. It is used in thermometers, mercury cell batteries, silent switches, barometers, and antiseptics. Unfortunately, although well acquainted with its uses, we are less familiar with the hazards mercury presents when it is used in industry, laboratories, and schools or when it contaminates the general environment.

There are three common forms of mercury: elemental mercury, aptly named quicksilver; inorganic salts such as mercuric chloride; and organic mercury compounds such as phenylmercuric acetate and methyl mercury. Both elemental mercury and mercuric chloride vaporize readily.

The form of mercury most dangerous to humans appears to be the group of organic mercury substances known as alkyl mercury compounds. This group includes methyl mercury and ethyl mercury. Methyl mercury is the main form in which mercury is found in foods. However, any of the three categories of mercury are potentially convertible to methyl mercury by microorganisms found in the bottom muds of aquatic environments.

EFFECTS ON HUMANS

Mercury can enter the body by inhalation or by ingestion or by absorbtion through the skin. Ingested mercury is absorbed into the bloodstream through the gastrointestinal tract. Inhaled mercury is absorbed into the bloodstream through the lungs. Probably as much as three-fourths of inhaled mercury vapor is absorbed.

The inorganic salts of mercury accumulate in the liver and kidneys. In the blood they associate with plasma proteins. Although they do not accumulate in the brain to a very large extent, loss from the brain tissue is slow.

Organic mercury compounds are absorbed through the skin to a greater degree than inorganic compounds and so are more likely to cause skin allergic reactions. It appears that organic mercury compounds can be divided into two groups on the basis of their toxicity to man. In one group are compounds such as phenyl mercury, dimethyl mercury and methoxyethylmercury. These are converted by the body to inorganic compounds which are excreted relatively rapidly. In the other group are the much more toxic monoalkyl mercury compounds such as methyl mercury and ethyl mercury. These associate with blood cells and are excreted extremely slowly from the body. The half-life of methyl mercury in the body, or the time required for half of the administered dose of methylmercuric nitrate to be eliminated, was determined in one experiment to be about 70 days. This rate of elimination corresponded to only 1 percent per day of the total body content.

Besides normal avenues of excretion, some forms of mercury such as methyl mercury penetrate the placenta to the fetus. Birth defects due to methyl mercury have been noted in babies born even three to four years after the mother was poisoned. Mercury may also be excreted in milk after the mother is exposed to a source of mercury. Methyl and ethyl mercury accumulate in the brain. Approximately 98 percent of mercury found in the brain is methyl mercury. Mercury fed to human volunteers as methylmercuric nitrate was found to accumulate 0.5 percent in the liver and 10 percent in the brain.

It is generally believed that mercury exerts its detrimental effects by inhibiting enzymes,[1] which are vital biological substances. Thus, mercury can stop chemical reactions necessary for growth and normal life functions. It may also attach itself to cell membranes and prevent their normal functioning. Mercurials are used medically as diuretics since they decrease membrane permeability and therefore increase water loss from the body. Mercuric chloride is used as an antiseptic since it inactivates bacteria by combining with their thiol groups.

Mercury has serious mutagenic effects; it has been shown to cause chromosomal breakage (damage to hereditary material) and to interfere with normal cell division. Much more study of the long-term mutagenic and carcinogenic effects of low concentrations is needed.

Acute poisoning by mercury is relatively rare. Early symptoms are gastrointestinal and pulmonary, including chest pains and difficulty in breathing. In the later stages of acute poisoning, insomnia and delirium develop. Mild cases will either recover in about two weeks or develop the symptoms of chronic poisoning, which include, in addition to the emotional instability described earlier, tremor and sore, bleeding gums. In severe cases death can result from exhaustion.

Susceptibility to mercury poisoning seems to depend on the individual. A dramatic example is provided by the case of the Huckelby family. This family shared meals of pork from a hog fed on floor sweepings from a New Mexico

[1] The chemical group with which mercury probably combines is the thiol or sulfhydryl (SH) group. (Amino-, aldehyde-, hydroxyl-, and phosphorus-containing groups are also possibilities.)

granary. The grain was later found to have been treated with a mercury-containing fungicide. The mother and father, four of their children, and a grandchild were unaffected. However, a 20-year-old daughter was paralyzed and could not speak. She has since recovered almost completely. Her 15-year-old brother remains blind and partly paralyzed, and her 10-year-old sister is blind and almost completely immobile. The mother was pregnant when she ate the contaminated food, and her baby was born blind and severely retarded.

Fish and shellfish containing mercury are responsible for two of the more spectacular incidences of mercury poisoning. Both occurred in Japan between 1953 and 1965. During this period, a strange disease was reported in the area around Minamata Bay. The disease caused a lack of coordination, narrowing of the field of vision, and sometimes death. Animals and birds seemed affected as well as humans. A total of 46 people died, and 120 were poisoned before it was established that the disease was actually mercury poisoning from eating contaminated fish and shellfish. The source of the mercury was waste effluent from a plastics factory upstream from the bay. Methyl mercury compounds were found in the effluent and were being concentrated by the fish and shellfish in Minamata Bay.

A similar incident occurred at Niigata, Japan, leaving 100 poisoned and 6 dead. Fish in the area were found to contain 5 to 20 parts per million mercury. Japan has now set a limit of 0.01 part per million methyl mercury in discharge waters.

SOURCES IN THE ENVIRONMENT

Mercury is released to the environment in a number of ways. Principal sources of environmental contamination include the mining and refining of mercury ores and the subsequent dissipative uses to which refined mercury is put. In addition, the burning of fossil fuels which contain mercury releases mercury to the atmosphere. Natural processes such as the weathering of rock, evaporation of mercury from the earth's crust, and volcanic action are further sources of environmental mercury.

Mercury is mined in this country mainly on the West Coast. It is not widely distributed in nature but is found in highly concentrated ores. The ore which is economically important is cinnebar, mercuric sulfide. Airborne concentrations of mercury in mines can reach 5,000 milligrams per cubic meter in the form of aerosols and vapors unless proper ventilation methods are used. (Acute poisoning occurs when the airborne concentration of mercury is 1.2 to 8.5 milligrams per cubic meter.)

Elemental mercury is obtained from cinnebar by heating the ore in the presence of oxygen or lime (calcium hydroxide). The procedure yields mercury vapor, which must then be condensed back to a liquid. Some loss of mercury vapor to the atmosphere can occur during this process, but stack losses normally do not exceed 2 to 3 percent. Using the lower figure, one can calculate that mercury-refining processes added 64,000 pounds of mercury to the environment in 1969. In addition, the refining of other metals such as gold may involve ores containing very small quantities of mercuric sulfide. If the amount of mercury vaporized is too small to be economically important, it may be allowed to escape.

Six million pounds of mercury were being used annually in the United States in the late 1960's. In 1968, 36 percent was mined in the United States, 22 percent was imported, 18 percent was recycled, 24 percent was released from government stockpiles. Most of this mercury is used in manufacturing. The Oak Ridge group [1] has divided the uses of mercury into potentially recyclable uses and nonrecyclable or dissipative uses (see Table M.1). Actually only 520 out of 1,180 tons of mercury were recycled

TABLE M.1
USES OF MERCURY

Use	Percentage
Recyclable	
Electrical devices (batteries, lamps)[a]	27
Measurement and control devices (switches, thermometers)[a]	11
Laboratory uses[a]	3
Chlorine–caustic soda plants	33
Subtotal	74
Nonrecyclable	
Agricultural (fungicidal seed dressings	5
Dental	4
Catalysts	2
Pulp and paper industry (prevention of slime formation)	1
Pharmaceuticals	1
Paints (preservative for latex paints)	12
Subtotal	25
Total[b]	99

From R. A. Wallace, W. Fulkerson, W. Shults, and W. Lyon, *Mercury in the Environment: The Human Element*, Oak Ridge National Laboratory Report No. ORNL-NSF-EP-1, Oak Ridge, Tenn., 1971.

[a] Only 520 of 1,180 tons of mercury in these categories were recycled. The remainder probably found its way into the environment in one way or another.

[b] Does not add to 100 because of rounding.

from the electrical, measurement, and laboratory use categories. The remainder probably found its way into the environment in one way or another.

In 1968 the largest single source of mercury contamination of water was the chlorine–caustic soda industry. During the summer of 1970, however, an astonishing change was effected by a combination of threatened lawsuits and the setting of tentative water quality standards. An 86 percent decrease in mercury discharges from chlorine–caustic soda plants was noted by government monitors in the period July to September 1970.

Mercury used as a seed dressing to prevent fungus diseases has resulted in numerous instances of bird and wildlife poisonings. Alkyl mercury compounds such as methyl mercury appear to be the main problem. There is also some evidence that mercury dressings can increase the concentration of mercury in certain plants grown from treated seed. In the spring of 1970, the U.S. Department of Agriculture suspended the interstate transport of 42 alkyl mercury compounds used as seed dressings. In October, 1971, the Environmental Protection Agency[2] confirmed the previous cancellation as well as canceling other uses of alkyl mercury compounds. The registration of use of any mercury compound as a treatment for rice seed was revoked in March 1972 because of the possible conversion of all mercury compounds to toxic alkyl mercury in the wet environment in which rice grows.

Mercury-treated wheat seed sold in Iraq was responsible for an epidemic of mercury poisonings in early 1972. The sacks of seed had evidently been labeled as poisonous and were marked for planting use only. Nevertheless, peasants who were unable to read fed the seed to livestock and, seeing no immediate ill effects, used the seed in the making of bread. The livestock which later fell sick were sold for slaughter so that the epidemic spread beyond the peasants. When the government decreed very strict penalties for persons dealing in the contaminated seed, some of the farmers dumped their seed into the Tigris River. Mercury pollution was thus spread to the river and its fish. One newspaper reported an unofficial death toll of 400. Many more people were reputedly blinded or paralyzed or suffered brain damage. This may have been the largest epidemic of mercury poisoning on record.

In Sweden, methoxyethyl mercury com-

[2] Authority to register pesticides was transferred to EPA in a governmental reorganization.

pounds have been substituted for alkyl mercury compounds as seed dressings since 1966. In addition, there has been an overall 70 percent drop in the use of dressed seed. No effect has been apparent on crop yields. The decline in several bird species in Sweden seems to have been reversed.

The use of mercury as a preservative in latex paints may present a little-recognized hazard if the paints are used with inadequate ventilation. These paints might otherwise be considered especially safe to use around children, because of their lack of **Lead.** A case has been recorded in which a fungicidal paint containing di-(phenylmercuric) dodecenyl succinate, used in painting a greenhouse, emitted mercury vapors at low levels. A concentration of 10 micrograms per cubic meter or less was sufficient to cause severe injury to certain species of roses growing in the greenhouse. Some house paints have been shown to cause mercury levels in air to rise above 100 micrograms per cubic meter for short periods of time. This concentration is far in excess of healthful mercury levels. Because of the possible conversion of any mercury compound to alkyl mercury, the registration for use of mercury compounds in marine paints was suspended by the Environmental Protection Agency in March 1972.

Although the pulp and paper industry is rapidly phasing out its use of mercury to prevent the formation of slime, the industry has contributed to mercury pollution in the past, as shown by the following data. Concentrations obtained by sampling aquatic organisms above and below a paper mill were *Isoperla* stone flies, 72 parts per million above and 2,400 parts per million below; *Asellus aquaticus* (a sowbug), 65 parts per million above and 1,900 parts per million below; *Sialis* (burrowing alderfly), 49 parts per million above and 5,500 parts per million below.

It was at one time believed that a good way to dispose of mercury was to dump it in lakes and rivers; being very heavy, mercury would sink to the bottom, presenting no hazard to wildlife or humans. Unfortunately, this has not proved to be the case. For instance, one year after a textile mill stopped using organic mercury compounds, sampling of the water below the plant revealed high concentrations of mercury in the layer of water just above the bottom sediments. Mercury was apparently redissolving in the water from compounds which had previously settled to the bottom.

Fish hatcheries have used mercury compounds to treat fish for parasites. Such use presents three possible problems. The fish which are treated directly with mercury accumulate it to varying extents in their tissues. Salmon and trout treated for one hour with 2 parts per million of Timsam, which contains 6.25 percent ethyl mercury phosphate, stored more mercury than is legally allowable in their gill tissue for 2 weeks, in their blood for 6 to 8 weeks, and in their liver for 20 weeks. The level in their kidneys had not declined to permissible levels after 28 weeks when the experiment ended. On the other hand, pyridylmercuric acetate, when used to treat the fish, was accumulated only in the kidney.

Fish which were fed on contaminated fish showed similar patterns of mercury accumulations at different concentrations. Thus, both legal-sized fish treated at hatcheries and fish which eat contaminated undersized fish may be potential hazards. In addition, the mercury-containing waters used to treat the fish would be a source of mercury contamination if released into rivers and lakes. The 1972 ruling of the Environmental Protection Agency which canceled registration of mercury compounds for such uses should prevent further pollution from this source.

Cement manufacture may contribute to the release of mercury to the atmosphere since it involves heating limestone and shale to 1,500° C. This temperature can drive off mercury contained in the rock. Using figures for cement production in 1960, one author estimates that hundreds of tons of mercury are released per year by cement manufacturing processes.

One of the largest sources of environmental contamination by mercury is probably the burning of the fossil fuels, coal and petroleum. Little is known about the concentration of mercury in fuels or how much is vaporized when they are burned. Analyses of fly ash, the particulate matter remaining after coal is burned, show 10 percent or less of the original mercury remaining. The assumption, then, is that the major portion of mercury in coal is released to the air. Coals have been reported to contain between 0.012 and 33 parts per million of mercury. Multiplying an average figure of 1 part per million by the world production of coal for 1970, one set of authors calculate the mercury release from the burning of coal to be 3,000 tons per year. Regarding coal used for power generation, another study determined that a 700-megawatt coal-fired power plant released 2.5 kilograms (about 5 pounds) of mercury per day.

Figures for the mercury content of petroleum do not appear to be available. If the content is estimated at 1 part per million by comparison to coal [4], one can calculate the mercury released to the atmosphere annually by petroleum combustion. The burning of the close to 2 billion tons of petroleum produced in 1970 would result in the addition of almost 2,000 tons of atmospheric mercury—assuming that all the mercury is released to the air.

It is difficult to compare the impact of man's activities on the addition of mercury to the environment with the contributions from natural sources. Estimates vary, depending on the assumptions made and the environmental compartment examined. One group of investigators, using the concentrations of mercury deposited in the Greenland ice sheet, estimate that man's activities have led to a doubling of the atmospheric mercury concentrations. They ascribe the doubling primarily to activities which disrupt the earth's crust (see Table M.2).

Other investigators [4] estimate that in 1970 man's activities were accelerating the release of mercury to the environment 18 times over natural processes. Even assuming all of this mercury reaches the oceans, however, they calculate that, because of absorption and sedimentation processes, the time necessary to effect a doubling of the oceans' mercury content would be about 3,450 years.

Thus, estimates of overall mercury contamination do not place the biosphere in imminent

TABLE M.2

MERCURY CONCENTRATIONS IN ICE SAMPLES (IN NANOGRAMS PER KILOGRAM OF WATER)

Time of Deposition	Concentration	
800 B.C.	62	
1724	75	
1815	75	Average = 60 ± 17
1881	30	
1892	66	
1946	53	
1952	153	
1960	89	
1964 (fall)	87	
1964 (winter)	125	Average = 125 ± 52
1965 (winter)	94	
1965 (spring)	230 ± 18	
1965 (summer)	98	

From Herbert V. Weiss, Minoru Koide, and Edward D. Goldberg, "Mercury in a Greenland Ice Sheet: Evidence of Recent Input by Man," *Science* 174 (November 12, 1972): 693. Copyright © 1972 by the American Association for the Advancement of Science.

NOTE: The sample deposited in 1724 was recovered from Antarctica; the other samples came from Greenland. The sample deposited in the spring of 1965 was analyzed in triplicate, and the error shown gives the average deviation from the mean.

danger. Locally, of course, specific contamination sources can pose serious problems.

IN THE AIR

Early studies in the felt hat[3] industry and the electronics and lamp industry are the basis for current limits set by the American Conference of Governmental Industrial Hygienists (ACGIH) for airborne mercury. The maximum allowable concentration of inorganic mercury was set at 100 micrograms per cubic meter for an 8-hour day. This may well be too high a level for elemental mercury. An international symposium has suggested values of 50 micrograms per cubic meter for mercury vapor and 100 micrograms per cubic meter for the inorganic salts. No limits are set for 24-hour exposures or general background levels. Long-term experiments with rats have shown that continuous exposure to concentrations of metallic mercury of 2 to 5 micrograms per cubic meter caused mercury deposition in the brain and organs and dysfunction of the higher nervous centers. Such studies have influenced the setting of the Russian limit for 24-hour exposure at 0.3 microgram per cubic meter.

Since organic mercury compounds are more toxic than inorganic ones, the ACGIH limit for eight hours of exposure to organic compounds is 10 micrograms per cubic meter. This limit may also be inappropriate in that toxicity varies greatly among the organic compounds. The limit for exposure to phenyl mercury compounds, for instance, may be unnecessarily strict since its effects are similar to those of inorganic mercury. The Russian limit for alkyl mercury compounds, on the other hand, is 5 micrograms per cubic meter for an eight-hour period. No limits are set for a 24-hour exposure.

[3] The Mad Hatter of *Alice in Wonderland* was, in fact, a representative of an occupational group which was exposed to mercury.

The data on mercury exposure of the general public seem to show correlations between a high mercury content in the air and a high index of industrialization. Measurements of mercury in the air taken in San Francisco average around 0.002 microgram per cubic meter if a sea breeze is blowing inland. If the winds are from industrial areas, values average 0.008 microgram per cubic meter and may go as high as 0.02 microgram per cubic meter. Rather rural areas yield airborne mercury levels ranging from 0.003 to 0.009 microgram per cubic meter. Chicago averages 0.01 microgram per cubic meter, Cincinnati 0.1, and Charleston, West Virginia, 0.17. New York City has between 1 and 14 micrograms per cubic meter. In March 1973 the Environmental Protection Agency set final air emission standards for mercury ore–processing facilities and chlorine–caustic soda plants. These facilities are stated to be the only sources identified at present as capable of raising the concentration of mercury in their immediate vicinity above 1 microgram per cubic meter. The standards specify that such sources may not emit more than 5 pounds of mercury into the atmosphere per day. This level is believed to be low enough to prevent ambient mercury concentrations from exceeding the aforementioned level of 1 microgram per cubic meter. Investigations are proceeding to identify other major sources of mercury in the air.

IN WATER

Mercury enters natural waters primarily from agricultural fungicides, from mining and smelting operations, and from industrial discharges. Rainfall brings down much of the airborne mercury, which therefore also ends up in the water.

Fish taken from waters located near nonindustrial areas generally contain less mercury than fish caught near industrialized areas, even though there are no direct mercury inputs. This may be the result of the presence, in industrial areas, of high levels of airborne mercury. The mercury would be washed down by rain and eventually absorbed by fish. Background levels in water are generally from a few hundredths to a few tenths of a part per billion. For instance, seawater has 0.03 part per billion mercury. Concentrations of mercury in water great enough to kill organisms range from 0.03 part per million for *Scenedesmus* (a species of algae) to 2 parts per million for bacteria to 3 parts per million for snails. Because their rate of mercury uptake may be high compared to their rate of excretion, aquatic organisms tend to magnify the mercury concentrations found in water. For instance, while seawater has 0.03 part per billion of mercury, marine plants might have 0.03 part per million, and the tissues of marine animals might have 60 parts per million. This phenomenon of biological magnification increases the difficulty of setting standards for water concentrations. It has been estimated that freshwater macrophytes and plankton will concentrate mercury 1,000 times; fish, 1,000 to 3,000 times; and freshwater invertebrates (i.e., shellfish), up to 100,000 times.

Bodies of water which have been contaminated with mercury require a variable period of time to recover. In Sweden some lakes are still considered contaminated 25 years after mercury discharges have stopped. A lake in Wisconsin is being polluted by mercury that is dissolving out of its sediments, even though the last mercury was added 12 to 15 years ago. A few Swedish lakes have been decontaminated by the natural deposition of mud which has covered mercury-containing sediments in as little as 10 years. Other lakes may take as much as 100 years to recover.

IN FOOD

It has been calculated from the Niigata incident that 1.5 milligrams of mercury per day as methyl mercury can be fatal. This corresponds to a diet of 250 grams (about half a pound) of fish per day containing 5 to 6 parts per million of mercury.[4]

The Swedish government has set its standard for the permissible levels of mercury in fish on the basis that 50 parts per million of mercury in fish eaten daily will cause symptoms of poisoning (as shown by the Minamata story). According to Japanese pharmacological data, decreasing this amount by a factor of 10 would remove the risk. Dividing by a further safety factor of 5, the level was set at 1 part per million, wet weight.

Unfortunately it was not realized at the time that the Japanese data were for dry-weight analysis of mercury content. Since fish are about 80 percent water, the Swedish standards are set five times higher than intended. The standard should actually have been set at 0.2 part per million. If this were the standard, however, most fishing areas in Sweden would probably have to be closed. The Swedish government has been understandably reluctant to change the standard. Instead, it has recommended that fish be eaten only once a week. In the United States, the Food and Drug Administration based its standard on the corrected data and the fact that people here eat about one-quarter the amount of fish consumed in Sweden. The level has been set at 0.5 part per million, wet weight.[5]

If one-half pound of fish at 0.5 part per million mercury were eaten per day, this would

[4] Mercury found in fish is exclusively methyl mercury. The mercury in egg yolks is 50 to 90 percent methyl mercury; in poultry it is 73 to 74 percent methyl mercury.

[5] As of September 1970 in the United States there were fishing restrictions set in parts of 18 states due to mercury contamination of the waters.

TABLE M.3
MERCURY CONCENTRATIONS IN TUNA AND SWORDFISH

DESCRIPTION	DATE CAUGHT	MEAN CONCENTRATION (PPM) AND STANDARD DEVIATION OF MEAN	
		Wet-Weight Basis	*Dry-Weight Basis*
	Museum Specimens		
Skipjack tuna, Mass.	1878	0.27 ± 0.02	0.91 ± 0.06
Skipjack tuna, Mass.	1878	0.64 ± 0.02	1.51 ± 0.04
Albacore tuna, Calif.	1880	0.27 ± 0.03	0.59 ± 0.06
Bluefin tuna, Woods Hole	1886	0.38 ± 0.01	1.14 ± 0.04
Skipjack tuna, San Diego	1890	0.45 ± 0.02	1.05 ± 0.04
Skipjack tuna, Hawaii	1901	0.42 ± 0.02	0.92 ± 0.01
Skipjack tuna, Philippines	1909	0.26 ± 0.01	0.53 ± 0.02
			0.95 ± 0.33[a]
Swordfish, Baja, Calif.	1946	0.52 ± 0.10	1.36 ± 0.31
			1.36 ± 0.31[a]
	Recent Specimens		
Albacore tuna, fresh, Calif.		0.13 ± 0.01	0.44 ± 0.05
Skipjack tuna, fresh, Pacific		0.18 ± 0.03	0.62 ± 0.10
Albacore tuna (A), canned in water		0.48 ± 0.04	1.53 ± 0.12
Albacore tuna, canned in oil		0.30 ± 0.02	0.66 ± 0.04
Albacore tuna (B), canned in water		0.38 ± 0.03	1.29 ± 0.11
			0.91 ± 0.47[a]
Six swordfish, fresh, California		0.23 to 1.27	0.94 to 5.08
			3.1 ± 1.5[a]

From G. E. Miller, P. M. Grant, R. Kishore, F. J. Steinkruger, F. S. Rowland, and V. P. Guinn, "Mercury Concentrations in Museum Specimens of Tuna and Swordfish," *Science* 175 (March 10, 1972): 1121.

[a] Mean mercury concentration.

yield 0.114 milligram of mercury daily. This is 7.4 percent of the intake rate which proved fatal at Niigata. Given that mercury is known to cause genetic damage, studies are urgently needed to determine whether such a standard is really safe.

The FDA has found mercury levels of over 0.5 part per million in some canned tuna fish samples. All of this tuna has now been recalled. The swordfish problem appears to be more widespread, with the possibility that as much as 90 percent of the available fish contains more than 0.5 part per million mercury.

Whether the mercury found in tuna and swordfish is due to man's activities in polluting the environment has, however, been seriously questioned. For instance, in one study the concentrations of mercury in the muscle tissue of preserved museum samples of tuna and swordfish were compared to the concentrations in recently caught specimens. The results are shown in Table M.3. This type of data makes it seem unlikely that mercury pollution as yet had any significant effect on the mercury content of such ocean fish as tuna. Apparently because of their position near the top of the food chain in the ocean, tuna and swordfish have always had high mercury concentrations. Unfortunately, this finding does not make them any more healthful to eat.

In contrast, by comparing mercury levels in the muscle tissue of preserved and fresh fish taken over a 50-year period from the Lake St. Clair–western Lake Erie area of the Great Lakes, other researchers have been able to show increases in the mercury levels of freshwater fish. These increases appear to parallel the increasing industrialization of the area.

Studies have shown that of the major types of food only fish contain appreciable amounts of mercury. Some samples of bread, flour, grains, and eggs were found with levels of up to 0.06 part per million mercury; meats and vegetables contained up to 0.04 part per million mercury. Other foods contained less than 0.02 part per million mercury.

Since alkyl mercury compounds are so toxic to man and since concentrations can be magnified by organisms in the food chain, the Environmental Protection Agency in March 1972 *suspended*[6] registration of all alkyl-mercury-containing products used as **Pesticides**. In addition, since other types of mercury compounds can be converted to alkyl mercury in the marine environment, the use of any mercury compound for rice treatment, laundering, or marine paints was suspended. These uses were considered to constitute an imminent hazard to the public health. Further, stating that "all pesticidal uses of mercury pose a substantial question of safety," the agency canceled registrations for such mercury products.

[6] *Suspension* immediately bars further interstate shipment of a product. Manufacturers challenging a *cancellation* order, on the other hand, may ship their product until the case is finally decided.

HANDLING

Mercury poisoning has long been recognized as an industrial hazard, but it is also a possible danger in laboratories, hospitals, and schools. When mercury is spilled, it is almost impossible to clean up completely, as anyone who has ever dropped and broken a thermometer is aware. The mercury breaks up into minute droplets and lodges in cracks and crevices. The increase in its surface area increases its vaporization. Dirt and grease collecting on the surface of the droplets can decrease vaporization. The gradual decrease, plus the fact that normal evaporation is retarded by vapors collecting near the mercury surface, is probably why more mercury poisonings don't occur. These control methods are suggested for spilled mercury. After the large particles have been swept up, a special sweeper designed to vaporize mercury out of small cracks and then suck it up may be used, or powdered sulfur may be dusted over the area to form nonvolatile mercury sulfides, or the area may be coated with commercial aerosol hairspray.

REFERENCES

1. R. A. Wallace, W. Fulkerson, W. Shults, and W. Lyon, *Mercury in the Environment: The Human Element*, Oak Ridge National Laboratory Report ORNL-NSF-EP-1, Oak Ridge, Tenn., 1971.
2. R. Q. Stahl, *Air Pollution Survey of Mercury and Its Compounds, Litton Systems Incorporated*, Air Pollution Technical Data Series, U.S. Environmental Protection Agency, Office of Air Programs, 1969.
3. T. A. Wojtalik, "Literature Review: The

Accumulation of Mercury and Its Compounds," prepublication copy, TVA, Division of Environmental Research and Development, Muscle Shoals, Ala., 1970.

4. J. Gavis and J. F. Ferguson, "The Cycling of Mercury Through the Environment," *Water Research* 6 (1972): 989.
5. TVA press release, Knoxville, Tenn., October 14, 1970.
6. M. A. Churchill, "Mercury in Flesh of Fish Taken from TVA Reservoirs," TVA information release, Chattanooga, Tenn., January 15, 1971.
7. Neville Grant, "Mercury in Man," *Environment* 13, no. 4 (1971).
8. Terri Aaronson, "Mercury in the Environment," *Environment* 13, no. 4 (1971).
9. G. E. Miller et al., "Mercury Concentrations in Museum Specimens of Tuna and Swordfish," *Science*, March 10, 1972, p. 1121.
10. H. V. Weiss et al., "Mercury in a Greenland Ice Sheet: Evidence of Recent Input by Man," *Science*, November 12, 1971, p. 692.
11. *Federal Register* 36, no. 234 (1971): 23239.
12. *Federal Register* 37, no. 61 (1972): 6419.
13. C. E. Billings and W. R. Matson, "Mercury Emissions from Coal Combustion," *Science*, June 16, 1972, p. 1232.
14. E. Arrhenius, Letters, *Science*, June 9, 1972, p. 1072.
15. J. Miettinen, Letters, *Science*, June 9, 1972, p. 1074.
16. *New York Times*, March 9, 1972, p. C-3.
17. R. J. Evans, J. D. Bails, and F. M. D'Itri, "Mercury Levels in Muscle Tissues of Preserved Museum Fish," *Environmental Science and Technology* 6, no. 10 (1972): 901.
18. J. T. Tanner et al., "Mercury Content of Common Foods Determined by Neutron Activation Analysis," *Science*, September 22, 1972, p. 1102.

N

Nitrates and Nitrites

Special care must be taken to assure that chemicals which are added to human foods or which may be present in drinking water are safe at the allowed concentrations. Concern has been expressed that this is not the case with nitrates and nitrites.

SOURCES IN THE HUMAN DIET

Humans encounter these chemicals from three main sources: (1) Nitrates and nitrites are food additives used in cured meat products such as frankfurters and corned beef. (2) Water supplies may be contaminated with nitrates from fertilizer and from the natural weathering of rocks and soil; nitrogen fixation by microorganisms may also furnish nitrates in water, and nitrates may leach into groundwater from sewage wastes disposed of in septic systems. (3) Certain vegetables, notably spinach, beets, radishes, eggplant, celery, lettuce, collards, and turnip greens, naturally contain large amounts of nitrates.[1] Nitrate concentrations in vegetables have been shown to be increased by the use of high-nitrate fertilizers. Nutritional deficiencies such as lack of molybdenum may increase plant nitrate concentrations, as may a shortage of sunlight during maturation. Insufficient water and chemical damage to plants may also raise nitrate levels. Vegetables probably contribute as much nitrate to the average diet as cured meat products.

There are two areas of concern with respect to the effects of nitrates and nitrites on human

[1] Values up to 3,000 parts per million have been found.

health. In the first place, *nitrites* can act on hemoglobin in the blood, causing a disease called methemoglobinemia. In this disease the iron present in hemoglobin is oxidized, rendering it incapable of carrying oxygen to the body's tissues. Symptoms of methemoglobinemia are those characteristic of a lack of oxygen. At low levels (10 to 15 percent of the total hemoglobin as methemoglobin) there is cyanosis, a bluish coloration of the lips and extremities. At higher levels (30 to 45 percent of the total hemoglobin as methemoglobin) breathing becomes labored. Recovery is possible if exposure to nitrites ceases. Death, which occurs at about 70 percent methemoglobin, is from asphyxiation.

The second possible danger is that nitrites may react with another group of chemicals, the secondary amines, to form nitrosoamines; the reaction may occur either in foods or in the human body. N-nitrosoamines and N-nitrosoamides are chemicals which have been proved to cause cancer in animal experiments. Some evidence also exists that they may cause birth defects and that they can cause genetic damage.

Nitrates, although poisonous in large quantities, are considered a potential problem because they can be converted to nitrites by microbial action. This conversion is likely to occur in the rumen of cattle or the cecum of horses or in fresh or prepared spinach if conditions are suitable for bacterial growth. Jars of baby food spinach left open and unrefrigerated have caused nitrite poisoning. Nitrates may also be converted to toxic nitrites in damp forage materials, causing instances of cattle poisoning. In adult humans, the stomach and upper intestinal tract are too acidic to allow the growth of microbes which could form nitrites from nitrates. Infants under four months of age, however, may have low enough stomach acidity to allow the conversion.

FOOD ADDITIVES

Nitrates and nitrites are used in the preparation of certain meat and fish products. Curing with nitrates was an early method of preserving meats before the era of refrigeration. It is now used in large part for the flavor it imparts to meat. Meat can be cured either by placing it in a brine or "pickling" solution or by pumping the curing solution through veins in the meat. Originally only nitrates were used for curing. After it was found that nitrites derived from the nitrates were actually responsible for the curing, nitrites were added directly to the product. Nitrites cause the characteristic pink color of cured meat,[2] as well as giving it the "cured" flavor.

From a public health standpoint, however, a third important function of nitrites is to prevent the growth of toxic bacteria like *Clostridium botulinum*, responsible for the deadly food poisoning botulism. This last function is complex and depends on the amount of heat, acid, and salt involved in the curing process. Botulism has occurred within the last few years from the consumption of improperly cured or improperly handled cured fish, such as lightly smoked salmon, tuna, halibut, cod, and chub. Federal law limits the residual concentration of nitrites in *meat* to 200 parts per million (as sodium nitrite). Salmon and shad have the same limit, while smoked chub *must not contain less than* 100 parts per million.

At the 200-parts-per-million level, 100 grams of meat could convert between 1.5 and 5.7 percent of an adult's hemoglobin to methemoglobin. This may be compared to normal blood concentrations of about 0.7 percent methemoglobin. The standards are based on the estimated amount of hemoglobin which would be inactivated in the worst case. The worst case is

[2] The color is due to the reaction of nitric oxide, formed from nitrites, with myoglobin in the meat, which results in the formation of nitrosyl myoglobin.

presumed to be the heavy smoker living in an urban area who eats one-quarter pound of corned beef daily (on either white or rye). His smoking may inactivate 7 to 10 percent of his hemoglobin by furnishing carbon monoxide. **Carbon Monoxide** from cigarettes and from urban air will combine with hemoglobin to form carboxyhemoglobin, which is incapable of transferring oxygen to the tissues. The total amount of hemoglobin "tied up" by carbon monoxide and nitrite in this individual would theoretically fall just short of causing the first symptom of oxygen starvation, headaches.

Because of their smaller blood volume, children may be more likely to be poisoned by excessive amounts of nitrite in foods. For example, a case has been recorded in which children were poisoned by eating frankfurters and bologna containing more than 200 parts per million of nitrite.

Thus, the situation with respect to nitrites and nitrates in food is complicated by the facts that, (1) although nitrite in high concentrations is toxic, especially to children, too little nitrite in certain foods leaves open the possibility of food poisoning due to bacterial growth, and (2) people seem to have become fond of the flavor imparted by a chemical used originally as a preservative.

NITRATES IN DRINKING WATER

The most common cause of methemoglobinemia is the presence of nitrites or nitrates in drinking waters. Nitrates from chemical plants or farm fertilizers can find their way into water supplies, most commonly into farm well waters. Nitrites are also sometimes added to private water supplies as anticorrosion chemicals.

Nitrate poisoning from water supplies occurs almost exclusively among infants during their first few months of life. They can be poisoned by the water itself, by their mother's milk if she drinks the water, or by the milk from cows drinking it. Infants appear to be the only susceptible group because of their previously mentioned low stomach acidity. Also, infant hemoglobin may be more readily converted to methemoglobin than adult hemoglobin, and infants suffer a decrease in hemoglobin after birth.

There is some controversy over the level of nitrate in water supplies which is likely to cause infantile methemoglobinemia. The main problem seems to be that waters are not always sampled until some time after a poisoning occurs. The level of nitrate may well have changed in the interval. In any case, no poisonings have occurred at levels less than 45 milligrams of nitrate per liter of water (45 parts per million). This is the recommended limit for public water supplies.

Nitrate does occur in waters at levels above this recommended limit.[3] Since nitrates currently are not being removed from drinking waters,[4] public health officials should warn residents in such areas of the possible danger to infants.

NITROGEN OXIDES IN THE AIR

Very high concentrations of nitric oxide in the air could theoretically cause methemoglobinemia (and have done so in animal experiments) but are unlikely to be encountered by the general public at the present time.

NITROSO COMPOUNDS

Not all the information necessary to assess the public health significance of carcinogenic nitrosoamines is yet available. The chemicals which could react to form N-nitrosoamines are present in food. Saltwater fish, for instance, are high in amines like trimethylamine and dimethylamine;

[3] Methods of removal are available. See **Water Treatment and Water Pollution Control, tertiary treatment.**
[4] See **Drinking Water Standards.**

the curing process for preserved fresh or lightly smoked fish then adds nitrites. Whether the reaction to form compounds like N-nitrosodimethylamine does take place, however, is not known. Reliable data on the concentrations of nitrosoamines in other foodstuffs are not available either.

It has been reported that tobacco smoke contains large quantities of both amines and nitrate. Further, the reaction which forms nitrosoamines is enhanced by a variety of agents, among which is the thiocyanate ion, known to be found in higher concentrations in the saliva of smokers than of nonsmokers. Whether these two factors act to increase smokers' exposure to nitrosoamines cannot be stated at this time. Smokers, however, might take note of the possibility.

It is also possible that nitrites and secondary amines once taken into the body combine to form nitrosoamines. Wolff and Wasserman [1] feel that the normal highly acid condition in the human stomach and upper gastrointestinal tract prevents this reaction. They argue that the growth of bacteria capable of either transforming nitrate into nitrite or forming nitrosoamines from nitrite and secondary amines is inhibited. They further point out that nitrates and nitrites rapidly move out of the stomach and upper gastrointestinal tract and are excreted in the urine. Thus, perhaps some danger of nitrosoamine formation in the bladder exists. This formation has, in fact, been demonstrated experimentally in rats.

Certain drugs may also provide the materials for nitrosoamine formation in the body. Oxytetracycline, antipyrine disulfiram and tolazamide are examples of drugs which might be suspect.

To summarize, then: High levels of nitrites derived from the nitrates in drinking water may lead to methemoglobinemia among infants. In addition, nitrosoamines are carcinogenic substances. Nitrates, nitrites, and secondary amines, the chemicals which can be converted in test tube experiments into nitrosoamines, are found in food, drugs, tobacco smoke, and drinking water. Nonetheless, information is not available on whether nitrosoamines are actually present in foods or tobacco smoke or whether they are formed from the contributing substances in the human body. This information is necessary in order to make intelligent decisions on the safety of nitrates and nitrites as food additives. It is also needed in order to set permissible nitrate levels in drinking waters. Therefore, it would be only prudent for the government to encourage and support further research in these areas. Moreover, as long as nitrates and nitrites are allowed to be food additives, the information available ought to be brought to the public's attention. The individual consumer would then be allowed to make an informed decision on whether the possible risks are balanced by sufficient gains to continue using such products.

REFERENCES

1. I. A. Wolff and A. E. Wasserman, "Nitrates, Nitrites, and Nitrosoamines," *Science* 177 (1972): 15.
2. *Public Health Service Drinking Water Standards*, U.S. Dept. of Health, Education and Welfare, Public Health Service, 1962, pp. 47–50.
3. Air Quality Criteria for Nitrogen Oxides, AP-84, U.S. Environmental Protection Agency, National Air Pollution Control Administration, 1971, p. 9-1.

Nitrogen Oxides

See Air Pollution, contaminants—photochemical pollution; Photochemical Air Pollution, nitrogen oxides.

Noise Pollution

One way to assess the degree to which people are bothered by a pollutant is to determine how much they are willing to spend to rid themselves of it. By this measure we must conclude that noise pollution is not a serious problem.

Advertisers rarely mention quietness in appliance operation and can show that purchasers have not responded well when they do. Many people, in fact, equate noise with power. Buyers of lawn mowers, motorcycles, and snowmobiles are suspicious of quiet models. A group of homemakers asked to test a quiet vacuum cleaner queried, "Is it really cleaning?"

Property values around freeways and airports may not be significantly lower than in quieter areas. One study found that the most common response to a noisy freeway environment is to keep the windows shut or to put up storm windows.

And perhaps it may seem that simple. Noise is a somewhat unusual form of pollution because if you ignore it it will go away. Unlike **Mercury** or some **Pesticides,** noise does not accumulate as an environmental contaminant. Rather it is a form of energy pollution. Another example of energy pollution is **Thermal Pollution.** Such types of pollution decay rapidly with time. It would be wrong, however, to assume that since the pollutant itself is short-lived there are not lasting effects.

A well-documented body of information exists showing that noise can adversely affect humans in both physiological and psychological ways. Hearing losses in particular occupations such as boilermaking and artillery manning are well known. In fact, however, we all find hearing more difficult as we age. Young ears can distinguish a wide range of sounds, from low to very high frequencies, while older ears lose the ability to distinguish high-pitched sounds. A comparison of some industrialized versus nonindustrialized peoples suggests that this hearing loss may not be a requisite accompaniment of old age.

The Mabaans, an African tribe from the southeastern Sudan, whose environment does not include even moderately loud sounds, do not experience old-age hearing loss. Mabaans who move to industrialized Khartoum, on the other hand, show both hearing loss with age and an increase in heart disease. Similar findings are reported for the Todas, who live in a pastoral region of India.

Furthermore, a closer inspection of other data reveals economic effects. For instance, an increased turnover in property has been observed in noisy areas near airports. Job performance can be adversely affected by loud noise, especially if accuracy and mental effort are involved. The use of outdoor areas for conversation is not possible for an estimated 5 to 10 million people who live or work in urban areas. When interference with television or speech or sleep is included, as many as 22 to 44 million people can be said to have lost part of the use of their homes and grounds because of noise.

Thus, contrary to our initial conclusion, noise pollution is a serious environmental concern. The apathetic attitude toward noise should be overcome; vigorous efforts should be made to alert people to the grave effects which may stem from an excessively noisy environment.

MEASURING SOUND

Sound is produced by the vibration of some object, such as vocal cords, the hood on a car or truck, or the tiny diaphragm inside a telephone receiver. The vibrating material alternately pushes against the adjacent air, compressing it, and then retreats, allowing it to expand. This compression-expansion phenomenon, propagated through the air, is called a sound wave. If it reaches a suitable surface, like an eardrum, it can set that vibrating.

A sound wave is characterized by its particular frequency, which is the number of times per second that the air is compressed and expands. The frequency of compression determines the pitch; the greater the frequency, the higher the pitch. If the frequency is between 20 and 20,000 hertz,[5] a normal young ear will hear it. Middle C on a piano corresponds to 260 hertz, high C to 1,000 hertz. The range of frequencies audible to other species can vary.

A sound wave also has its own particular strength or intensity, a characteristic which is related to loudness. There is a tremendous range, over a billionfold, between the softest sound to which the human ear responds, such as the whisper of a leaf, and the loudest sound to which it has been exposed, probably the liftoff of a Saturn rocket. In order to reduce this large range to a more manageable size, a logarithmic scale is used. Each unit on the scale is called a bel, which is further divided into 10 smaller parts called decibels.[6]

[5] Frequency is measured in hertz, which equal cycles per second, or the number of alternate compression-expansions per second.

[6] The mathematical relationship by which decibel levels are calculated can be expressed in two ways. In one formula the sound pressure level is related to a reference pressure of 0.0002 microbar. In the other formula the energy level of the sound, in picowatts, is related to a reference energy level of 1 picowatt—that is, $db = 10 \log_{10} I/I_o$, where I equals sound energy level and I_o is the reference energy level of 1 picowatt.

Logarithmic scales respond differently to changes in quantities from the way the more familiar linear scales do. Thus, if the intensity of a sound is doubled, there is a 3-decibel increase on the scale. Suppose one small plane gives a reading of 76 decibels on a sound-level meter; then, two small planes will cause a doubling of the intensity of the sound, but the sound will register 79 decibels on the meter. If the sound from 10 small planes is measured, intensity is multiplied 10 times, and there is a 10-decibel increase on the scale. However, the ear usually judges a 10-decibel increase to be only an approximate doubling of the noise source. In order to invest the concept of decibels with some semblance of reality, Table N.1 lists familiar sounds and their approximate sound intensities in decibels.

Pure tones, composed of a single frequency, are the exception rather than the rule. A flute, played softly, will give almost pure tones, but most sounds have a number of different frequency components. Each component can have its own decibel level. Sound meters equipped with filters can focus on the individual frequency components, or, more usually, on small groups of frequencies, and measure their decibel level. A graph of the frequency groups or bands in a sound versus their decibel levels is like the fingerprint of a sound—a unique picture of the sound, called its spectrum.

While a sound-level meter can respond to the actual sound level at all frequencies, the human ear edits what it receives. High-frequency sounds are judged louder than lower-frequency sounds when both are at the same level of intensity. For this reason a system for weighting frequencies has been devised, called the A scale. The A scale is an attempt to measure sounds in a manner analogous to the way the human ear hears them. A sound-level meter equipped with the proper filters will weight the frequencies, as the ear would hear

TABLE N.1

NOISE LEVEL OF COMMON SOUNDS

Decibels[a]	Sound
	Physically painful to hear
140	
130	
	Maximum in a boiler shop
120	
	Unmuffled motorcycle Jet engine test control room
110	
	Power mower Rock and roll band
100	
	Inside a subway car Electric blender
90	
	10-horsepower outboard motor 20 to 50 feet from heavy traffic
80	
	Tabulating machines in an office 20 feet from a passing automobile
70	
	Dishwasher Accounting office
60	
	Conversation from a distance of three feet Private business office
50	
	Average home
40	
30	
	Broadcasting studio
20	
10	
0	

From "Noise—Sound Without Value," Committee on Environmental Quality, Federal Council for Science and Technology, September 1968.

[a] Measured on the C or unweighted scale, re 0.0002 microbar.

them, will sum their weighted decibel levels, and will register one number, the A-weighted sound level: db(A). Such a measure has been shown to correspond well to the degree that specific sounds annoy people. Other weighting schemes have been proposed and given letter designations, but the A scale is most commonly used.

A final parameter necessary to describe a sound is its variation in time. Sounds are either steady state—that is, continuous, like a siren—or impulsive, like a gunshot. Steady state sounds can fluctuate in intensity. All of these characteristics influence how a sound affects us, whether it is acceptable or whether it is a nuisance.

EFFECTS OF NOISE ON PEOPLE

Research has shown that noise affects people in several different ways. Hearing damage is the most well-known effect. Hearing damage involves some injury to the receptor cells, or the structure containing them, in the inner ear. The eardrum and middle ear are very rarely damaged by noise. Only an extremely loud sound pressure could burst the eardrum. Noise can also have more subtle effects. It may influence task performance, deteriorate one's mental or physical state, interfere with sleep, and increase the rate of accidents.

Sounds in the 60- to 80-decibel range may initiate what is known as a threshold shift. A threshold is the decibel level at which one begins to hear a particular frequency. Threshold shifts mean that the required decibel levels are higher. That is, the sound must be louder before it becomes audible. Shifts can be temporary (the ear will recover in a few hours to a few weeks) or permanent (in which case they represent hearing loss). In general, as sound levels increase and as the time of exposure increases (up to a maximum of 8 to 16 hours), the risk of hearing damage rises. The recovery time from a temporary shift also varies, being longer for loud noises and lengthy exposures. These facts mainly apply to continuous noises. Impulsive noises pose an additional risk of actual mechanical damage to the eardrums or receptor cells, since one may be exposed to very high levels of sound in impulsive forms, like gunshots. A person would not of his own volition stay in an area with such noise levels for any *continuous* period.

In the age range 5 to 10 years, significant hearing damage is mainly caused by toy caps. In the range 10 to 18 years, loss is caused by toy caps and firearms. The main problem in the 18 and over group is occupational exposure. It is estimated that 6 to 16 million workers in such occupations as construction, heavy industry, flying, printing, mechanized farming, and truck driving are exposed regularly to hazardous noise levels. In addition, some recent studies contain disturbing evidence of hearing loss among nonoccupationally exposed young people. Tests given to 7,000 school children show the following relationship between hearing loss and age:

Grade	*Some Hearing Impairment*
6th	3.8%
9th and 10th	10.0%
College freshmen	30.0%

In the same study, when the original college freshmen were tested in their second year of college, 61 percent showed some impairment. It is suspected that listening to amplified rock and roll music may be causing hearing loss among young people on an epidemic scale. At one subcommittee hearing it was averred that certain rock groups contract that they will produce 120 decibels or forfeit payment.

Frequent exposure to noises of 80 db(A) or above and even exposure of a sufficient duration to sounds in the 70- to 80-db(A) range probably lead to diminished hearing ability. It has been suggested that noises in the range of 70 db(A) or greater should be reduced or avoided to prevent gradual hearing loss with age. At least one author reports, however, that the sound of one's own voice at one's own ear is between 72 and 82 db(A). Most people understandably feel this is a comfortable range for sound levels; some find it enjoyable.

Noise causes three types of physiological responses in addition to threshold shifts. First, there is a voluntary muscle response in which the head and eyes turn to the noise source and the body prepares for action. This is exemplified by a startle response to a sudden loud noise. Such reactions decrease very little with repetition. Steady loud noises above 90 decibels cause an overall increase in muscle tension.

Second, the involuntary muscles respond to noises over 70 decibels by general reduction in blood flow to the peripheral body parts (fingers, toes, ears) and changes in such functions as heart rate, breathing, and intestinal motility. Some of these responses, but not all, disappear on continued stimulation.

Third, loud noises can cause neuroendocrine responses which, from animal experiments, appear to affect sexual and reproductive functions and which cause overgrowth of the adrenal glands, among other effects. These results, of course, cannot be directly extrapolated to humans. Nonetheless, workers exposed to high noise levels exhibit an increased incidence of cardiovascular disease, ear, nose, and throat problems, and equilibrium disorders. For instance, German steelworkers have been shown to have a higher than normal incidence of abnormal heart rhythms.

High noise levels obviously cause stress situations. Whether the induced stress can cause organic diseases like ulcers and hypertension has still not been proved conclusively, although it is suspected that it may do so.

Adverse psychological responses to noise have been noted in several areas. Task performance is hurt by noise that is intermittent, although steady state noise even up to 90 db(A) does not seem to have much effect. Complex tasks are affected more than simple ones, and accuracy is reduced more than quantity. Even when good performance is maintained, excess noise is fatiguing. People working in noisy environments are much less tolerant of noise in their nonoccupational environment than those working in quiet surroundings. Certain types of noise can of course be helpful, such as rhythmic noises or those which mask distracting sounds. Time judgment is another response upset by noise. Further, noise has been shown to produce anxiety symptoms such as headache, nausea, and irritability. One study indicated that mental hospital admissions are more numerous in noisy areas than in comparable quiet ones.

It is generally felt that noise interferes with restful sleep. Steady sounds below 33 decibels usually cause no complaints. At levels of 33 to 38 decibels some people are bothered, while at levels between 38 and 48 decibels most people complain. Sounds above 48 decibels will cause extremely large numbers of protests. Intermittent or fluctuating noises such as an occasional passing car tend to be more disturbing than steady noises. People do not seem as able to adapt to noise during sleep as is commonly believed. Laboratory measures of the soundness of sleep—heart rate, blood flow, and electroencephalograph tracings—show that the body responds to repeated noise exposures at virtually the same levels as initially. In general, the effect of loss of sleep on health is not known.

Despite the fairly extensive body of evidence concerning the effects of noise on humans, there is some uncertainty as to how noise exposure standards should be set. Not everyone reacts in the same way to noise, either physiologically or psychologically. The same noise will produce threshold shifts of different sizes in different people. As is pointed out in "Noise Assessment Guidelines" [11], 10 percent of the population can be expected to object to any noise not of their own making, while another 25 percent are disinclined to object to any noise, however loud. The situation is further complicated by the question of how one should define hearing loss. An acceptable definition might be that a hearing impairment indicates difficulty in understanding everyday speech. But what the critical frequencies are for everyday speech is still undecided.

A proposed set of standards for both occupational and nonoccupational exposure is given in Table N.2. The occupational limits would protect 80 to 90 percent of exposed workers. They assume, however, an 8-hour workday followed by 16 quiet hours away from the place of work. This second condition can certainly not always be met. The nonoccupational limits are set some 15 decibels below occupational limits in an attempt to protect 100 percent of the population at all frequencies.

TABLE N.2
RECOMMENDED LIMITATIONS ON EXPOSURE TO NOISE

db(A)	Occupational Limit	Nonoccupational Limit
115	15 minutes or less	2 minutes
110	30 minutes	4 minutes
105	1 hour	8 minutes
102	1½ hours	
100	2 hours	15 minutes
97	3 hours	
95	4 hours	30 minutes
92	6 hours	
90	8 hours	1 hour
85		2 hours
80		4 hours
75		8 hours
70		16–24 hours

From A. Cohen, J. Aticaglia, and H. H. Jones, "Sociocusis," *Sound and Vibration* 4, no. 11 (1970): 12.

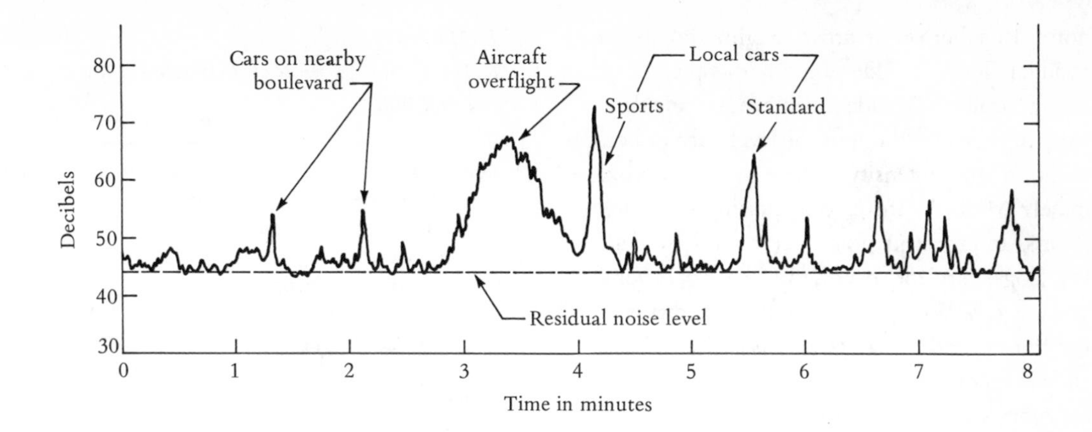

FIGURE N.1

NOISE-TIME PROFILE OF A SUBURBAN NEIGHBORHOOD

From "Community Noise," NTID 300.3, U.S. Environmental Protection Agency, December 31, 1971.

EFFECTS OF NOISE ON WILDLIFE

Noise, like so many other pollutants, affects not only humans but wildlife and other animals. This is an area which has so far been only minimally explored. Evidence from laboratory studies indicates that if noise is great enough to cause hearing loss among animals, predator-prey relationships may change—especially where the prey depends on auditory danger signals or the predator expects to hear its victims. Mating failures could occur where mating calls are necessary. Similarly, community survival may be endangered if distress or warning cries go unheard.

Noise can also cause sexual and reproductive problems, possibly by interfering with neuroendocrine balances. Other effects of noise are behavioral; it can cause huddling, piling up, decreased exploration, or avoidance of certain territory. These may well be maladaptive responses in terms of species survival. Sound in general is classed as an adverse stimulus for mammals, birds, and insects. This, of course, is the basis for commercially available devices which scare away birds and animals from farms or residential areas.

Documented effects of noise on wildlife in field situations are scarce. One of the best known is the hatching failure of sooty terns in Dry Tortugas, Florida. Sonic booms scared the terns off their nests and caused a massive hatching failure in 1969.

MEASURING NOISE

Defining noise is not as easy as defining sound. Sound can be described in purely physical terms while noise demands a value judgment. A simple definition, used in the December 1971 Environmental Protection Agency report on noise, is that "noise is unwanted sound." This is probably true, but it does not help to make noise a quantifiable entity. Nor does it account for individual preference.

A first approach to the problem is to examine the various environmental noise levels in which people now live. The noise level during an afternoon period in a normal suburban neighborhood is pictured in Figure N.1. Several general features of noise profiles are apparent. In the first place, there is a residual noise level or background of steady, unidentified noise. In suburban or urban neighborhoods the residual level is due almost completely to distant traffic. The identifiable noise incidents superimposed on the residual level vary in both duration and intensity. There is an approximately 33-db(A) increase in noisiness at some points. Intermittent noises are caused by barking dogs, airplane overflights, and occasional passing cars. Steady state noises in addition to the hum of traffic may be produced by industry and air-conditioning units. City locations, which are more subject to the steady state noises, may easily be 20 db(A) noisier, day and night, than suburban residential areas.

Three figures are commonly used to indicate the noise profile of a community: L_{90}, L_{50}, and L_{10} or possibly L_1. L_{90} is the noise level exceeded 90 percent of the time. This corresponds to the residual or background level of noise. In Figure N.1, L_{90} or residual level is about 44 db(A). L_{50} is the noise level exceeded 50 percent of the time; half the time the noise is above this level and half the time it is below it. L_{50} may be thought of as roughly the average noise level. L_{10} or L_1 is the noise level exceeded 10 percent or 1 percent of the time, respectively. This figure should represent the level of high-intensity, short-duration noise in an area. In fact, it is the least sensitive of the three measures since there may be very loud noises of very short duration. These can be extremely annoying and yet, because of their short-lived character, would not contribute significantly to the L_1 or, especially, to the L_{10} figures.

Table N.3 gives L_{90}, L_{50}, and L_{10} values for an assortment of environments. They are values, however, for single locations and thus must be viewed as typical rather than average values. The noisiest environment sampled in this particular study was a third-floor apartment next to an eight-lane freeway in Los Angeles. The noise level was so great inside the apartment that even with the windows shut it was not possible to converse in normal tones. The landlord reported that he was unable to rent the apartment for more than a month at a time. By contrast, on the north rim of the Grand Canyon, the average and residual noise levels were over 50 db(A) lower. The major noise problem was chirping crickets. Even the noisiest moments at the Grand Canyon site, due to crickets or sightseeing aircraft, were apparently quieter than the residual noise levels in the Los Angeles apartment and New York City tenement.

TABLE N.3
NOISE PROFILES OF VARIOUS LOCATIONS (IN DECIBELS)

Location	L_{90}[a]	L_{50}[a]	L_1[a]
Third-floor apartment next to freeway	73	77	87
Second-floor tenement in New York City	63	68	81
Urban residential near airport	51	55	92
Urban residential	46	51	66
Suburban residential near city	39	43	62
Small-town residential cul-de-sac	38	42	55
Farm valley	33	36	48
North Rim of Grand Canyon	19	23	43

From "Community Noise," NTID 300.3, U.S. Environmental Protection Agency, December 31, 1971.

[a] Arithmetic averages of the 24-hour values in the entire day.

One criterion for a suitable environment in terms of sound levels is that normal, relaxed conversation should be possible. A certain amount of background noise can be tolerated when listening to someone speak because speech is actually redundant; that is, it can be understood even if some of it is not heard with complete clarity. The amount of noise which can be tolerated depends on how far away the listener is.

At 15 to 20 feet, the length of an average classroom or living room, the background noise level must be below 50 db(A) for normal communication. When two people converse, a comfortable distance is 5 feet. In this case the background level can be as high as 66 db(A). (Compare this to the residual noise level in the Los Angeles apartment cited previously.) Downtown in a large city, where background outdoor noise levels can reach 76 db(A), one cannot converse on the sidewalk without raising one's voice. It has been shown that small children and older people need even lower backgrounds to understand speech at these same distances. There are obvious implications for the acoustic design and location of schools.

A number of studies have been performed to determine what it is about noise that actually disturbs people. It is generally accepted that sound intensity or simple loudness is a necessary but not sufficient measure of the perceived noisiness of a sound. Dr. Karl Kryter of the Stanford Research Institute has found several physical characteristics of sound that affect people's perception of noisiness: the frequencies a noise contains and their sound intensity; the complexity of the frequency organization—i.e., how the frequencies are concentrated into bands; the duration of the noise; and the length of time necessary for the noise to reach its maximum intensity. These factors are felt to control the masking, loudness, noisiness, and startle effects of noise. It is possible, using a weighting scheme for these factors, to calculate a perceived noisiness level (PNL) for sounds. Perceived noisiness is measured in decibel units, which because of the weighting factors are now given the designation PNdb.

Other studies have shown that annoyance may be governed by feelings about the need for a particular noise. Is the noise preventable, and if so are the noise producers attempting to do something about the problem? Is the noise source an important activity (e.g. a military airport might be felt to be necessary for national defense)? Is the noise under the hearer's control (e.g., vacuum cleaner noises are less annoying than airplane overflights)?

Feelings about noise may be affected by the level of noise people are accustomed to, by the time of day (night noises are more annoying than day noises), and even by the age, education, and economic level of the hearer.

There are numerous schemes which attempt to predict people's reactions to noise. Obviously the psychological factors mentioned tend to be extremely difficult to quantify. As pointed out in "Effects of Noise on People" [4], rating schemes will generally have a number of the following features in common:

1. A method of identifying noise events.
2. A method of measuring the intensity and duration of the noises.
3. A tabulation of the number of events in a given time period.
4 An account of the background noise levels.
5. An account of the variation in intensity of the noise events.
6. Special weighting factors such as season (window open or closed), time of day, type of community, pure tone content of the noise, or previous noise exposure of the area.

Noise perception and prediction schemes are usually referred to by their initials. A few of the most common ones are listed in Table N.4. How well these measures correlate with field studies, to determine the annoyance of people who actually are exposed to noise, varies. Some schemes are better than others for particular types of noise.

One fact seems clear: Annoyance is easier to

TABLE N.4

NOISE PERCEPTION AND PREDICTION SCHEMES

Abbreviation	Scheme	Special Features
PNL	Perceived noise level[a]	Used in calculation of CNR
EPNL	Effective perceived noise level[b]	Corrects PNL for noise due to jet aircraft; used to calculate NEF
CNR	Composite noise rating	Original method used to forecast noise exposures around airports
NEF	Noise exposure forecast	Improves CNR by correcting for pure tones and duration; proposed by FAA for prediction of community reaction to airport noise; takes into account factors such as types of aircraft expected to use an airport, subjective noise levels,[b] flight paths, the number of flights during the day and night periods

From "Community Noise," NTID 300.3, and "Effects of Noise on People," NTID 300.7, U.S. Environmental Protection Agency, December 31, 1971.

[a] Measured in PNdb.
[b] Measured in EPNdb.

predict than community reaction. That is, most of the schemes show correlation with how annoyed people actually are, as judged by questionnaires or interviews. Whether they will take action, ranging from informal complaints to legal steps, depends on the presence of leadership in the community, the prevailing attitudes toward regulatory agencies, and other as yet unidentified factors. There is a general correlation between degree of noise exposure and community reaction, but in any particular case prediction of community reaction is not easy. Nor can a typical profile of the person who will complain be drawn.

SOURCES OF NOISE

The major noise sources to which the American public is exposed are: transportation equipment, including aircraft, trucks, cars, and buses; construction equipment; recreational vehicles such as powerboats and snowmobiles; other internal-combustion engines including generators, chain saws, and lawn care equipment; heavy industrial machinery; and office or home appliances such as washers, food blenders, air conditioners, and calculators.

Transportation noise is probably the best-studied noise source, and it also probably has an impact on the largest number of people. In most communities distant traffic provides the background upon which other noises are superimposed. Table N.5 contrasts the growth in the United States population from 1950 to 1970 with the growth in the number of transportation vehicles in the country. Whereas the population increased approximately 30 percent, the number of vehicles in all categories more than doubled during this 20-year period.

AIRCRAFT

The commercial aircraft fleet is made up of a small percentage of nonjet airplanes and a large proportion of turboprop and turbofan jets. The major noise-producing feature of turboprop planes is the high-velocity gas stream exhausted

TABLE N.5

INCREASE IN POPULATION AND TRANSPORTATION VEHICLES IN THE UNITED STATES, 1950–1970

	1950	1970
Population[a]	151	204
Passenger cars[a]	40.4	87.0
Trucks and buses[a]	8.8	19.3
Motorcycles (highway)[a]	0.45	1.2
Turbofan aircraft (2 or 3 engines)	0	1,989
Helicopters	85	2,800

From "Transportation Noise and Noise from Equipment Powered by Internal Combustion Engines," NTID 300.13, U.S. Environmental Protection Agency, December 31, 1971.

[a] In millions.

from the jets, which produces turbulence as it mixes with air. Turbofan jets have less mixing turbulence noise, but the whine of the fan is a significant noise source. The newer fan jets such as the DC-10 are quieter than older types but are in a minority in airline fleets.

On takeoff and landing, turbofan jets deliver from 84 to 105 decibels to a listener 1,000 feet from the craft. Noise levels inside the cabins vary from 72 to 92 decibels for commercial jets to 75 to 100 decibels for commercial propeller planes. A major source of noise inside the aircraft cabin is the vibration of the airplane walls.

One must also consider the problem of the sonic boom. Planes flying at supersonic speeds compress the surrounding air, creating a wake

Jet plane approaching Logan Airport in Boston, Massachusetts.

of increased air pressure, much as a boat does in the water. Waves of this type are formed along the length of the plane, wherever there are discontinuities. The waves interact with each other at large distances away from the plane to cause a bow and tail shock. Sonic booms are judged equal to a subsonic noise of 107 db(A). Because this is an intolerable level, initial plans for the development of supersonic transports were defeated in Congress.

General awareness of the noise problem in the United States came at about the time the commercial jet aircraft was introduced. The number of planes has increased tremendously since then. Although the number of airports is only slightly higher, the size of the population living near airports has grown significantly. Commercial jets are generally derived from military jets, in whose development considerations other than noise were given priority. Thus, the basic concepts of jet aircraft design were formed with no view to producing a quiet product. Newer engines like the CF6 in the DC-10s are much quieter than those in the older jets. However, most existing planes do not meet the current noise standards, FAR-36, which were established by the Federal Aviation Agency in November 1969 because these regulations apply only to new acquisitions. Reductions of up to 14 EPNdb at designated takeoff and landing points can be expected as older planes are replaced.

Although a great deal of research is being done on retrofit measures to quiet existing planes, there is nothing in sight for the near future. On the other hand, it is possible to reduce noise levels by altering takeoff and landing procedures. On takeoff the pilot can climb for an initial period at full thrust. He then reduces thrust for some distance over heavily populated areas and later resumes climbing. Reductions of 3 to 10 EPNdb are possible for the older-model turbofan jets. A landing procedure involving an initial steep descent followed by a normal (3 percent) slope to the runway can reduce noise levels, but it has not yet been proved as safe as a conventional approach.

A promising class of future noise sources is vertical takeoff or short takeoff and landing craft, abbreviated V/STOL. The helicopter is an example of vertical takeoff and landing craft. There are also conventional planes still in the development stage which will require a much shorter (about 2,000 feet) runway than is needed by current aircraft. Because such planes will be able to use small airfields or even land in present urban centers, they possess a large noise pollution potential. Helicopter noise, composed of low-frequency throbbing sounds, is especially difficult to insulate against and is often underrated by common noise rating scales. Helicopters also usually fly at low altitudes because they are used for short hops, surveillance, and sightseeing. It should be possible to shield heliports and urban runways by appropriate placement of nonresidential high-rise buildings and by careful choice of sites, e.g., industrial areas.

HIGHWAY VEHICLES

The problem of vehicle noise was given official recognition as early as the beginning of the century, when mufflers were required on cars to prevent their scaring horses. Today, highway vehicles continue to have a tremendous environmental noise impact. In most urban and suburban communities the residual noise level, that continuous background to which urban man is subject, is due to highway traffic sounds. The average four-lane freeway is traveled by 6,000 to 10,000 vehicles per hour. At distances beyond 100 feet from the highway, their noise blends into a continuous sound whose level depends on the traffic flow. In general terms, there is a 3-decibel increase in level for every doubling of the traffic flow, and a 6-decibel increase for every doubling of vehicle speed.

Up to speeds of about 60 miles per hour, noise increases with speed. This increase is primarily due to tire noise, which for cars first becomes noticeable at about 35 mph. At highway speeds of 50 to 60 miles per hour, tire noise is the major noise source from automobiles, reaching about 65 to 75 db(A). The control of high-speed tire noise is not well advanced at present.

Truck tires are even noisier than passenger car tires. The state of wear of the tire, the type of road surface, and the vehicle's load can cause variations of up to 20 decibels in the noise caused by truck tires. The main problem, however, is their crossbar tread design, which causes noise levels up to 25 decibels higher than automobile tires at 50 to 60 mph. This tread type, also used in passenger car snow tires, wears twice as long as continual-rib automobile tires and also provides better dry and wet traction. Crossbar tread tires are thus preferred from both economic and safety standpoints. When new, the tires generate 80 to 85 db(A) at 55 mph and at a distance of 50 feet, but the figure rises by 10 decibels when they are half worn. There are newer tread designs which are less noisy. However, because of the lower cost of retreads one-half of the tires on the road in 1971 were retreads, mainly from older tread design tires.

Other noise sources from automobiles and trucks include exhaust noises, induced and radiated noise from the engine, aerodynamic turbulence, and body rattles. Cars become noisier with age as mufflers deteriorate and rattles increase. Trucks also become noisier as they age and as they are fitted with replacement mufflers and retreaded tires.

In the early 1970's, 97.5 percent of the trucks on the road were powered by gasoline engines and 2.5 percent by diesel engines. Diesel engine trucks are generally 10 decibels noisier than gasoline engine trucks and 12 to 18 decibels noisier than gasoline engine cars. As tire noise becomes more dominant at high speeds, however, the noise levels of the two truck types converge.

Because buses have their motors located in an enclosed compartment and are usually fitted with better mufflers, they are generally quieter than trucks. At 50 mph, buses reach about 80 to 87 db(A), the noise stemming primarily from tires. But buses generate noise levels of over 90 db(A) at the engine intake grill while accelerating away from a bus stop.

Interior noise levels in transportation vehicles vary from acceptable to somewhat over recommended noise exposure levels. Decibel levels in autos vary among manufacturers. Some economy imports may be 10 db(A) noisier than domestic cars because they have stiffer suspensions, less resilient tires, and less sound treatment of the body. Predictably, expensive models are quieter than economy models among United States brands. Speed is of course important; in one study, on an asphalt road, with the windows closed, automobile interior noise levels ranged from 64 to 73 db(A) at 35 mph to 63 to 82 db(A) at 60 mph. Traveling with the windows open adds 5 to 15 db(A) to these figures, while air conditioning adds something over 5 db(A). Sound levels inside buses range from 72 to 80 db(A), but inside trucks, noise levels may reach 100 db(A), a situation which poses a risk of hearing damage.

To reduce noise levels at lower speeds, better engine and muffler designs as well as the introduction of gas turbine engines for trucks may be necessary. New auto exhaust standards (see **The Automobile and Pollution**) requiring catalytic converters in exhaust systems will require larger mufflers as well, because the higher gas temperature produced will affect noise levels. Dual exhaust systems, which are quieter than single exhausts, will need dual catalytic converters, representing a significant cost increase. Larger mufflers will mean redesigning automobile underbodies, larger radiators, larger fans and fan shroudings. These changes will all be necessary to maintain even the present levels of noise.

RAILROADS

The two general types of rail systems in this country are long-distance freight and passenger trains and rapid transit systems such as subways. In 1971, there were 15 rapid transit systems in the United States. Wheel-rail noise is the most significant problem because the rails tend to develop areas of severe wear both where the train brakes on entering a station and on small-radius curves. The cars used in rapid transit are generally inferior in insulation to long-distance railway cars and thus are noisier inside. Subways suffer additionally from noise reflected off tunnel walls. On the other hand, elevated systems are noisy because their track beds have a less rigid support than ground-based systems.

United States rapid transit systems are among the noisiest in the world, possibly because of the division of responsibility in which transit authorities are responsible for the tracks while manufacturers are in charge of car design. The use of mufflers and engine insulation techniques could be expected to decrease noise levels. For rapid transit systems, welded tracks, ballasted track beds, and improved door seals seem to promise 5 to 15 db(A) reductions in noise levels.

RECREATIONAL VEHICLES

Included among recreational vehicles are snowmobiles, motorcycles, pleasure boats, and all-terrain vehicles.

One of the phenomena of recent times is the explosive growth in the number of snowmobiles operated in the United States. From a level of zero in 1950, the number in existence grew to 1.6 million in 1971; by 1970–1971, sales reached 600,000 per year. About 30 percent are operated by farmers, foresters, and utility servicemen in their occupations. The remainder are pleasure vehicles, of which 80 percent are situated in rural communities of 25,000 or less.

The noise level generated by a snowmobile depends on its age and type. A 1971 model may reach 77 to 86 db(A) at 50 feet while an older, poorly muffled one may generate 90 to 95 db(A). The operator of a snowmobile is in serious danger of permanently losing part of his hearing ability unless he wears a helmet or other ear protection device. Levels at the operator's position are generally about 105 to 115 db(A). Some operators remove factory-installed mufflers in the mistaken belief that they can achieve more power. Actually all they get is more noise since improved muffler and exhaust systems currently prolong engine life and increase power. In a number of areas, laws are pending which would lower permissible levels to 73 db(A) at 50 feet within two to three years.

In 1971 there were 2.6 million motorcycles in the United States, but the number is expected to grow to 9 million by 1985. The noise from a motorcycle increases with its speed. Four-cycle engine models suffer decreases in both economy and performance with silencing, but in two-cycle engine vehicles exhaust muffling can actually increase power. As with snowmobiles, noise levels present a definite hearing damage hazard to the operator. Helmets are recommended both for accident protection and to prevent hearing loss. Motorcycles which operate on roads are legally required to be muffled. The major manufacturers of motorcycles have agreed to reduce noise levels from motorcycles that operate off roads to 92 db(A) at 50 feet. They are also attempting to convince their customers that noise is not synonymous with power. They then intend to produce more heavily muffled vehicles which will weigh more and be less powerful.

Pleasure boats in the United States numbered about 6.5 million in 1971. Of these about 500,000 had inboard motors and the rest had outboards. Noise levels from outboards range from 65 to 90 db(A) at 50 feet while inboards generate 75 to 105 db(A). The noisiest boats are inboard water ski towing craft with dry-stack exhausts (that is, they exhaust to the air as opposed to under the water). Pleasure boat operators are subjected to noise levels of 73 to 115 db(A). There is again risk of hearing damage from the noisier types.

IMPACT OF TRANSPORTATION NOISE

It is estimated that in 1971, 5 to 10 million people in urban areas could not converse in reasonable voices outside their homes or offices, largely because of vehicle noise. Another 2.5 to 5 million Americans were bothered at home by freeway noises, and 7 to 14 million were annoyed with traffic noises in some other way or to a lesser extent. Thus, 7 to 14 percent of the population suffered from poor freeway planning and the lack of effective vehicle silencing devices.

For those living near airports there is the additional stress of air traffic noises. Some 7.5 million people in 1971 were living in "impacted" areas around airports. They are, that is, exposed to noise exposure forecast (NEF) levels of 30 or greater; these levels are considered unsuitable for residential living. Actually an NEF level of 30 is the level at which 30 percent of the exposed population is *very* annoyed. Even at NEF values of 20, 20 percent of the population will be very annoyed. If one includes all those who experience interference with sleep, television, or radio, the number could be as high as 15 million people bothered.

Another serious focus of aircraft noise disturbance is impacted schools. Near John F. Kennedy Airport in New York City alone there are 220 schools serving 280,000 pupils. Teaching is reported to be frequently interrupted by jet flyovers, giving rise to the cynical term *jet-pause teaching.* Similarly, aircraft noise in Inglewood, California, near the Los Angeles International Airport, has rendered some school buildings unusable. Such instances can be expected to increase with further growth in the number of planes in operation unless legislative guidelines for noise reduction are set.

CONSTRUCTION

A typical high-rise building requires about two years of construction time from site preparation to cleanup and finishing. People who live or work in the area are therefore subjected to construction noise in the range 40 to 70 db(A) for that period of time. If the number of people exposed to this type of construction is added to the number exposed to city road construction, public works construction (sewers, water mains) and the construction of non-high-rise buildings, the following estimates of construction noise exposure can be made:

1. Some 34 million people are subjected to noise levels which interfere with normal speech communication on a long-term basis. A total of 25 million probably are exposed briefly to such noise levels as they pass by construction sites.
2. About 3 million adults and perhaps 2.5 million children under four years old suffer sleep disturbance due to construction noise.[7]

It can be seen that although actual hearing damage risk is probably limited to equipment operators, construction noise is still a problem

[7] Data are from [1].

which affects a large segment of the population.

At 50 feet from earthmoving equipment such as backhoes, noise levels range from 73 to 96 db(A). These levels could be reduced by 5 to 10 decibels immediately by better engine muffling. Semi-movable equipment like cement mixers generate 75 to 90 db(A) at 50 feet. Better engine muffling could again achieve noise reductions. Noise from stationary equipment such as pumps and generators would be the easiest to attenuate since this type of equipment can be enclosed. Present levels are 70 to 80 db(A). A long-term reduction of 15 to 20 db(A) in the noise from all types of construction equipment powered by internal-combustion engines is achievable with design changes or a switch to gas turbine or electric power.

In another category of construction equipment—pile drivers, rock hammers, pneumatic wrenches, etc.—the main noise comes from impact. Noise levels may easily reach 100 db(A) at 50 feet. It is possible to use sonic pile drivers instead of impact pile drivers and thus almost eliminate one noise source. The sonic pile driver uses barely audible vibrations to sink piles.

In 1971 it was stated that *operators* of construction equipment generally were making no effort to use existing noise reduction information. Although equipment *manufacturers* were concerned, they did not utilize all of the noise reduction techniques they had available because of the increased cost involved. The concern which was shown, however, existed in part because such countries as Switzerland and Belgium had designed strict noise codes to protect equipment operators. Companies desiring to export equipment to these countries must meet their standards. Recent occupational health and safety legislation in the United States is expected to force the use of the same noise reduction techniques on equipment produced for domestic use. The feeling within the construction industry is generally that federal legislation would be preferable to a patchwork of local ordinances.

Jackhammer in city street.

A number of noise reduction procedures may be implemented in the construction industry. Quieter construction techniques may be substituted where they are available. For instance, welding may be utilized instead of riveting. Concrete may be mixed offsite, and prefabricated modules may be used. Quieter machines can be substituted for noisy ones; electric power could replace the diesel engines, and hydraulic power tools may be used in place of pneumatic power tools. Noisy operations could be scheduled for relatively noisy times of day to take advantage of greater masking, although one wonders whether the possibility of hearing damage might not increase as a consequence. The positioning of equipment might be an important phase in noise reduction. Noisy equipment could be stationed as near the center of the site as possible, and barriers could be erected around stationary equipment or even around the entire site. If present trends continue and no noise reduction techniques are employed in the construction industry, a 50 percent increase in potential exposure to such noises can be expected in the next 30 years.

HOME APPLIANCES

One of the hazards associated with living in a highly industrialized, consumer-oriented society is noise radiated from timesaving home appliances. Lawn mowers, food blenders, garbage disposal units and other appliances can be more than just noisy conveniences, they can actually be dangerous to one's hearing. Home appliances can be separated into four groups, according to noise level and size; the ranking is adjusted to take account of the usual distance which separates the operator from the appliance:

1. *Quiet Major Equipment*
(*Under 60 decibels*)

Refrigerator
Freezer
Electric heater
Humidifier
Floor fan
Dehumidifier
Window fan
Clothes dryer
Air conditioner

2. *Quiet Equipment and Small Appliances*
(*60–70 decibels*)

Hair clippers
Clothes washer
Stove hood exhaust fan
Electric toothbrush
Water closet
Dishwasher
Electric can opener
Food mixer
Hair dryer
Faucet
Vacuum cleaner
Electric knife

3. *Noisy Small Appliances*
(*70–80 decibels*)

Electric knife sharpener
Sewing machine
Oral lavage
Food blender
Electric shaver
Electric lawn mower
Food disposer (grinder)

4. *Noisy Electric Tools*
(*Over 80 decibels*)

Electric edger and trimmer
Hedge clippers
Home shop tools[8]

The first group, which comprises equipment subjecting the user to less than 60 db(A), presents no risk of hearing damage, very little annoyance, and, because noise levels are steady, no startle effects. Some sleep interference is possible from, for instance, air conditioners, but they may also mask other annoying sounds. These appliances are generally the source of home background noise levels. A 10-db(A) reduction in noise levels would probably be desirable to reduce the possibility of sleep interference or interference with speech at short distances from the appliance.

The second group, consisting of quiet equipment and small appliances which subject the user to 60 to 70 db(A), contributes significantly to speech interference and annoyance effects. Because exposures tend to be brief and infrequent, the risk of hearing damage seems negligible. A 10-db(A) reduction in noise levels would allow more normal speech during opera-

[8] Data are from [1].

tion and reduce annoyance of those not operating the equipment.

Equipment in the third group exposes people to noise levels of 70 to 80 db(A). Here there is a risk of hearing damage, especially for those who use the appliances for long periods of time—home seamstresses or yard care workers, for example. Severe to moderate speech interference results from the use of appliances in this group. A 10-db(A) reduction would reduce the risk of hearing damage and ameliorate the annoyance potential of such devices.

The last group produces noise levels at the operator's ear of over 80 db(A), posing a significant risk of hearing loss. Those who use the machines for long periods are in greatest danger. Speech interference and annoyance are effects to be expected for those not making direct use of the equipment. It is conceivable that neighbors' sleep could be disturbed. The operator is also working at levels which have been shown to produce "task interference," especially by masking of warning or danger signals. A 10-db(A) reduction would be beneficial, but a larger reduction is really needed to reduce noise effects to reasonable levels.

Almost all of these appliances could be quieted by existing technology. The reason they are at present so noisy is that manufacturers do not believe people will pay for "quietness" in appliances. Quiet vacuum cleaners, blenders, and hair dryers have not been successfully marketed. However, quiet models of other appliances such as air conditioners and dishwashers have attracted customers, especially second-time customers. As mentioned previously, some individuals equate noise with power. Also, manufacturers apparently feel that the public is willing to accept noise as the price of convenience. Clearly, they need not.

The major noise sources in home appliances are the fans, motors, and cutting blades. Noise is radiated as well from appliance housings and the adjacent walls, floors, cabinets, and sinks. Wrapping and damping, the use of flexible connections and vibration insulation, better balancing, and smoother mechanical connections could all reduce noise.

Reductions of 10 to 15 db(A) in air-conditioner noise could be achieved by better motor and compressor mounting. The noise from dishwashers could be reduced by 10 to 15 db(A) if splash curtains were used to prevent water from drumming against the tubs, if the pump and motor were better insulated, and if rubber hose connections were employed. To quiet garbage disposal units, manufacturers could utilize motor enclosures. The unit could be insulated from the sink, and the sink could be damped to prevent its resonating. Flexible electric and water connections and a cover for the disposer mouth could also be employed. Vacuum cleaner noise could be reduced by better blower design combined with vibration insulations and damping and sealing of the canister. These steps could result in a reduction of 10 db(A) for a canister model and 5 db(A) for an upright model.

Since the main stumbling block to further noise reduction in the home appliance industry appears to be consumer apathy toward appliance noise, or at least the belief of manufacturers in such apathy, a program to educate the consumer about the choices he could have would be helpful. This, along with the labeling of appliances as to their noise levels, which should soon become mandatory in accordance with the Noise Control Act of 1972, would allow consumers to make reasonable choices about the value of noise reduction.

SMALL INTERNAL-COMBUSTION ENGINES

Besides the appliances discussed above, the average homeowner is likely to have or live next door to an appliance powered by a small internal-combustion engine, most probably in the form of a lawn mower. Also in this group are leaf blowers, tillers, edge trimmers, snow blowers, and garden tractors. Chain saws and the small generators used in mobile homes and boats may also be powered by internal-combustion engines. These devices usually have air-cooled, single-cylinder, four-cycle engines.

In the early 1970's there were 17 million lawnmowers of this type in use in the United States. A typical rotary lawn mower will expose the operator to noise levels of 92 db(A). Riding mowers tend to be even noisier (low to mid 90's) and consequently more dangerous to one's hearing. A suburban homeowner might accumulate four to five hours of noise exposure at an 80- to 90-db(A) level on a weekend morning if he cuts and edges the grass, picks up the clippings, and works his garden. His neighbors would probably suffer speech interference if they tried to use their yard at the same time or left their windows open. Lawn mowers could probably be quieted by up to 20 db(A). In the absence of consumer demand, however, the industry does not currently feel pressured to make the change. The use of a two-cycle engine with two cylinders would reduce noise levels but increase both initial costs and fuel costs, the former by 30 to 50 percent.

Chain saws are a particular problem because of their high noise level, up to 115 db(A) at the operator's position. Since they must be portable and lightweight, they are very difficult to muffle. Even at 50 feet, noise from chain saws has been measured at 83 db(A). Hearing can be damaged up to 25 feet away, and speech will be interrupted even several hundred feet from the chain saw. Used in suburban communities for tree removal, these devices will obviously be annoying to residents. The engine noise can probably be reduced somewhat, but the problem of blade noise, which is significant, is not easy to solve. Reductions of 5 db(A) are possible. Table N.6 indicates noise from equipment powered by small internal-combustion engines.

TABLE N.6

NOISE FROM SMALL INTERNAL-COMBUSTION ENGINES AT DISTANCE OF 50 FEET

Equipment	Decibels[a]
Lawn mowers	74
Garden tractors	78
Chain saws	82
Snow blowers	84
Lawn edgers	78
Model aircraft	78
Leaf blowers	76
Tillers	70

From "Transportation Noise and Noise from Equipment Powered by Internal Combustion Engines," NTID 300.13, U.S. Environmental Protection Agency, December 31, 1971.

[a] Levels at the operator's position will generally be higher.

BUILDING EQUIPMENT AND TRANSMISSION IN BUILDINGS

Multistory buildings require mechanical and electrical devices to maintain them at the proper temperature and otherwise serve the occupants: pumps, boilers, steam valves, elevators, air compressors, fans, and emergency generators. Although noise levels from these pieces of equipment may reach 90 or 100 db(A) at 3 feet from the source, insulation techniques for the equipment and the building can reduce noise to almost any desired level. One of the main factors in noise reduction is cost, as the insulation may add significantly to the cost of the building. Another factor is that certain noise levels may be desirable to maintain privacy.

Building equipment manufacturers are aware of noise problems from their equipment and recognize that they will profit from producing quiet machinery. Advantage has been taken of easily incorporated techniques but the large amounts of money essential for more costly research have not been allocated. The Walsh-Healy Public Contracts Act requires that machine noise in public buildings be no greater than 80 db(A) at 3 feet, and local ordinances are often patterned on this federal legislation. All equipment manufacturers are not able to produce equipment meeting the specifications. In this area, in contrast to others, however, there are pressures for developing quieter equipment. The public expects noise levels to be low in large buildings, and architects want to use open space designs. Furthermore, there are legal requirements to be met. Finally, building owners can successfully use a quiet environment as a sales feature.

Noise is transmitted within buildings in two ways, as airborne noises which require a continuous air path, and as structure-borne noise, illustrated by the impact noises of footsteps or door slamming. Structure-borne noises can be especially irritating since they tend to be both loud and intermittent.

Building codes which specify noise transmission limits were passed in Europe as early as 1938. At the present time, such codes are in force in Austria, Belgium, Bulgaria, Canada, Czechoslovakia, Denmark, England, Finland, France, Germany, the Netherlands, Norway, Scotland, Sweden, Switzerland, and the USSR. In the United States, the Walsh-Healy Public Contracts Act specifies noise levels in public buildings constructed under federal contract. In addition, there are Minimum Property Standards for Multifamily Housing, which are the responsibility of the Federal Housing Authority. The FHA specifies limits on airborne noise and recommends impact noise limits. These standards, however, have not always been enforced. The Department of Defense uses the FHA standards, and there are local ordinances of varying sorts in other parts of the United States.

Noise transmission within buildings can be reduced in several ways. Accoustical materials such as ceiling tile, sound deadening board, or blankets all absorb sound. Wall, floor, and door assemblies can be constructed so they are soundproof. Rooms can be insulated from each other if baffles and traps are used in air-conditioning duct systems. Heating and ventilation system equipment can be chosen for its low noise emission. Plumbing noises can be reduced by avoiding high-velocity or pressure systems and thin-walled, small-diameter piping. Plastic pipes and lightweight bath and shower stalls tend to radiate more noise than their conventional counterparts. A great deal more research is needed on reducing plumbing noises.

ECONOMICS OF NOISE

The economic costs to society of noise are several. Airports are currently operating at less than capacity because of noise regulations which restrict their hours of operation. For instance, at Washington's National Airport no jet traffic is allowed from 11 P.M. to 7 A.M. Other airports restrict the use of certain runways. One estimate is that noise restrictions reduce possible airport use by 20 percent. The profitable cargo trade is especially affected by night restrictions.

Whether property values suffer in noisy locations is still open to question. In one study, a 62 percent smaller changeover was found for a quiet area compared to a noisy one near an airport, although no actual difference in property values was noted. In this case the increased number of sales may have hidden a possible drop in property values. In another comparative study, however, a decline in property values was found in the noisier area. No decrease in value has been found in areas studied near freeways.

In the case of airports, jet engines may be modified to reduce their noise level, or insulation from air traffic noise may be provided by the purchase of land around airports or the insulation of buildings. One estimate is that $5.7 billion would be required to equip all existing jet engines with noise control devices. However, considering the current state of the art, even taking this step will not reduce noise levels at all points to acceptable values. Some combination of methods is probably necessary.

If all aircraft were quieted by existing methods, there would be a number of economic benefits. A decrease in the right-of-way needed for airports would be possible, and an increase in airport capacity would occur. Litigation costs would fall, and property values near airports might rise. Transportation costs to and from airports could be diminished since the airports now could be located closer to population centers. It should be noted that the location of airports at a distance from large population centers does not always solve the problem of airport noise. Dulles Airport was built at some distance from Washington, D.C., but for economic reasons a great deal of industry has grown up around the airport, producing many new jobs. The people want to live near where they work. Consequently, builders are agitating for zoning changes to allow them to build housing in areas previously ruled unsuitable because of noise levels.

As far as freeways are concerned, it seems that improved freeway design and barriers are likely to be less costly (by a ratio of 2 to 1) than buying up buffer land zones. Reducing vehicle noise is not as economical as the first two measures in terms of reducing freeway noise, but, since it would reduce the noise level in the total urban environment, it would still be a valuable area of endeavor.

Cases of workmen's compensation involving noise injuries are few. In 1966 there were only 500. However, it has been estimated that many more workers are actually eligible under present laws. In order to protect all workers who are now exposed to levels producing hearing damage, industry might have to spend $12 million per year. On the other hand, noise reduction controls, if included in the original construction of buildings, could reduce this cost.

Even in the area of household noise, the cost of accidents or fatalities due to noise fatigue, annoyance, and the missing of warning signals is probably significant, perhaps in the billions of dollars per year. The cost of quieting household appliances varies with the type of appliance; estimates to decrease noise from such devices range from 0.1 to 20 percent of the cost of the item. The cost of noise control could place an unfair burden on less affluent members of society, who are limited to purchasing the lowest-cost models of most appliances. On the other hand, the cost to society of medical treatment and research, errors and loss of manpower can be expected to increase if noise pollution continues to grow unchecked.

Much research still needs to be done on the economic aspects of noise abatement and noise effects. Although some of the effects of noise pollution are known, more must be discovered about its effects on health, productivity, property values, and the quality of life. Furthermore, the cost of noise pollution control to the economy as a whole needs to be illuminated.

CONCLUSION

Comparisons of noise level studies carried out in the 1930's, 1940's, and 1950's, with data from 1971 seem to show no change in the residual noise levels where land use patterns have not changed. That is to say, rural areas are still quiet and cities have always been noisy. Yet where rural areas have become suburban or suburban areas have become more urbanized, there has been as much as a 10-db(A) rise in residual noise level. Since the land area occupied by urban areas, freeways, and airports has increased over this period, the United States population as a whole has become exposed to higher residual noise levels.

The 1971 report to the President and Congress by the Environmental Protection Agency estimated that residual noise levels could rise from the 1970 levels of about 46 db(A) to almost 50 db(A) by the year 2000 if there were no aggressive efforts at control. Noise from highway vehicles will increase as the number of vehicles multiplies during this time period. Land areas bordering airports or freeways are expected to grow from the present level of 2,000 square miles to 3,300 square miles. On the positive side, the employment of existing and projected noise control technology will make it possible to decrease the residual level to 42 db(A) by the year 2000, even though the number of noise sources will increase.

The Clean Air Act of 1970 established an Office of Noise Abatement and Control within the Environmental Protection Agency to study noise pollution. The National Environmental Policy Act authorizing environmental impact statements includes noise as one of the aspects to be investigated. However, the Noise Control Act of 1972 provides the authority for what will, it is hoped, be effective measures for reduction of noise. The act states that EPA can set noise emission standards for construction and transportation equipment, including recreational vehicles,[9] as well as for other types of motors, engines, and electrical devices. In addition, EPA is given the authority to label consumer products both as to noise levels and as to their noise-reducing effectiveness. Finally, EPA is to conduct research itself, as well as coordinate other government programs dealing with noise.

[9] But not including aircraft. Authority in this sphere will remain with the Federal Aviation Agency.

It still appears that a combined program of public education and federal legislation is necessary to achieve control of noise pollution. The public must be alerted to the dangers and economic costs of noise pollution so that people may make intelligent choices and exert appropriate pressures.

REFERENCES

The first nine references are reports from the U.S. Environmental Protection Agency, December 31, 1971:

1. "Noise from Construction Equipment and Operations, Building Equipment and Home Appliances," NTID 300.1.
2. "Community Noise," NTID 300.3.
3. "Effects of Noise on Wildlife and Other Animals," NTID 300.5.
4. "Effects of Noise on People," NTID 300.7.
5. "Social Impact of Noise," NTID 300.11.
6. Transportation Noise and Noise from Equipment Powered by Internal Combustion Engines," NTID 300.13.
7. "Economic Impact of Noise," NTID 300.14.
8. "Fundamentals of Noise: Measurement, Rating Schemes and Standards," NTID 300.15.
9. "Summary, Conclusions and Recommendations from Report to the President and Congress on Noise," NRC 500.1.
10. Karl D. Kryter, *The Effects of Noise on Man*, Academic Press, New York, 1970.
11. "Noise Assessment Guidelines, Technical Background," U.S. Dept of Housing and Urban Development, TE/NA 172, 1972.
12. "Noise—Sound Without Value," Committee on Environmental Quality, Federal Council for Science and Technology, September 1968.
13. "The Noise Around Us—Findings and

Recommendations," Report of the Panel on Noise Abatement to the Commerce Technical Advisory Board, U.S. Dept. of Commerce, September 1970.

Nuclear Power Plants

See also Electric Power; Fossil Fuel Power Plants; Radiation Standards; Thermal Pollution.

The Atomic Energy Commission has predicted that by the year 2000, 60 to 70 percent of all electric power generated in the United States will originate from nuclear power plants. In 1971 there were only 22 operable plants, but more than 100 more were either under construction or had reactor orders already placed with manufacturers (Figure N.2). Most of these will be ready before the decade is over. This is literally a tide.

It was only 1956 when the first experimental boiling water reactor went into operation at the Argonne National Laboratory, and it was only about a year later when the first pressurized water reactor began producing electricity at Shippingport, Pennsylvania. The phenomenal growth in reliance on nuclear power for electricity may be attributed to a recognition that the direct economic cost of nuclear power generation will not be far different from the costs associated with conventional **Fossil Fuel Power Plants.** It is also related to the scarcity of sites for future hydroelectric power plants.

FIGURE N.2

NUCLEAR REACTORS IN THE UNITED STATES AND PUERTO RICO, SEPTEMBER 30, 1971

From *The Nuclear Power Industry*, U.S. Atomic Energy Commission, 1971.

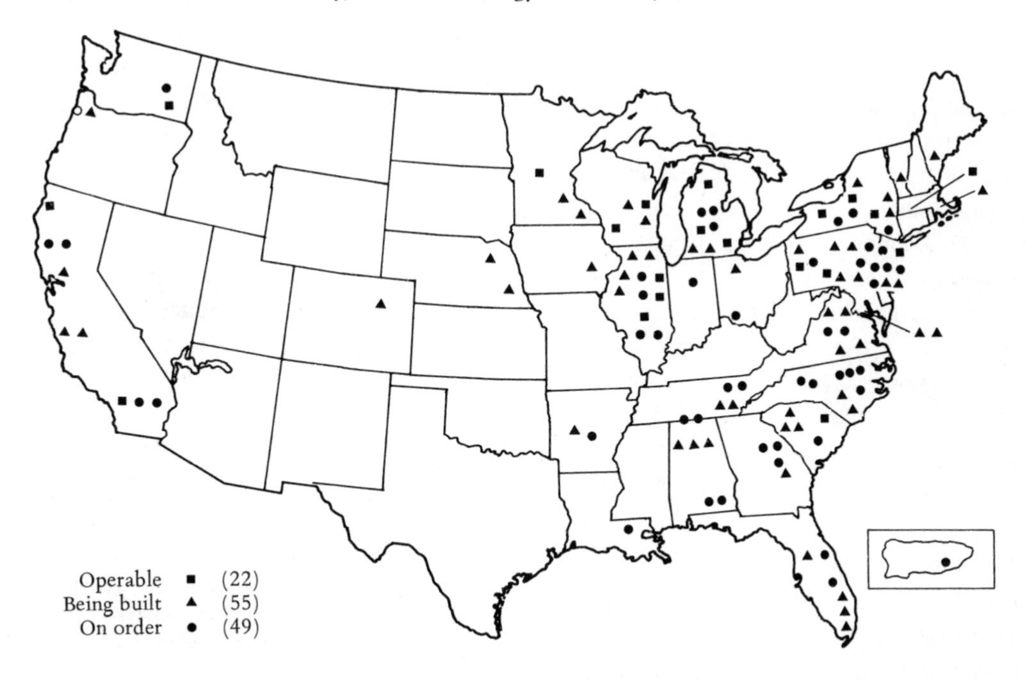

Nuclear power plant at Plymouth, Massachusetts. Heat is produced by fission rather than by combustion. Steam turns the turbines to generate electricity.

The advent of nuclear power does not solve the problem of **Thermal Pollution** which plagues **Fossil Fuel Power Plants.** For today's nuclear reactors the problem is even more severe than for fossil fuel power plants. As for the problem of air pollution by **Sulfur Oxides, Particulate Matter,** nitrogen oxides, and the like, these substances will indeed be absent from the emissions of the nuclear plant. In their place, however, are more exotic substances which are radioactive. The cooling water leaving the plant also carries radioactive substances. We hasten to add that the control of air and water discharges of radioactivity from nuclear plants are probably the most closely controlled of any industrial wastes *ever.*

TWO TYPES OF REACTOR

All but a few of the nuclear power installations in the United States which are currently in place, under construction, or planned are of one of two types: the boiling water reactor (BWR) or the pressurized water reactor (PWR). Common to both is the "core," in which controlled fission of uranium-235 takes place,[10] releasing heat energy, which is captured by a coolant. In both types of reactor the coolant is water, which also serves to slow down neutrons.

The core is made up of basic elements:

1. Enriched uranium dioxide is encased in cylindrical rods of stainless steel or zircaloy (alloy of zirconium) to protect against the reaction of fuel with water and against the

[10] A neutron strikes the U-235 nucleus, creating the unstable element uranium-236. The unstable U-236 breaks apart with a release of energy and neutrons into a light element (mass number 85 to 104) and a heavy element (mass number 130 to 149). There are a number of possible combinations of light and heavy elements. The neutrons strike other uranium-235 nuclei, perpetuating the fission process. The successive fission reactions are termed a *chain reaction.*

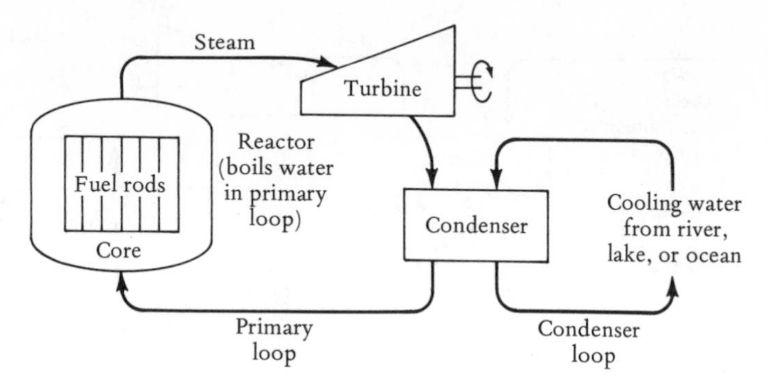

FIGURE N.3
OPERATION OF A BOILING WATER REACTOR

escape of fission products. There is a trend toward the use of zircaloy.

2. Control rods, which absorb neutrons, are used for two purposes, coarse control (start-up or shut-down) and fine control (to regulate operation). The rods, which are held by electromagnets, may be dropped into the reactor to shut down the reactor. Lowering the rods decreases neutron flux and decreases power. Raising rods has the opposite effect.
3. A shielding of concrete and iron or steel layers attenuates gamma rays and neutrons.

Although the components of the cores of the two types of reactor are similar, the operation of the core differs in the BWR and PWR. In the BWR, the water is allowed to boil to steam at about 540°F, and the steam turns a turbine directly for power generation. In the PWR, the water is held under 2,000 pounds per square inch pressure to prevent boiling. The heat which is absorbed from the fission raises the temperature of the water to about 600°F. The heated water in this primary loop passes through a heat exchanger,[11] which boils a parallel stream of water to about 540°F. The steam thus generated is used to turn a turbine.

The BWR operation is referred to as "direct cycle," and the PWR operation is called "indirect cycle." Figures N.3 and N.4 illustrate the two concepts and also show the condenser loop common to both types of plant; the condenser loop carries cool water from river, lake, or ocean to condense the spent steam from the turbines. Not shown in Figure N.4 is the fact that four loops of pressurized water extend from the reactor to four different steam-generating heat exchangers.

ORIGIN AND CONTROL OF RADIOACTIVE WASTES

Radioactive wastes are discharged into the air and water by both the BWR and the PWR plants, although their waste outputs do differ. Before we detail these differences, it is useful to

[11] The heat exchanger finds wide use in the chemical industry and in the power industry. Its purpose is to exchange heat between two streams of liquid or gas. A frequent design consists of a number of parallel pipes inside a large cylinder. One fluid runs in the pipes, the other in the space around the pipes within the cylinder. Ideally, the two streams, which may run in the same or opposite directions, never mix. The stream which was originally warm is cooled during passage (even condensed), and the stream which was cool is warmed (even vaporized or boiled) during passage. Inevitably, in heat exchangers with water streams, corrosion causes eventual leakage and failure.

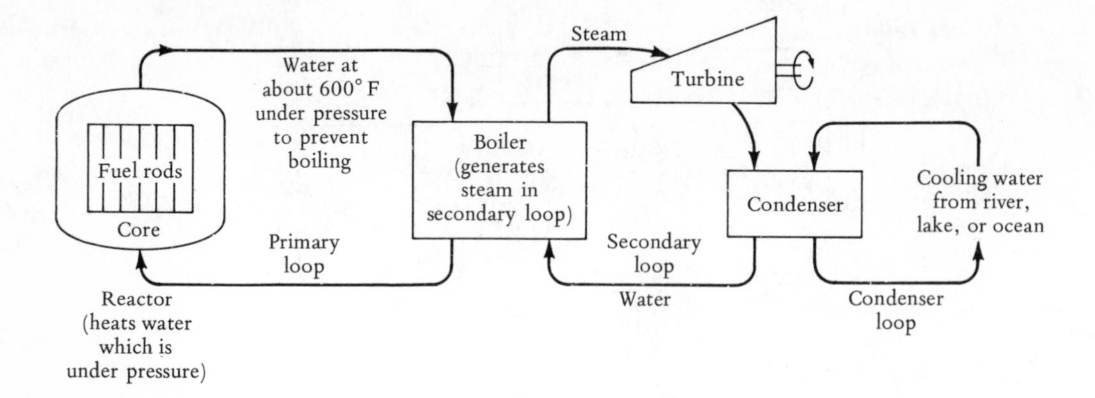

FIGURE N.4
OPERATION OF A PRESSURIZED WATER REACTOR

distinguish the two processes by which wastes are generated: fission and activation.

In the fission process, uranium-235 is converted to uranium-236, which subsequently separates into two large fragments, plus neutrons and occasionally tritium; the process is accompanied by a release of energy. Fission is supposed to be confined to the interior of the fuel rod, but even from there tritium,[12] which may be produced in the fission process, may diffuse out of the rod. The zirconium in zircaloy, the currently favored cladding material for the fuel elements, combines with hydrogen and hence may limit diffusion in newer plants. Not only may tritium diffuse out of the rods, but leaks can develop on account of defective cladding. One percent defects is a figure mentioned in preparing design estimates. The leaks provide the fission products such as radioiodines (I-131 and I-133), xenons (Xe-133 and Xe-135), krypton-85, strontium-89, cesium-137, etc., which reach the coolant. Besides diffusion and leakage, there is another way in which fission contributes wastes in the coolant. In the fabrication of the fuel rods, some uranium is left adhering to the exterior of the rods, and a small amount of fission may take place on the outer surface of the rods themselves instead of inside the rods. The uranium adhering to the surface of the rods is called "tramp uranium" (presumably because it rides the rods).

The second process by which radioactive elements occur in the primary coolant water is called activation. Neutrons from fission may strike and be captured by the nuclei of substances in solution or by substances in the cladding or for that matter by the elements in water itself. The new compounds are frequently radioactive. The iron and other elements in the cladding may be activated, and the activation products may enter the coolant via corrosion. Alternatively, elements which have entered solution via corrosion may be activated while in the coolant. Oxygen in the coolant may also be activated. One typically may find the radioelements cobalt-58, cobalt-60, manganese-54, iron-59, chromium-51, molybdenum-99, and zinc-65 in the coolant. The maximum permissible concentrations (MPCs) of these elements in water discharged to the environment are listed in **Radiation Standards.** Whereas tritium

[12] Tritium is a form of radioactive hydrogen.

production may differ depending on the cladding, it does not appear that the corrosion products will differ significantly with the two cladding materials.

Much has been written about tritium production and escape. Tritium is produced in the PWR and BWR by neutron activation of boron in the coolant and subsequent decay of the induced radioelement, by neutron activation of lithium in the coolant, by neutron capture by deuterium in the coolant, by fission of the uranium, and by other processes. Fission is thought to produce more than all the other processes combined. However, most of the tritium is retained within the fuel rod, so that the quantity of tritium in the primary coolant is derived primarily from the reactions other than fission. Because the tritium is largely retained in the rods, the processing of spent fuel may present a more challenging situation than the power reactor with regard to tritium management. Fuel reprocessing plants are taken up in the next section.

With this background on radiochemical origins of wastes, we can describe the actual technological sources. The PWR plant, which vents gases periodically from the system, discharges much less gaseous wastes than the BWR. Because krypton-85 and xenon-133 (products of uranium fission) from the PWR are very long in decaying, the gaseous wastes which include radioiodines may be stored for 60 to 120 days prior to release through a final filter and venting through a tall stack. The BWR, on the other hand, produces gases (in addition to fission products) which decay rapidly. Nitrogen-13 is the predominant component of the gases which are vented from the condenser; the gases are delayed 15 to 30 minutes in the line prior to final filtering and ejection through a tall stack. We contrast the total activity released in gaseous wastes for the two types of facilities. Of the currently operating PWRs, the maximum total activity is about 55 curies[13] per year; of the BWRs, the smallest total activity is about 145,000 curies per year.

Liquid wastes from leaks and drains in the primary coolant system may be treated and in some cases returned to the system. Treatment of BWR leaks consists of detention and filtration and is followed by dilution and discharge of the wastes with condenser water. If the waste stream in the BWR is to be reused in the reactor loop, the filtered stream is demineralized with an ion-exchange resin and is stored. The waste stream from the PWR is detained and then evaporated. The slurry from the evaporator is mixed with cement and stored in a 55-gallon drum for disposal offsite. The distilled liquid from the evaporator is diluted and discharged with the condenser cooling water. If the waste stream in the PWR is to be recycled into the reactor loop, it is demineralized following evaporation.

The filters and ion-exchange resins eventually become solid wastes, which are generally encased in cement in 55-gallon drums. This is the same fate noted for evaporator slurries, which were mixed in cement and drummed. The solid wastes are shipped offsite to special burial grounds set aside for disposal of radioactive wastes; the grounds are overseen by the Atomic Energy Commission (AEC).

Knowing the origins of the wastes, we are now in a position to compare the performance of operating plants with the standards which were set for them. The 12 plants for which we have data (in operation before 1970) are mostly small in size, only three producing more than 200 megawatts of electric power. The newer nuclear plants will have a much greater capacity; among those now being planned, few are rated at less than 500 megawatts of electric

[13] Curies represent the number of disintegrations per second of radioelements.

power and many will have nearly double that power rating.

The discharge standards set for the plants were derived from AEC regulations for licensing, which required that the exposure above background be less than 5 rems per year on the facility site and less than 0.17 rem per year off the site. However, since meteorological and physical factors vary between locations, different discharge levels are allowable at each plant. It should be noted at this point that the standards for plant performance are undergoing change, but the topic is delayed to later in the discussion.

Up to this time (1972), there has been no record of a plant's exceeding its allowed discharge of liquid and gaseous wastes. In fact, most plants in the period from 1967 to 1970 produced less than a few percent of their allowable discharges of noble gases (such as xenons and kryptons) and activation gases and less than 1 percent of their allowable halogens (radioiodines principally). In the same period, the average annual discharge concentrations of liquid wastes from PWRs were generally less than 10 percent of the allowable levels. Of the BWRs, four of the six plants considered produced average annual concentrations of liquid wastes between 15 and 25 percent of the allowable discharge levels in 1970. A complicating issue is whether simply the total activity of the discharge was recorded and compared to the level specified for combinations of unidentified radioactive elements, or whether the activities of individual radioelements were measured against their respective maximum permissible concentrations. Some plants chose the latter procedure, which seemed to provide a greater cushion for release levels. Several early plants had higher percentages of the allowable release than the others.

It is indeed difficult to fault the performance of nuclear power reactors in meeting the specifications set by the Atomic Energy Commission. Those specifications were based on guidelines recommended by the Federal Radiation Council (FRC). However, as pointed out in **Radiation Standards,** the guidelines set by the FRC were challenged in the late 1960's, and the FRC has gone out of existence in a government reorganization. Further, the role of the AEC as a promoter of nuclear power and the protector of the public welfare with regard to radiation has come under heavy attack. Recently an adversary organization (the Environmental Protection Agency) has been given responsibility to set environmental radiation standards. At the same time, the AEC has set new plant *performance* standards far more stringent than the old.

CAUSES FOR CONCERN

THERMAL POLLUTION

One of the most substantial objections to individual nuclear power plants is founded on the low thermal efficiencies of the plants. Because of these low efficiencies, a great deal of waste heat is deposited in the water, leading to a phenomenon known as **Thermal Pollution.** Less negative people than we simply call them thermal discharges. Optimists can find several positive effects and may call the effect thermal enrichment.

Both the BWR and PWR plants have thermal efficiencies in the neighborhood of about 30 to 32 percent whereas the best conventional **Fossil Fuel Power Plants** have an efficiency near 40 percent. By thermal efficiency is meant the fraction of the heat energy generated which is converted to electric power. Thus, nuclear reactors waste 68 to 70 percent of the heat generated; most of the waste heat goes into raising the temperature of the discharged condenser water. The best conventional plants waste about 60 percent of the energy, part being seen as stack gases at elevated temperatures and the larger portion going into increasing the temperature of discharged condenser water. The conclusion is that more waste heat is discharged into water by the nuclear reactor per unit or watt-hour of electrical energy generated.

In 1971 the U.S. District Court of Appeals in the District of Columbia handed down an important decision. With regard to the nuclear power plant at Calvert Cliffs on the Chesapeake Bay, the court criticized the AEC's implementation of the National Environmental Policy Act (NEPA) of 1970. The AEC, said the court, "should consider very seriously the requirement of a temporary halt in construction pending its [the AEC's] review and the backfitting of technological innovations." One result of the decision may be that new equipment for controlling thermal pollution will be required on many plants. The AEC cited a national power crisis as the reason for its position on not delaying new construction [8].

A recent concept which may eventually alleviate some siting problems associated with thermal pollution and distance from population centers is the offshore station. Both fixed and floating islands in the ocean for power generation are being considered. In 1972 the Public Service Electric and Gas Company announced plans to build a pair of floating stations off the New Jersey coast. The company indicated that it hoped to begin supplying electricity from the two stations by 1980 [6, 10, 11].

REACTOR INCIDENTS

On the basis of experience, the possibility of power reactor incidents which are hazardous to the public appears to be small—in fact, remote. Although several early reactors had serious incidents, the precautions incorporated in newer plants are more sophisticated. This is not to say that minor incidents no longer occur; they are written up regularly in *Nuclear Safety*, a journal devoted to disseminating information on safety in the nuclear field.

However, very little experience is available with the plants now in the building and planning stages, which are much larger than the early plants. Instead of building several of the larger-size plants in a staged program of testing, the AEC has in the works applications for over 20 plants with electric power ratings of 1,000 megawatts or more. Advocates of nuclear power argue that simple increases in scale should produce no insoluble problems. As of January 1971, the largest operating water reactor had an electric power rating of 715 megawatts. This was Dresden-2 (a BWR) in Illinois, and it was the site of an incident (not hazardous to the public) in June 1970. The report of the incident in *Nuclear Safety* (vol. 12, no. 5) concluded with the comment, "It is unfortunate that procedural, mechanical and control inadequacies can be recognized only upon the occurrence of some incident that puts them to the real test."

Although the operating experience of nuclear plants imparts confidence of their safety, one issue has recently been raised which casts a shadow over the wisdom of allowing further growth in the size of nuclear power plants. If a rupture occurred in the piping that supplies the primary coolant (water) to the reactor, the reactor would expel the water in the core as steam because of the intense heat in the core. With little water remaining in the core to carry away the heat from the fission reaction, the rods would heat quickly and melt. The molten core could melt through the reactor vessel and fall to the floor of the containment vessel. A steam explosion could occur if water were present at the base of the containment vessel. If the containment were damaged, fission products such as the radioiodines and noble gases could be released to the environment.

To guard against meltdown, in the event of a

failure of the primary system, power reactors have an "emergency core cooling system." The system responds to a loss-of-coolant situation by injecting water into the reactor vessel to cool the rapidly heating rods. It is the reliability of this safeguard system that has been called into question; some of the reservations originate from members of the staff of the AEC itself [13]. Two engineers within the AEC have indicated their belief that there is a need for a moratorium on power level increases and reactor design changes and that research on core cooling systems should be expanded. In May 1973 the issue was unresolved, and no emergency system had yet been tried in a realistic test.

RADIOACTIVE EFFLUENTS

The Atomic Energy Commission from its inception in 1946 has had an unusual double responsibility: to foster the growth of nuclear power and to protect the public and atomic energy workers from the effects of atomic radiation. In the decade past, the AEC took guidance on standards from the Federal Radiation Council (FRC). With the government reorganization of October, 1970, the functions of the FRC including the formulation of radiation standards and the responsibilities of the AEC in setting standards were transferred to the newly created Environmental Protection Agency. In early 1973 EPA had made no moves to alter the standards.

With regard to radiation discharges, we noted earlier that all nuclear power plants are well under their prescribed release limits. Nevertheless, a controversy arose in the late 1960's concerning **Radiation Standards** and nuclear power plants. To its credit, the Atomic Energy Commission has responded to its detractors and has issued tentative new design objectives for the light-water-cooled reactors [15].

Earlier the AEC had adhered to the environmental radiation standards promulgated by the Federal Radiation Council. The standards called for no more than 500 millirems per year[14] exposure to the most exposed individual in the offsite area and no more than an exposure of 170 millirems per year to the average individual in the offsite area. The release limits or allowable discharge levels originally prescribed by the AEC for new plants were based on these environmental standards. As pointed out above, most plants were able to achieve discharges which were only a few percent of their allowable releases. Now, the AEC plans to change the design objectives to reflect new and much stricter environmental standards. The AEC itself has chosen the environmental standards on which it will base the design objectives.

Given that the setting of radiation standards is the responsibility of the Environmental Protection Agency, one wonders how the AEC can preempt this activity. The answer is simply that the AEC is not setting standards; it is setting design objectives for nuclear power plants. In addition, the environmental standards which the design objectives are meant to achieve are so stringent that EPA is unlikely to take issue with them. The design objectives do not differ greatly from the actual operating experience of nuclear power plants. They represent a formal statement on the part of the AEC that the previous excellent operating experience can be, to a degree, assured in new plants. As of early 1973 the design objectives were *proposed;* they had not been issued yet as rules. The standards and objectives which will now be discussed may yet be altered in the final stage of rule making by the AEC.

The design objectives are meant to provide "reasonable assurance" that annual exposures from plant releases to individuals on the site boundary are less than 5 percent of natural

[14] One rem = 1,000 millirems.

background. Since the AEC assumes that the typical natural background is about 100 millirems per year, the objectives will lead to exposures, above background, of less than about 5 millirems per year. The individual on the site boundary corresponds to the "maximum individual"—that is, the person most exposed. Since the FRC environmental standard for the maximum individual was 500 millirems per year, the design objectives reflect a hundred-fold reduction in the allowable exposure of the maximum individual.

The new design objectives are also meant to provide "reasonable assurance" that the annual exposure of members of a sizable population is less than 1 percent of natural background or less than about 1 millirem per year. This exposure level may be compared to the FRC's environmental standard for exposure, averaged over the population, of 170 millirems per year.

In addition, the allowable radioiodine and particulate concentrations in air are reduced by a factor of 100,000. The reduction is prompted by the recognition that the radioiodines may be deposited on grasses eaten by cows and may thus occur in cows' milk. The human body concentrates iodine in the thyroid gland without regard to whether or not it is radioactive. Finally, the guidelines are intended to provide reasonable assurance that the exposure of any population would be less than 400 man-rems per year per 1,000 megawatts of installed nuclear electrical capacity. The number of man-rems of exposure is the number in the population multiplied by the average exposure of members of the population.

The guides provide for some flexibility in the operation of the plant; this flexibility is built into the guides "to assure that the public is provided a dependable source of power." Two levels of action are specified, depending on the actual experience of the plant. If the release averaged over a calendar quarter when converted to annual releases (by multiplying by four) is likely to exceed twice the design objectives (which are specified on an annual basis), the plant is required to investigate the causes and begin a program to reduce the release rates. If the release averaged over a calendar quarter is likely to lead to annual emissions greater than four times the design objectives, the AEC itself promises to take action to assure a reduction.

It must be noted again, however, that the above objectives and plans are those *proposed* in June 1971 and may differ from the ones finally chosen. It is worthwhile to indicate, though, that the AEC evidently believes these goals to be within reach. The AEC has estimated an increase in nuclear power generation costs of as much as 5 percent if the design objectives were implemented [16].

FUEL REPROCESSING PLANTS

Another area in which concern has been expressed deals not with the releases from individual power plants but with the operation of fuel reprocessing plants.

When the fuel rod is spent, the rod is shipped to a fuel reprocessing plant. There the fuel is dissolved, and the fission products are separated from the remaining fuel. All of the fission products mentioned earlier are treated or disposed of to prevent environmental contamination, but two of those products, tritium and krypton-85, are not at present treatable.

We noted earlier that tritium management is a potential problem for fuel reprocessing plants. Tritium was produced in the fission process and was largely retained in the fuel rods. Even though it is a radioactive element, it is much less hazardous from a radiological point of view than most radioelements, because of the very small energy associated with its beta particle emission.

Additionally, because it is a form of hydrogen, tritium is dispersed in the environment and distributed in organisms almost exactly as hydrogen is (mainly as water). Thus, there is no tendency for tritium to concentrate in particular tissues of the body or to be magnified in aquatic organisms. However, the large quantities to be produced ultimately, as well as tritium's stability (half of an amount of tritium will decay in about 12 years), make it important to consider tritium management. Projections on tritium accumulation from power generation for the year 2000 indicate that the increase in worldwide background radiation due to tritium's presence is likely to be insignificant. It is assumed that tritium will be uniformly distributed as water in the seas and atmosphere. That exposure might amount to 0.002 millirem per year; total background ranges from 120 to 250 millirems in most of the United States. Tritium was introduced to the atmosphere during weapons testing in the 1950's and 1960's. By the year 2000, the quantity remaining in the atmosphere from those tests will have decreased to a level comparable to the level projected for the tritium from nuclear power production for that time. Thus, the total tritium projected will not be of much importance. Nevertheless, local effects near fuel reprocessing plants may someday become significant.

Krypton-85 is a potential problem. The gas is essentially unreactive and will be expected to disperse rather uniformly in the atmosphere. Because it takes about 10 years for half of an original quantity to decay, it will be accumulating in the environment. Although an estimate of its level at the turn of the century is not ominous, the projection for later years gives impetus to plans and research to remove the gas from reprocessing plant effluents. Its lack of reactivity is the principal reason that there is as yet no process for its removal from plant stack gases. However, a removal process based on its physical properties had been proposed. Thus, krypton is not simply a potential hazard associated with reprocessing plants; it requires attention because of its possible global levels.

REFERENCES

1. S. Glasstone and A. Sesonske, *Nuclear Reactor Engineering*, Van Nostrand, New York, 1967.
2. S. Glasstone, *Sourcebook of Atomic Energy*, Van Nostrand, New York, 1967.
3. M. Eisenbud, *Environmental Radioactivity*, McGraw-Hill, New York, 1963.
4. "The Environmental Effects of Producing Electric Power," *Hearings Before the Joint Committee on Atomic Energy*, Parts 1 and 2, 91st Congress, 1st Session, October–February 1969–1970.
5. W. H. Oates, Jr., *Radiation Exposure Overview, Nuclear Power Reactors and the Environment*, U.S. Public Health Service, 1969.
6. T. Anderson, "Offshore Siting of Nuclear Energy Stations," *Nuclear Safety* 12, no. 1 (January–February 1971).
7. M. Eisenbud, "Radiation Standards and the Public Health," *Nuclear Safety* 12, no. 1 (January–February 1971).
8. *Science*, August 27, 1971.
9. "Radioactive Waste Discharges to the Environment from Nuclear Power Facilities," Addendum 1 to BRH/DER 70-2, U.S. Environmental Protection Agency, October 1971.
10. *New York Times*, March 16, 1972.
11. *New York Times*, March 23, 1972.
12. J. Blomeke and F. Harrington, "Waste Management at Nuclear Power Stations," *Nuclear Safety* 9, no. 3 (May–June 1968).
13. *Science*, May 5, 1972.
14. "Water Cooled Reactor Safety," European Nuclear Energy Agency, Organization for Economic Cooperation and Development, 1970.

15. *Federal Register* 36, no. 111 (June 9, 1971): 11113.
16. *Draft Environmental Statement Concerning Proposed Rule Making Action: Numerical Guides for Design Objectives and Limiting Conditions for Operation to Meet the Criterion "As Low as Practicable" for Radioactive Material in Light-Water-Cooled Nuclear Power Reactor Effluents*, U.S. Atomic Energy Commission, January 1973. This book deals with the impact of regulations proposed in [15].
17. A. Weinberg, "Social Institutions and Nuclear Energy," *Science*, July 7, 1972, p. 27.
18. *The Nuclear Power Industry*, U.S. Atomic Energy Commission, 1971.
19. *Science*, July 28, 1972, p. 330.
20. *Science*, November 3, 1972, p. 482.
21. "Nuclear Safety: 1. The Roots of Dissent," *Science*, September 1, 1972, p. 771.
22. "Nuclear Safety: 2. The Years of Delay," *Science*, September 8, 1972, p. 867.
23. "Nuclear Safety: 3. Critics Charge Conflicts of Interest," *Science*, September 15, 1972, p. 970.
24. "Nuclear Safety: 4. Barriers to Communication," *Science*, September 22, 1972, p. 1080.
25. "Calvert Cliffs Court Decision," *Hearings Before the Committee on Interior and Insular Affairs*, Parts 1 and 2, Serial no. 92-14, U.S. Senate, 92nd Congress, November 1971.

O

Oil Pollution

On January 28, 1969, at about 11:00 A.M. during and following normal well-drilling operations at Union Oil Company's Platform A (Lease 402) in federal waters off Santa Barbara, California, an oil leak occurred. Federal and state regulatory agencies were promptly notified, and company executives and management personnel were alerted.

Observations made at 7:30 A.M. on January 29 in a flight over the area by helicopter showed a major violently agitated area about 800 feet east of Platform A. . . . Several small emissions were observable along a west to east line with one small emission west of the platform. . . . An oil slick approximately 25 sq. miles in area extended easterly from the platform. Patches of oil were breaking off from the main body in windrows. These patches were tending toward the Santa Barbara, Montecito, Summerland, and Carpenteria beaches.[1]

This terse and unemotional introduction by an employee of Union Oil describes the beginning of the Santa Barbara oil spill, which, perhaps more than any other in recent years, has caused people to question the policies and procedures which are leading to pollution of the oceans with oil.

Great efforts were made to prevent the oil released at Platform A from coming ashore on the beautiful Santa Barbara beaches. However, a lack of effective preventive devices and a series of storms in the area foiled the attempts at protection. On February 4 and 5, beaches, seawalls, cliffs, homes, and boats were covered

[1] T. H. Gaines, "Pollution Control at a Major Oil Spill," *Journal of the Water Pollution Control Federation* 43 (1971): 651.

by the oil. Dead and dying birds, soaked with oil, littered the beaches.

As dramatic as this episode was, it is actually only a small part of the total problem. There are a number of other sources of oil pollution. Incidents are numerous but mostly small; many are of a chronic nature. It is estimated that 7,500 spills occur each year in United States waters alone, including both accidents and deliberate discharges such as dumping ballast. What should give us pause is that we do not know the effects of this continual addition of oil to the seas. Furthermore, when an oil spill occurs, we are woefully unprepared to minimize its impact.

TABLE O.1

SOURCES OF OIL POLLUTION OF THE WORLD'S WATERS, 1970

Source	Metric Tons per Year[a]	Percent of Total
Used motor and industrial oil	3,300,000	67.2
Tankers[b]	530,000	10.7
Other ships (bilges)	500,000	10.1
Refineries, petrochemical plants	300,000	6.0
Tankers and ship accidents	100,000	2.0
Nonship accidents	100,000	2.0
Offshore production[b]	100,000	2.0
Total	4,930,000	100.0

[a] 1 metric ton equals 1.2 short tons.

[b] Normal operations.

SOURCES OF OIL IN THE ENVIRONMENT

The Environmental Protection Agency attributes worldwide oil pollution of water to the sources listed in Table O.1. The total of almost 5 million tons per year is comparable to the amount lost during the whole of World War II.[2] It is easy to see that, although accidents involving offshore drilling operations or tankers are the most noted oil pollution incidents, they do not add the major share of oil to the environment.

Used oil, both industrial and nonindustrial, apparently constitutes the largest part of the world's oil pollution problem. This is oil which has, for example, been used to lubricate cars, trucks, and other vehicles as well as industrial machinery. Thus, the oil drained from your car on the occasion of an oil change may become a pollutant. Although such oil could be rerefined, the number of companies specializing in this process has been decreasing in recent years. It is not actually known what happens to a good deal of the oil used in transportation and industry each year. Portions are certainly added to the atmosphere by the combustion process, but the bulk of the oil probably reaches the oceans in one way or another.

Some 300,000 tons of oil or oil products are contributed to the environment by refineries and petrochemical plants each year. There are a number of steps in the initial refining of oil where accidental or deliberate spills can occur. For instance, evidence exists that minor spills accompany the loading and unloading of tankers at refineries. During the refining process itself, water is used in several steps. Desalting of the oil and desulfurization both result in the contamination with oil of water used by the refinery; so does the steam-assisted separation of the components of the crude oil.[3] The technol-

[2] A ton of oil varies in volume according to its composition (i.e., gasolines are lighter than diesel oils); however, in general, 7.0 barrels equals 300 gallons equals 1 ton.

[3] Crude oil is oil in the natural state in which it is found in the earth. It is not a single substance but a mixture of many (possibly thousands) of compounds. In order to separate partially the various components of the mixture, oil is refined. Fuel oils which are used for heating, transportation (i.e., gasolines), and the running of indus-

ogy for the closed-cycle operation of refineries and petrochemical plants is available. The amount of water used can be kept to a minimum and then purified of oil for reuse or release. Such operation could reduce the level of oil pollution almost to zero. Systems of this type are, however, not commonly encountered. For instance, a refinery at Southampton, England, discharges 120,000 gallons of contaminated water per minute into a salt marsh community. The water contains 25 to 30 parts per million oil. This effluent has caused an ecosystem collapse in the marsh by killing off an essential type of grass, *Spartina.*

Normal sea transport operations by tankers and other types of ships add a million tons of oil to the aquatic environment yearly. Ships accumulate oil in the bilge water, which is often pumped directly into the ocean. Tankers, carrying oil, have special problems. Their storage tanks must be filled with either oil or water in order to provide stability and maneuverability. Thus, as it unloads its oil, a tanker must fill its tanks with water. In this way, the water becomes contaminated with oil. The tanks must also be periodically cleaned, a process which generates more oily water.

These problems can be circumvented by a load-on-top technique.[4] This procedure is more complex than merely discharging oily ballast into the ocean, but, besides having the desirable result of not polluting the oceans, it can even be profitable. If the refinery to which the oil is delivered already has certain equipment (for desalting the recovered oil), the cost of the procedure is less than the value of the oil saved. On the other hand, load-on-top is suitable for use only with crude oils which are to be refined. Specialty crude oils which cannot be mixed with other types of oil, and already refined fuel oils cannot be loaded by this method. Neither can the technique be used to prepare a tanker for entry into dry dock. In these cases oily residues must be pumped into a slop tank, thence to be siphoned into shore facilities for processing. The load-on-top technique requires that the tanker have storage facilities for oily water and that shore facilities be available to receive the water. Unfortunately, since few areas have such essential facilities, even tankers properly equipped may find themselves forced to dump oily water into the ocean. It is estimated that 80 percent of the world's tanker fleets use the load-on-top technique. The other 20 percent are responsible for some 145 million gallons of discharged oil each year.

The world production of oil was estimated in 1971 at 600 billion gallons per year. Of this total, 60 percent is moved at sea. Oil also constitutes 60 percent of all the goods transported at sea. Figure O.1 compares the size of seaborne shipments of oil with that of other products. Thus, 360 billion gallons of oil must be loaded and unloaded annually and most move through restricted shipping lanes. It is not surprising, therefore, that tanker accidents account for a significant portion of the oil pollution of the seas. Between June 1964 and April 1967, 19 tanker groundings were recorded, resulting in 17 spills. In the same period, 238 collisions occasioned 22 spills.

TANKER INCIDENTS

The wreck of the *Torrey Canyon*, in 1967, while en route from the Persian Gulf to Milford Haven, was a widely publicized event; it was carrying a cargo of 117,000 tons of crude oil. In

---

trial machinery are some of the fractions derived from crude oil; other fractions are lubricating oils, waxes, and asphalts. Each is itself a mixture, but a less complex one than the original crude oil. Crude oils vary in composition and are often named for the area in which they were found (e.g., Kuwaiti crude oil).

[4] A complete description can be found in the article by M. P. Holdworth in [18].

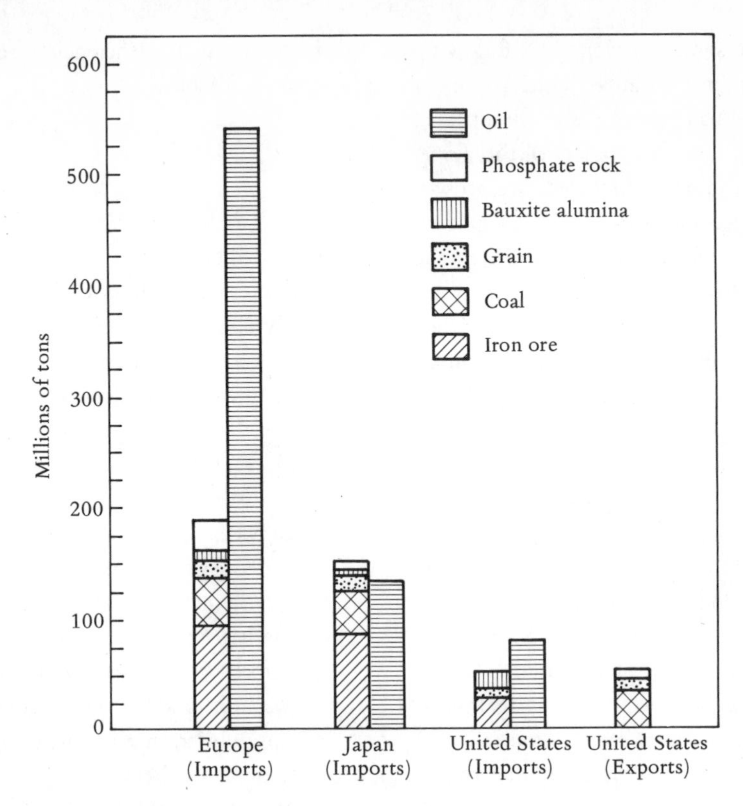

FIGURE 0.1

SEABORNE TRADE OF MAJOR BULK COMMODITIES, 1969

From *Deep Water Port Developments,* IWR Report 71-11, U.S. Army Corps of Engineers, December 1971. Prepared by Arthur D. Little Co., Cambridge, Mass.

an apparent attempt to save time, the captain cut too close to Land's End and ran aground on Pollard Rock of the Seven Stones, 15 miles west of Cornwall. This was the same place the *Thos. W. Lawson* went aground in 1907, spilling 8,000 tons of crude oil, in one of the earliest tanker spills recorded. The *Torrey Canyon* was hung up on the rocks for six weeks despite efforts to set it afire with bombs; it is thought that almost all the oil the ship carried was lost during that period. A large portion was carried out to sea by winds, but 36,000 tons reached the beaches of Brittany, Land's End, and the north and south Cornwalls.

The infamy of the *Torrey Canyon* accident is not attributed to the oil itself. Although the oil caused great harm to marine birds, the slick would apparently have posed only a small acute hazard to marine life. The incident is important because great quantities of detergents were used to clean the beaches. An estimated 2 tons of detergent per ton of oil were consumed in preparing the beaches for the summer holiday season which was due to start in a few weeks. The detergents and the detergent-oil mixture were shown to be the cause of a massive kill of coastal life forms. Fish, crabs, lobsters, shellfish, starfish, and shrimp were destroyed in waters as deep as 7 fathoms and up to a quarter-mile from shore. In the intertidal zone all organisms were killed except for some anemones. A follow-up study has shown that in areas which were *not* treated the oil had completely disappeared within two years. Long-term effect of the *Torrey Canyon* spill and cleanup included overgrowth of seaweeds due to the loss of normal grazing sea creatures. The balance of marine life in the area is apparently being slowly restored.

If the location of an oil spill can be described as in any way fortunate, the one that occurred off West Falmouth, Massachusetts, in 1969 might be so designated. Scientists from the nearby Woods Hole Oceanographic Institute were able to do a detailed analysis of the effects of the spill with sophisticated equipment. The accident involved the coastal barge *Florida,* which went aground in Buzzards Bay off West Falmouth on Cape Cod. Although the spill was relatively small compared to that of the *Torrey Canyon* (168,000 gallons were lost, as against 30 million from the *Torrey Canyon*), mortality of marine life was high. This is attributed to the fact that No. 2 fuel oil was being transported by the barge. This fuel is the type used in home heating, and the refining process makes it richer in the more toxic aromatic and phenolic fractions than is crude oil. Winds were onshore for two days after the spill, and heavy seas mixed the oil into the water.

The wreck of the tanker *Arrow* in Chedabucto Bay off the coast of Nova Scotia in 1970 illustrates the extreme lack of enforceable standards for tankers and their crews. A member of the study group that reported on the disaster pointed out that the radar, depth sounder, and gyrocompass of the tanker were not working, and the standard compass had a variable error. No proper coastal charts were being carried, and the crew had been poorly trained.

An accident in San Francisco Bay in 1971, involving two tankers from the same company, is reputed to have been caused by a lack of ability to use modern navigational equipment. As the number and capacity of tankers afloat increase, such deficiencies become more and more intolerable.

OFFSHORE DRILLING

Oil drilled from the United States Outer Continental Shelf is expected to supply eventually about 10 percent of the nation's demand. The quantity of oil recovered by offshore drilling could amount to 63 million gallons per day by 1975 and 100 million gallons per day by 1985. Figure O.2 shows various types of offshore drilling rigs.

Between 1954 and 1969, 8,000 wells were drilled offshore. One estimate suggests that by 1980, 3,000 to 5,000 wells will be drilled annually. Accidents, especially blowouts, may therefore be expected to increase. When a well blows, oil and/or gas under pressure burst explosively from the earth. Twenty-five blowouts occurred between 1954 and 1969, two of them causing serious oil pollution.

Normal drilling operations are accompanied by chronic spilling of small amounts of oil. Pipelines carrying oil to shore are subject to rupture and may pose a particular problem in areas, such as Santa Barbara, which frequently experience earthquakes. Other forms of marine pollution result from drilling operations; barium sulfate may be dropped overboard with drilling mud and drill cuttings, as may acids which are sometimes used to increase the flow of oil.

The Santa Barbara spill added between 1 and 4 million gallons of oil to the ocean. Confusion exists as to the cause of the leak; human error and negligence have both been suggested as causes. It is not clear, moreover, whether the channel presents particular geological hazards to well drilling. There are natural oil seeps in several portions of the channel, notably at Coal Oil Point, where tar covers the rocks.

The major damage caused by the spill seems

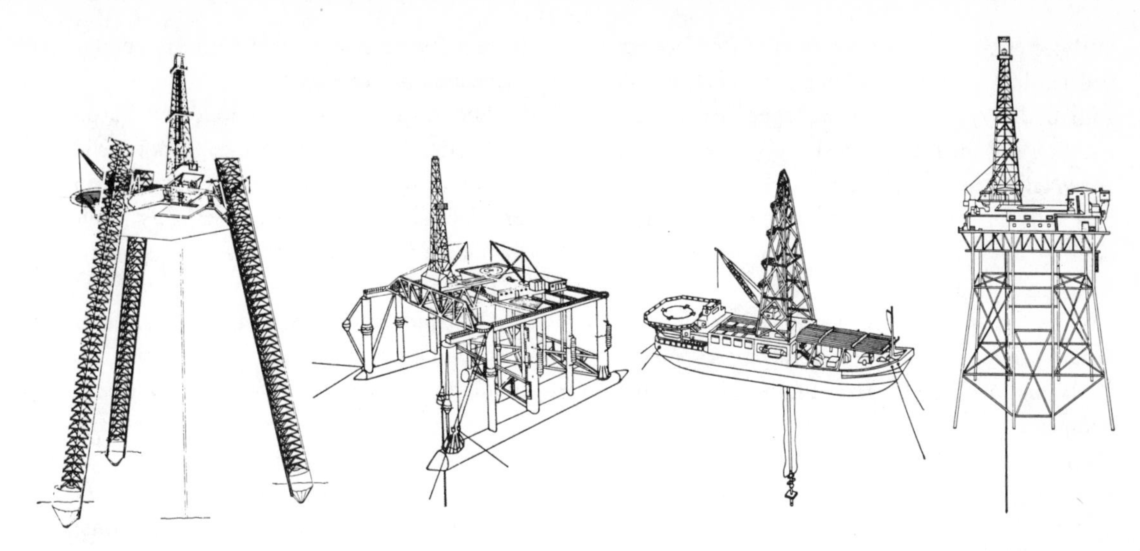

FIGURE O.2
OFFSHORE DRILLING RIGS
Used by permission of Standard Oil Company (New Jersey).

to have involved birds and beaches; a minimum of 3,600 birds perished. The oil leak had slowed to between 100 and 1,260 gallons per day by August 1969, and most of this quantity was being collected by a tentlike underwater device. Nevertheless, the slick washed in and out many times requiring repeated cleanup of the beaches. The oil also sank into the sand and may reappear as the sands shift.

Compared with the Santa Barbara disaster, the blowout on Platform Charlie in the Gulf Coast off Louisiana apparently caused much less environmental damage. Eight wells on the platform, belonging to the Chevron Oil Company, were out of control from February 10 to March 31, 1970. For the first month the oil and gas gushing from the well were aflame. High-pressure water jets containing dispersants were sprayed at the plumes of oil and gas in the air while the fire was being brought under control. The fire was finally brought under control by March 10, and the oil then started to flow into the Gulf waters.

Both the season and weather conditions were extremely favorable from the point of view of preventing biological damage. Although about 1.3 million gallons of oil were lost into the waters of the Gulf, the oil evaporated or was dispersed before it could reach the biologically rich Mississippi Delta area and nearby bird refuges. Only once was a beach seriously contaminated. In addition, the spring flood stage of the Mississippi acted to prevent oil from damaging the delta. Major bird migrations were already over; the oysters and crabs had not spawned yet, and the menhaden, a major commercial fish, had not arrived at its spring feeding grounds. Finally, the commercial fishermen for menhaden and inshore shrimp had not started their seasons. If they had begun, their nets would have been fouled with oil.

For almost a month the fire consumed the oil released from the platform, and during this time a force of experts and equipment were gathered to combat the ensuing oil spill. Nonetheless, the major conclusion drawn from an assessment of those efforts was that the existing technology and equipment were inadequate to deal with a spill in heavy waters on the open sea. Equipment which had performed well in protected areas failed to provide appreciable control of this spill for much of its history. One estimate is that only 15 percent of the oil spilled was recovered.

Chevron was subsequently charged with failure to have adequate safety devices on its wells and was fined $1 million. Eight other companies—Humble, Continental, Union, Gulf, Tenneco, Kerr-McGee, Mobil, and Shell—in the Gulf were inspected and charged with violations. It is speculated that the companies were attempting to make up for time lost during Hurricane Camille and had removed safety chokes to speed up production.

EFFECTS IN THE ENVIRONMENT

In contrast to more subtle and less visible pollutants such as **Pesticides** and **Mercury**, spilled oil is responsible for several readily identifiable phenomena. No one needs to be told that beaches covered with oil are a pollution problem. Anyone who has sailed through an oil slick and had to clean off his boat afterward will regard oil as an undesirable addition to natural waters. Aesthetics aside, one is concerned about the plight of birds which are soaked and blackened with oil from these incidents.

Moreover, oil spills cause major economic disasters in holiday resort areas and may disrupt commercial fishing ventures. A spill of 3 million gallons from the *Ocean Eagle* covered the San Juan beaches with oil in 1968, and the Santa Barbara spill caused great economic hardship to an area dependent in large part on its tourist trade. Beaches can be superficially cleaned, but oil that sinks into the sand may reappear as the sands shift. People who walk on such a beach may find their feet smudged or blackened by the oil.

The major philosophy followed in dealing with oil spills has been to get rid of visible oil as fast as possible. If the oil can't be seen, the problem is solved. Whether this policy is justifiable has been challenged by several lines of inquiry.

Oil is not a single substance but a mixture of many different chemicals. The refining process involves partial separation of the various components of crude oil; the products are less complex, but still varied, mixtures. In general, those fractions which vaporize at low temperatures, including aromatic hydrocarbons and phenolic compounds, are the most toxic to life. Thus, No. 2 fuel oil, rich in the low-boiling fractions, is extremely toxic to marine organisms. Fractions which boil at lower temperatures do, however, tend to evaporate readily at sea or dissolve into the water. Because they eventually become too dilute to cause much harm, the longer an oil slick is kept away from the biologically rich coastal zone, the less damage it will do to life forms.

Oil and water do not mix well. Oil does not adhere readily to submerged aquatic life such as fish. The addition of dispersants or detergents to oil slicks is based on their ability to emulsify oil. That is to say, they allow the oil to mix into the water and thus disappear. But another action of dispersants is to allow the oil to stick to wet surfaces such as fish gills. This property of emulsified oil makes the toxicity of the mixture of oil and dispersant much greater than that of oil alone. Some experiments have shown that oil-dispersant mixtures can be 1.3 to 248 times more toxic than oil alone. Furthermore, the formulation of most dispersants includes not only a surfactant, which does the actual emulsifying, but also stabilizers and a solvent. The solvents may be the same as those components of crude oil known to be most toxic to marine life.

Aquatic birds seem especially vulnerable to the harmful effects of oil pollution. When a bird's feathers are soaked with oil, the creature is unable to fly and loses the insulation which enables it to withstand cold. Oiled birds often drown or die of exposure. Diving birds, which continually hit the surface of the water in search of food, are most severely affected. The birds do not seem to avoid oil slicks and may even be attracted to them, perhaps by the presence of oil-sickened fish or shellfish.

Oil also blinds birds, destroys their feeding grounds, kills their food supply, and may be directly toxic to them, depending on the type of oil spilled. Contaminated birds preen themselves and swallow large quantities of oil in the process. It has been estimated that hundreds of thousands to millions of birds may be killed each year by oil pollution. The numbers *actually* reported may be much too low, perhaps only 10 percent of the total since counts generally reflect the number of birds washed ashore. Birds on the open sea that are unable to fly, however, will simply sink into the water.

An attempt to save an oil-soaked duck, victim of an oil spill in Tampa Bay, Florida.

Effects on whole bird populations are difficult to determine because many factors may be involved. Nevertheless, the decline of alcids in Britain is believed to be attributable to oil pollution. The wintering population of common eiders off the Massachusetts coast is known to have been reduced from 500,000 in 1952 to 150,000 in 1953 as a result of the wrecking of two tankers.

Attempts to save oiled birds have been widely publicized but in fact have failed dismally. A general recovery rate of 10 percent prevails with a few exceptions. The main problem, if the oil itself is not too toxic to the birds, is in rewaterproofing the feathers. Compounds which effectively clean oil off the feathers also remove their natural wax coating, and no substitute has yet been found for this coating. Handling also seems to hurt the feather structure. The successes which have been achieved generally depended on keeping the birds in captivity until they molted and grew new sets of feathers. Maintaining the birds alive under conditions of captivity is not easy, for they are subject to disease, especially if crowded together. Furthermore, they often refuse to eat.

Detergents too may be directly toxic to birds or cause subsequent breeding failure. It is speculated that dispersants may affect birds' taste buds, making them unable to distinguish appropriate foodstuffs.

Crude oil, fuel oil, and waste oil are known to contain carcinogens. Studies have revealed an excessive incidence of cancers in tar workers and men in engineering professions in which contamination with oil is chronic. The higher-boiling fractions are implicated in skin cancers. Such alleged carcinogens as 3,4 benzo(a)-pyrene are found in tar, one of the higher-boiling fractions. The carcinogens found in oil have been detected in bottom sediments and in the tissues of barnacles and oysters. In some areas polluted by oil, investigators have ascertained that fish have a high frequency of possibly cancerous lesions.

The possible carcinogenic effect of oil is one of the problems to be considered in determining the chronic effects of oil pollution. That is, concern should be directed not only toward acute effects due to large and well-publicized spills but toward the long-term results of the small continual additions to water.

Reports of the West Falmouth spill, which was intensively investigated by the Woods Hole group, provide some insight into what those chronic effects might be. Floating oil disappeared soon after the spill occurred. The massive kill of fish, bottom dwellers, and organisms that live in the intertidal zone was seen for only a few days. The disappearance of the oil was not the end of the story, however.

Nine months after the spill, the oil was still found in the bottom sediments. The area covered by oil had increased beyond the area of the original spill to a ratio of one acre for each barrel of oil initially spilled. Not only was the original site of the spill underpopulated, but there was evidence of damage to bottom organisms in the section to which the oil had spread. The destruction of animal and plant life on the ocean floor was thought to be allowing erosion of the bottom and fostering the further spread of the oil slick.

The statement has been made that no permanent damage is apparent from the *Torrey Canyon* spill, and no oil can be seen in that area. One member of the Woods Hole group has pointed out that oil is not directly visible at West Falmouth either. More sophisticated techniques are needed to show that the oil, which sinks into sediments and apparently disappears, may still be present.

At West Falmouth, effects were also found on organisms not directly killed. Some species of mussels failed to reproduce, while oysters and clams incorporated oil into their tissues. Oysters experimentally removed to clean waters for periods up to six months were not able to cleanse themselves of the oil components. Filter feeders like oysters and clams concentrate many pollutants besides oil—**Mercury** and **Pesticides,** for example. Since these organisms are part of man's food supply as well as part of the food chain in general, the possibility that they are concentrating carcinogenic substances from oil pollutants should not be overlooked.

A number of studies of the Santa Barbara spill were carried out shortly after it occurred. Short-term effects on fish and most other marine life were minor. Many of the volatile components had apparently evaporated from the oil before it reached the beaches, lowering the mortality of species in the intertidal zones. Intertidal plants and algae were, however, injured when they were coated with oil; damage occurred to barnacles as well, presumably because they seined the oily water in search of food. On the other hand, many organisms were able to burrow into the sand and escape initial effects of the spill.[5] Whereas short-term consequences may be of minor significance, the study of the West Falmouth spill showed that the long-term effects and the nonvisible effects may be of greater importance. Unfortunately, these are more difficult to detect.

Perhaps the primary difficulty in determining the effect of chronic oil pollution is that so many factors other than oil pollution influence marine population size and the health of individual organisms. Other pollutants such as mercury, pesticides, **Arsenic,** and heat,[6] as well as climatic changes, disease, and natural cycles,

[5] The use of detergents in the *Torrey Canyon* incident allowed the oil to reach and kill even these creatures.

[6] See **Thermal Pollution.**

combine to complicate the picture. Some trends are nonetheless becoming discernible.

In 1964–1965 there was a decrease in the shrimp catch in the Gulf of Mexico, despite the fact that an increased number of post-larval-stage shrimp reached the nursery grounds near the Main Pass oil field. Shrimp and fish kills were noted in the area in 1967 and poor menhaden catches in 1964–1965. In addition, oyster grass in the area is dying back. These events suggest a common pollution influence, and oil is a likely candidate.

The natural oil seeps in the Santa Barbara Channel give evidence for one of the effects which chronic oil pollution would have on the environment. Rocks in the Coal Oil Point area are covered with tar, the substance which remains after many of the components of crude oil have evaporated. Animals and plants seem unable to attach to this tar. Thus, one effect would be to drastically decrease the living area available to marine intertidal organisms.

Some doubt exists as to the eventual fate of the more persistent fractions in oil. These high-boiling, tarry substances are only very slowly degraded by natural processes. A number of recent reports have noted the presence of large collections of tar lumps in the oceans, far from land. In the Mediterranean Sea and the North Atlantic quantities of lumps as great as 0.5 milliliter per square meter have been found. The lumps vary from small particles to pieces the size of oranges. Some investigators report that part of the low-boiling fractions is still present in the lumps. A bacterial film covers the lumps, and barnacles may be seen growing on them. One of the surface-feeding fish, the saury, has been found to contain the tar lumps in its stomach. Since porpoises, tuna, and other commercial fish feed on the saury, the components of the tar lumps are finding their way into a portion of the food chain which leads to man.

CLEANUP TECHNOLOGY

It is appropriate at this point to take a brief look at the technology available for containing and cleaning up oil spills. Methods for detecting spills are fairly sophisticated. Optical methods include photographic records and involve correlation of the color of the oil slick with its thickness. From the latter, the volume of oil can be calculated. Ultraviolet imaging and radar have been used in detecting oil spills. Once a spill is located or reported, much damage can be avoided if the spill can be contained. Currently, this is attempted with booms, air curtains, and chemical gels or absorbents.

Booms are floating devices which are anchored to boats or shore structures and provide a physical barrier to the movement of the oil slick. They consist of (1) a flotation device which may be styrofoam blocks or even empty oil drums and (2) the barrier itself. Booms were used in both the Santa Barbara and Platform Charlie spills but did not perform very well in either case. Although an area of shellfish beds was probably protected by booms in the Platform Charlie incident, heavy weather in both cases made handling the booms extremely difficult.

Since the original oiling, the Santa Barbara harbor has been protected by an air curtain. This can be produced simply by blowing air through a submerged hose or pipe in which a series of holes has been punched. The air bubbles to the surface, forming a bubble curtain or air curtain. Currents formed at the surface by the air repel the oil slick. Air curtains might be especially advantageous for use in an oil spill from a ship in port. The spill could be contained, but emergency equipment would still be able to reach the injured vessel.

A number of chemicals have been developed to help contain slicks. Gels, absorbents, congealants, and trapping agents all solidify the oil

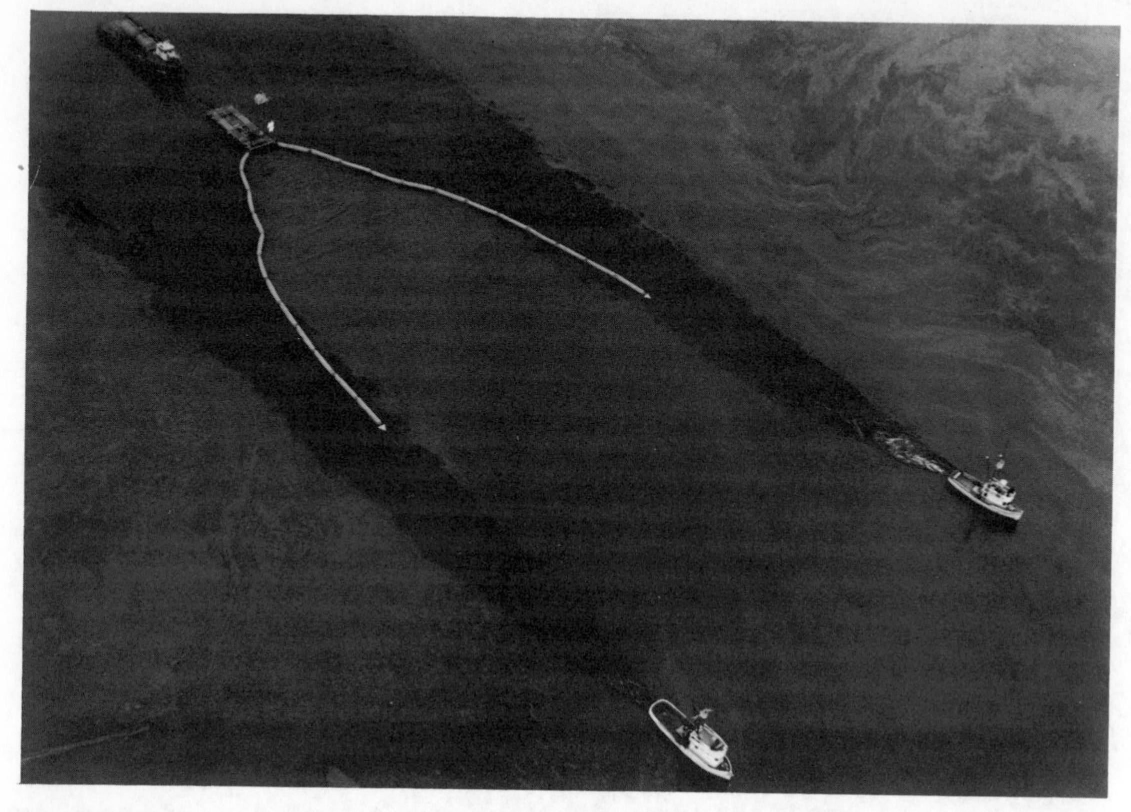

An oil spill collection system in the Santa Barbara Channel. Oil from a slick is directed by floating booms into a skimmer (at the point of the V). Suction pumps then move the oil and water mixture to the vessel immediately behind, where the water is removed. The system cuts a dark swath of clean open water (at left) in an oil slick (lighter areas at right).

and make it easier to skim from the water surface. Sinking agents are also available. The French navy used large quantities of chalk to sink oil from the *Torrey Canyon* spill. The use of sinking agents is not recommended for waters less than 100 meters deep. Experience at West Falmouth, moreover, indicates that this method of disposal may not be wise.

The best dispersal methods for oil spills probably involve some form of mechanical collection of the oil. This would produce the least harmful effect on the environment and also possibly lead to salvage of the oil.[7] In the 1970 spill in the Gulf of Mexico, skimmers were able to keep up with the flow of oil during good weather but did not perform adequately in bad weather.

Despite the proved danger posed to marine life by dispersants, these substances may be used to remove visible oil pollution if a fire hazard exists. For this reason, 1,000 barrels of Corexit

[7] See [15] for descriptions of devices for skimming oil from the surface of the water.

and 500 barrels of Cold Clean were used at Platform Charlie while the fire was being brought under control. Some use of Corexit was also made during the initial part of the Santa Barbara spill. Other justifications for the use of dispersants include hazard to human life from the oil or provable hazard to a major segment of a wildfowl population. In the United States, dispersants may not be used for any reason in the following circumstances: on distillate fuel oil spills; on spills of less than 200 barrels; on shorelines; in waters less than 100 feet deep; where danger to fish or other marine life exists; in waters where currents might carry the dispersants ashore within 24 hours; and where the dispersants might enter surface water supplies.

Once oil is deposited on beaches, technology is of little help in removing it. The most common method of disposal is to spread straw on the sand to absorb the oil. The straw is then raked up by hand. However, a new problem is created by the disposal of the oil-soaked straw. Burning the straw adds to air pollution, and quantities of straw such as were used at Santa Barbara can prove too large to be handled at nearby dumps. It has been reported that a good deal of the Santa Barbara straw was dumped in a canyon. Areas of the shore which can be left alone may well cleanse themselves. Certain remote coves did so within two years after the *Torrey Canyon* spill. In another incident, however, the natural processes seem to be taking five or more years.

The best method of dealing with pollution, in the final analysis, is almost always prevention. Tankers can be designed to withstand collisions and for emergency off-loading at sea. Certain sea-lanes could be specifically set aside for tanker traffic, and controls similar to those in air traffic could be instituted. Offshore drilling regulations should be constantly updated to require use of the best technology available. Certain areas of the continental shelf which both contain oil and border on valuable recreational or scenic areas might be set aside as reserves or "escrow" lands, perhaps to be drilled in the future when the technology of offshore drilling has made the possibility of another Santa Barbara disaster remote. These last two suggestions are at present being implemented in the Santa Barbara Channel.

LAWS

Several United States laws pertain to the area of oil pollution. The Water Quality Improvement Act of 1970 prohibits the discharge of oil in harmful quantities into the navigable waters of the United States, onto shorelines, and in contiguous zones. It makes the perpetrator of spills liable for cleanup costs.

The Federal Water Pollution Control Act Amendments of 1972 require that a permit be obtained by anyone desiring to discharge harmful pollutants (such as oil) into the United States inland or coastal waters. Although a permit system has been in operation by authority of the Refuse Act of 1899, under the new law not only has authority for the granting of permits been given to the Environmental Protection Agency, but also a series of steps has been specified whereby no discharge of pollutants into U.S. waters will be allowed after 1985. The amendments provide severe penalties for violators.

The Oil Pollution Act of 1961 was designed to implement the International Convention of 1954 for the Prevention of Pollution of the Sea by Oil. The convention required that no oil be dumped within 50 miles of a country's shoreline and that oil dumped beyond that limit be recorded. The convention was amended in 1962 to extend the zone of prohibition and to require new ships over 20,000 tons to have oily-water storage facilities.

The convention was further amended in 1969 (1) to limit permissible area discharges to 20 gallons per mile, (2) to allow coastal states to take appropriate action "to prevent, mitigate or eliminate" oil pollution after an accident (previously nations were not legally allowed to interfere with salvage operations), and (3) to make the owner of a tanker liable for oil pollution he caused (war or acts of God are excepted). As of May 1973, these last amendments had been signed but not ratified by the United States.

A serious problem with the 1969 amendments convention is that a number of nations which have large tanker fleets or provide registration for fleets either did not sign or were not even present at the convention: Liberia, Greece, Japan, Panama, and Honduras. Twenty-two percent of the world's tonnage was registered by Liberia as of 1968, including the *Torrey Canyon* and the *Arrow*.

Drilling in the Outer Continental Shelf was authorized when the Submerged Land Act and the Outer Continental Shelf Lands Act established that the states own the land and mineral rights for three miles out from the mean high tide line. The federal government owns the land and mineral rights beyond that distance. The first leases off California were offered for sale in February 1968. Seventy-one leases were bought by oil companies for a total of $600 million. Part of the leasing revenue currently goes into the Land-Water Conservation Fund.

The leases allow the oil companies to drill for oil and produce it. If no oil is found within a specified period of time, the lease reverts to the government. If oil is found, royalties are paid on it to the government. By May 1969, an additional $1.3 million had been paid to the government as royalties on three leases. After the Santa Barbara disaster the Secretary of the Interior ruled that oil companies are responsible for pollution, for any reason, from their drilling operations. A National Oil and Hazardous Materials Contingency Plan has been devised and is to be reviewed each year. This attempts to provide for the organization and mobilization of manpower and equipment to combat large oil spills in United States waters.

PROGNOSIS

The oil pollution problem can be seen to have many facets. Oil is added to the oceans from many sources. Some spills, such as occur with tanker accidents and from offshore drilling accidents, are both huge and blatant. Small and chronic additions may, however, be the most significant source of oil in the sea.

The burgeoning demand for oil in the United States should be considered. An average of 13.1 million barrels was consumed each day in the United States in 1968, when we imported about 20 percent of our total needs. The consumption rate will rise to an estimated 18.6 million barrels per day by 1980, and the percentage provided from foreign sources does not give evidence of decreasing in this decade. The quantity of oil obtained from the outer continental shelf is expected to expand so that by the middle of the decade about 10 percent of the nation's demand will be flowing from this source.

Major studies of the long-term effects of oil on aquatic life have only recently begun. Visible effects may be reported with alarm, but the yet unnoted long-term effects may finally prove to be more serious than is now supposed.

REFERENCES

1. T. A. Murphy, "Environmental Effects of Oil Pollution," *Journal of the Sanitary Engineering Division*, Proceedings of the American Society of Civil Engineers, June 1971, p. 361.
2. "The Oil Spill Problem," First Report of the President's Panel on Oil Spills, Execu-

tive Office of the President, Office of Science and Technology, 1969.

3. "Offshore Mineral Resources: A Challenge and an Opportunity," Second Report of the President's Panel on Oil Spills, Executive Office of the President, Office of Science and Technology, 1969.
4. *The Role of Petroleum and Natural Gas from the Outer Continental Shelf in the National Supply of Petroleum and Natural Gas*, Technical Bulletin 5, U.S. Dept. of the Interior, Bureau of Land Management, 1970.
5. T. H. Gaines, "Pollution Control at a Major Oil Spill," *Journal of the Water Pollution Control Federation* 43 (1971): 651.
6. *Santa Barbara Oil Pollution, 1969,* U.S. Dept. of the Interior, Program 15080 DZR 10/70, Federal Water Quality Administration, 1970.
7. *Santa Barbara Oil Spill: Short-term Analysis of Macroplankton and Fish*, U.S. Environmental Protection Agency, Office for Water Quality Research, February 1971.
8. "Santa Barbara Oil Pollution," *Hearings Before the Subcommittee on Minerals, Materials and Fuels of the Committee on Interior and Insular Affairs*, U.S. Senate, 91st Congress, 2nd Session, July 21–22, 1970, March 13–14, 1970, May 19–20, 1969.
9. *Oil Pollution Incident, Platform Charlie, Main Pass Block 41 Field, Louisiana*, U.S. Environmental Protection Agency, Water Quality Office, May 1971.
10. J. E. Smith, ed., *Torrey Canyon Pollution and Marine Life*, Marine Biological Association of the United Kingdom, Cambridge at the University Press, 1968.
11. L. A. Griner and R. Herdman, *Effects of Oil Pollution on Waterfowl: A Study of Salvage Methods*, U.S. Environmental Protection Agency, Water Quality Office, December 1970.
12. *Review of the Problem of Birds Contaminated by Oil and Their Rehabilitation*, U.S. Dept. of the Interior, Fish and Wildlife Service, Bureau of Sport Fisheries and Wildlife, May 1970.
13. M. H. Horn et al., "Petroleum Lumps on the Surface of the Sea," *Science*, April 10, 1970 (p. 245).
14. "National Oil and Hazardous Materials Pollution Contingency Plan," *Federal Register* 35, no. 106 (June 2, 1970): 8508.
15. *Testing and Evaluation of Oil Spill Recovery Equipment*, Program 15080 DOZ, U.S. Environmental Protection Agency, Water Quality Office, December 1970.
16. "Proposed Regulations of the Department of Interior on Oil Pollution Under the Water Quality Improvement Act of 1970," *Hearings Before the Subcommittee on Air and Water Pollution of the Committee on Public Works*, U.S. Senate, 91st Congress, 2nd Session, August 4, 1970.
17. "Conventions and Amendments Relating to Pollution of the Sea by Oil," *Hearings Before the Subcommittee on Oceans and International Environment of the Committee on Foreign Relations*, U.S. Senate, 92nd Congress, 1st Session, May 20, 1971.
18. "Water Pollution by Oil," *Proceedings* of the Seminar at Aviemore, Inverness-shire, Scotland, May 4–8, 1970, ed. P. Hepple, Institute of Petroleum, London, 1971.
19. M. J. Cerame-Vivas, *The Ocean Eagle Oil Spill*, NTIS AD 681 062, Puerto Rico University at Mayagüez, 1968.
20. "The Oil Import Question," Cabinet Task Force on Oil Import Control, February 1970.
21. A. Nelson-Smith, "Effects of the Oil Industry on Shore Life in Estuaries," *Proceedings* of the Royal Society of London, Series B, 180 (1972): 487.
22. "Deep Water Port Policy Issues," *Hearings Before the Committee on Interior and Insular Affairs*, Serial no. 92-26, U.S. Senate, 92nd Congress, April 1972.

Organic Water Pollution

In this section the emphasis will be on dissolved oxygen. Why study dissolved oxygen? Because it is one of the key parameters for measuring the health of a stream. High concentrations of oxygen indicate a stream with balanced and varied populations. Low concentrations (near zero) are indicative of the presence of the few miserable species which thrive under such conditions. The degradation of organic wastes by aquatic organisms is the typical cause of the depletion of dissolved oxygen.

EFFECTS OF ORGANIC WASTES ON DISSOLVED OXYGEN IN A STREAM

When organic wastes such as those discharged from either a chemical plant or a city enter a watercourse, bacteria and protozoa utilize them as food (for growth and to provide energy for maintenance) in the same way that animals use food. Just as higher animals use oxygen to release energy, so, too, do the bacteria and protozoa. The demand for oxygen by these organisms is the mechanism by which oxygen is removed from the water. A stream free of organic and bacterial pollution will ordinarily have a relatively high concentration of dissolved oxygen, typically around 8 milligrams per liter of water. To describe the quantity of organic matter in polluted water, water chemists commonly use the measure referred to as BOD—biochemical oxygen demand.

The concept of biochemical oxygen demand is an old one, having originated around the turn of the century. In a sense, it is a tribute to the originators of the concept that it remains today one of the fundamental methods of describing water pollution. When polluting material is added to the water, bacteria and protozoa oxidize the pollutant. The complete biological oxidation by bacteria and protozoa of the soluble organics in 1 liter of polluted water would require a certain weight in milligrams of molecular oxygen. This value, the milligrams demanded per liter of sample for complete biological oxidation, is the BOD. Biological oxidation is quite slow; frequently the quantity of oxygen consumed in the first five days is reported since this represents nearly 90 percent of the final demand. Alternatively, an estimate of the final or ultimate demand is presented. Specification of the ultimate demand is equivalent to stating the oxygen-depleting capacity of a sample of polluted water. This is a valuable indicator of pollution since the lack of oxygen is responsible for fish kills, foul odors, and undesirable "pollution-tolerant" organisms.

However, BOD is more than a measure of the capacity to deplete oxygen. It is a surrogate measure of pollution itself. The organic substances in water are many, and their identification and measurement are a time-consuming task. BOD summarizes the quantity of *all* organics in a single number, the milligrams of oxygen demanded for complete biological oxidation of the organics.

With this measure of pollution one can describe the effects of wastes on the dissolved oxygen in a natural watercourse such as a stream or river. When microorganisms are present in abundance, the rate of oxygen removal is roughly proportional to the BOD in the stream. Thus, if the amount of BOD is large, the rate of oxygen removal is large. If there is only a small concentration of organics, the rate of oxygen removal is small.

However, as oxygen is removed to oxidize these wastes, the wastes themselves are converted to substances which cannot be further utilized by the bacteria and protozoa—typically, carbon dioxide and water. Thus, the more oxygen is removed from the water to oxidize wastes, the lower is the concentration of organics, and the slower is the further rate of oxygen removal.

The gradual lowering of dissolved oxygen is the pattern that would be seen if the stream were unable to replenish its oxygen. But the stream can replenish the oxygen by simply dissolving more from the atmosphere. The process is called "reaeration," and it is influenced by two main factors: the turbulence of the water and the oxygen deficit of the water.

Turbulence is a term which indicates the activity of water—how choppy the waves are, how rapidly the stream is mixing. If one were to stir a beaker of water, the more rapidly he stirred, the more turbulent the water would become. Turbulence depends on the velocity with which the water in a stream is moving. The greater the velocity, the greater is the turbulence. Turbulence causes the rapid incorporation of oxygen from the atmosphere into stream waters. Higher turbulence causes more rapid reaeration of the stream.

Oxygen deficit introduces a new concept. There is a limit to the amount of oxygen that water can dissolve, just as there is a limit to the amount of salt or sugar it can dissolve. The maximum concentration of oxygen in water is called the *saturation concentration.* If the water contains a concentration less than the saturation value, the difference between the saturation value and the actual concentration is called the *oxygen deficit.* It is a measure of how much more oxygen the water can dissolve before it can hold no more.

If the water were saturated with oxygen, the rate of reaeration would be zero since no additional oxygen from the air could be dissolved. If the concentration were less than the saturation level, there would be reaeration and its rate would be proportional to the difference between saturation and actual concentration, i.e., proportional to the *deficit.* Therefore, the higher the oxygen concentration, the lower the deficit will be and the slower the rate of reaeration. The lower the concentration, the higher the deficit will be and the greater the rate of reaeration.

With this background, we can describe the effect of reaeration on the oxygen in a polluted stream. The oxidation of organic pollutants

FIGURE O.3
RATES OF REMOVAL AND REPLENISHMENT OF OXYGEN

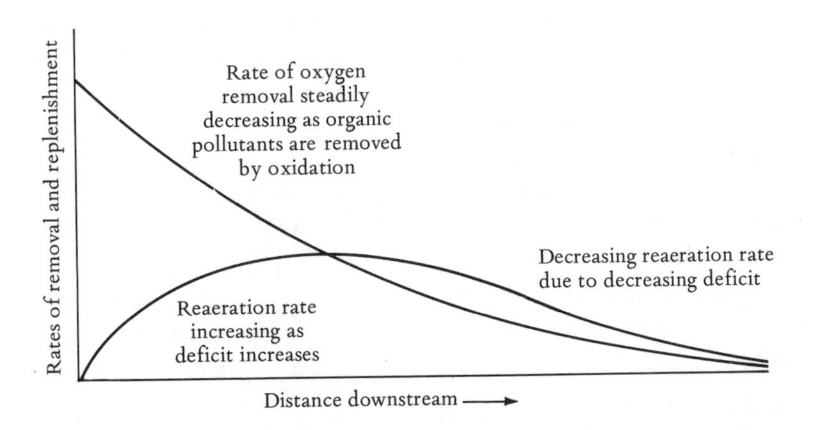

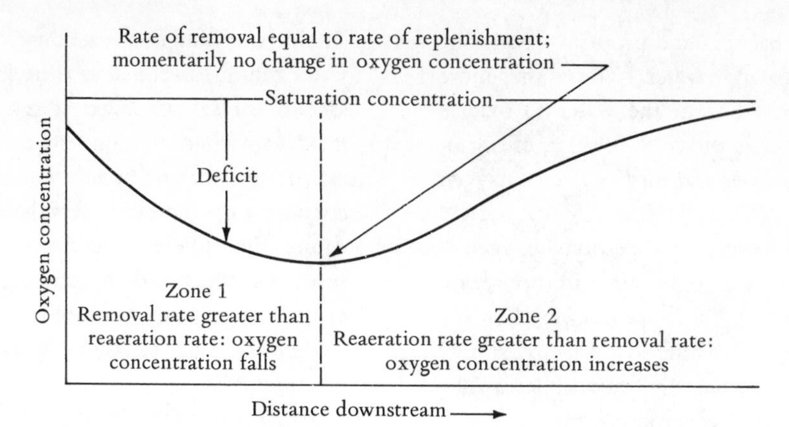

FIGURE O.4
THE OXYGEN SAG CURVE

withdraws oxygen from the water, decreasing the oxygen concentration. But the process of reaeration tends to restore the oxygen, the rate of reaeration increasing as the oxygen concentration falls. Or, speaking in terms of deficit, as the deficit increases because of the oxidation of organics, the reaeration rate rises. Thus, there are two competing mechanisms: oxidation, removing oxygen from the stream; and reaeration, restoring it. These mechanisms, when shown graphically, give rise to a characteristic curve of the dissolved oxygen level with distance. Because the curve "sags" prior to recovery, it is known as the *oxygen sag curve* (see Figures O.3 and O.4).

If the concentration of organics were initially large and the oxygen concentration near saturation, the removal rate would be expected at first to exceed the rate of reaeration. This would result in a decrease in the concentration of oxygen. But the more the oxygen concentration decreases, the more the rate of reaeration increases. Further, the rate of oxygen removal is falling because of the declining concentration of organics. Simultaneously, the rate of reaeration is increasing because of the decreasing oxygen concentration (rising deficit).

Soon a point is reached at which the two rates are balanced. Oxygen removal equals oxygen replenishment. For a moment, the oxygen concentration remains unchanged. But in the next instant the concentration of organics has fallen sufficiently so that the rate of oxygen removal has become less than the rate of replenishment. The oxygen concentration must then rise since reaeration exceeds removal, and as the water proceeds downstream (i.e., as time goes by) the rate of removal decreases still more. Since the oxygen concentration has risen, the deficit has decreased and the reaeration rate must simultaneously decrease. But the reaeration rate continues to exceed the removal rate, so that gradually the oxygen is restored to the stream. Eventually both rates fall to zero, and the oxygen content of the stream returns to its original level.

ECOLOGICAL RESULTS OF ORGANIC POLLUTION

The organisms downstream from a pollution discharge are the result of the presence of both organic solids and the high concentration of soluble organic wastes in the discharge. The

are responsible for a
oncentration in the

by the organic solids
st. Where the pollution
eam, the solids in the waste
ming a layer of sludge whose
noted for many miles down-
strea… the sludge, various microorganisms and higher forms act to decompose the solid substance. In general, dissolved oxygen is low in the sludge layer. Thus decomposition must take place in the absence of oxygen, yielding products which are not notable for their appeal to the senses, such as the gas hydrogen sulfide. The stream itself at this early stage may have a reasonable level of dissolved oxygen; the sludge layer, however, is nearly devoid of utilizable oxygen resources.

Both fine and coarse solid particles contribute to a turbidity which gradually lessens as the solids settle to the stream bottom. As oxygen decreases downstream from the pollution discharge, the combination of the low dissolved oxygen and the turbidity may prevent algal growth. The turbidity acts in this respect as a barrier to sunlight, which is essential for algae. With the clearing of the water and recovery of dissolved oxygen, algae begin to prosper. Their presence may result in a fluctuation in the oxygen content in the water. During the hours of daylight, the algae produce oxygen as a by-product of photosynthesis; an elevated oxygen level is thus obtained. But at night the respiration and decomposition of algae remove oxygen from the water, depressing oxygen concentration. As can well be imagined, the presence of algae severely complicates the picture described in the first portion of this section. The algae-caused fluctuations in dissolved oxygen may even deplete the stream's oxygen to such an extent that the normal aquatic life is harmed. The presence of a sludge layer and the hindrance of normal algal growth are the two events attributed to the presence of organic solids in a discharge.

Soluble organic wastes in high concentrations are responsible for the growth of the bacterial population in the stream since they serve as nutrients for the bacteria. Bacteria whose ideal environment is man will, on the other hand, gradually die away as the water proceeds downstream, despite the large nutrient supply. The new bacterial population also eventually declines, falling prey to higher organisms such as ciliated protozoans, rotifers, and crustaceans.

The condition of a low dissolved oxygen concentration, which results from high concentrations of soluble organic waste, gives rise to severe ecological changes. Upstream from the pollution discharge, where dissolved oxygen may be as high as 8 or 9 milligrams per liter, a large array of species exist. Sports fish, minnows, mayflies, snails, and the like are characteristic of the clean stream, prior to pollution. The entrance of organic wastes and subsequent lowering of the dissolved oxygen creates an environment suitable for "pollution-tolerant" organisms. One such organism is the sludgeworm, which consumes sludge as its source of nutrients and thrives in water with as little as half a milligram per liter of oxygen. Another is the rat-tailed maggot, a resident in the sludge of stream bottoms which breathes via a long tubular appendage reaching to the water surface. The replenishment of dissolved oxygen restores the characteristic populace of the clean stream.

REFERENCES

1. A. F. Bartsch and W. M. Ingram, "Stream Life and the Pollution Environment," *Public Works* 90 (1959): 104–10; reprinted in *Biology of Water Pollution*, U.S. Dept. of the

Interior, Federal Water Pollution Control Administration, 1967. (The FWPCA later became the Federal Water Quality Administration and still more recently has become the Water Quality Office of the Environmental Protection Agency.)

2. G. Fair, J. Geyer, and D. Okun, *Elements of Water Supply and Waste Water Disposal*, Wiley, New York, 1968.
3. H. B. Hynes, *The Biology of Polluted Water*, Liverpool University Press, Liverpool, 1960.

Organophosphorus Compounds

Organophosphorus compounds are organic molecules containing the element phosphorus. Included among the organophosphorus pesticides are compounds which serve as insecticides, miticides, fungicides, **Herbicides,** and defoliants.

Organophosphorus compounds poison insects and people by combining with the enzyme cholinesterase, a chemical which is necessary for proper functioning of the nervous system. Compounds similar to these were developed as nerve gases for chemical warfare. The combination of an organophosphorus compound and cholinesterase is a fairly stable one and is not broken down immediately by the body. For this reason, several nontoxic doses in a short period of time can produce symptoms of poisoning. Organophosphorus poisoning can be diagnosed by measuring the level of cholinesterase in the blood. Levels of enzyme 20 to 25 percent below normal are usually associated with the typical symptoms of poisoning, which include, in the early stages, respiratory distress, cough and running nose, shortness of breath, and tightness in the chest. In addition, there may be visual problems such as blurring and tearing of the eyes or dimming of vision. Tiredness and psychological symptoms such as an inability to concentrate, dreaming, and disturbed sleep have

TABLE 0.2

ORGANOPHOSPHORUS PESTICIDES

Chemical Group	Examples	Comments
Phosphoric acid derivatives[a]	DDVP	Active ingredient in Shell No Pest Strip
Thiophosphoric acid derivatives	Parathion; methyl-parathion	Widely used; very toxic to man; other members of this group generally of low toxicity
Dithiophosphoric acid derivatives	Phorate	Extremely toxic to man
Pyrophosphoric acid derivatives	TEPP	Extremely toxic to man

[a] Phosphoric acid = H_3PO_4.

in food chains. Their long-term hazard is thus very small. On the other hand, they include some of the more toxic chemicals produced. As little as 0.015 ounce of parathion might poison a man. This amount could be inhaled during one spraying session if safety procedures were ignored. The acute hazard of organophosphorus compounds is thus very great. In 1966 a study was made in California of accidents attributed to pesticides and agricultural chemicals. Organophosphorus compounds were involved in more accidents than any other group of chemicals, and parathion accounted for 40 percent of the organophosphorus accidents. However, not all of the organophosphorus compounds which have been prepared are as toxic to man as parathion.

The lack of persistence of organophosphorus compounds can also be a disadvantage. The compounds may need to be applied several times during a season to control pests. Furthermore, organophosphorus chemicals are often more costly on a weight basis than chlorinated hydrocarbons. Because of their greater toxicity, though, they may be used in smaller amounts.

The most commonly used organophosphates can be generally grouped into the categories mentioned in Table O.2. A discussion of the environmental effects of organophosphorus compounds is in **Pesticides.**

REFERENCES

1. "Occupational Diseases in California Attributed to Pesticides and Other Agricultural

P

Particulate Matter

Solid or liquid particles dispersed in air (sometimes called aerosols) are widespread and apparently deleterious to health. Smokestacks are a visible source of atmospheric particles, but particles are dispersed from many sources including the automobile. Particle concentrations in urban environments have been linked to a number of diseases including lung cancer and stomach cancer. Technology exists to abate most particulate pollution; its application to the problem waits on demand and on the public's willingness to bear the cost. Industrial exposures to particles will not be treated here. Some of the substances which are associated with occupational exposures such as **Asbestos, Beryllium,** and coal dust (see **Black Lung Disease**) are discussed separately.

THE FAMILY OF SUBSTANCES

Particulate matter, unlike other specific air pollutants, is not one substance or even a group of related substances. It consists of many components: sulfate salts and sulfuric acid droplets (see **Sulfur Oxides**), lead salts (see **Lead**), **Asbestos,** iron oxides, silica, carbon particles (soot), and other substances. Carbon particles are seen most commonly, but one cannot see the organic compounds adsorbed on the surface of the carbon particles. These organic compounds—particularly the polynuclear aromatic hydrocarbons, of which benzo(a)pyrene is an example—have been implicated as potential cancer-causing substances in air pollution.

SOURCES

Particles arise from technological processes such as combustion, grinding, and spraying and from natural sources such as wind erosion, the action of wind on the sea, and the sea's turbulence.

Cities, as centers of man's activities, may experience particle concentrations in the range of 60 to 220 micrograms per cubic meter; suburbs can be expected to have about half the city levels, and rural concentrations may be as low as 10 micrograms per cubic meter. Public awareness of particulate pollution has been noted at concentrations as low as 50 micrograms per cubic meter. On a daily basis, peaks in particle concentrations may occur in the early morning when atmospheric inversions (see **Air Pollution**) conspire with increased traffic to produce higher particulate levels. There is also a peak concentration in winter when fuel combustion for heating is at a maximum. In air pollution episodes, such as the one which occurred in the late fall of 1962 in the eastern United States, particulate concentrations in several cities rose to two to three times the usual level. The levels of dispersed particles are listed for a number of U.S. cities for the period 1961–1965 in Table P.1.

In order to set the discussion on control in perspective, we need to note the relative contributions of the principal sources of particulate matter. In 1966, technological sources discharged 11.5 million tons of particles into the air. About 25 percent of this total, or 3 million tons, stemmed from the combustion of fossil fuel for **Electric Power,** coal combustion accounting for 95 percent of this quantity or 2.85 million tons. Had there been no control on emissions from coal-fired power plants, about 21 million tons of particles would have been released. Thus, coal burning is at once the major source of particles in the atmosphere and the process to which the most controls have already been applied.

TABLE P.1

AVERAGE CONCENTRATION OF SUSPENDED PARTICLES IN U.S. URBAN AREAS, 1960–1965

Standard Metropolitan Statistical Area	Concentration ($\mu g/m^3$)[a]	Rank
Chattanooga	180	1
Chicago–Gary–Hammond–East Chicago	177	2
Philadelphia	170	3
St. Louis	168	4
Canton	165	5
Pittsburgh	163	6
Indianapolis	158	7
Wilmington	154	8
Louisville	152	9
Youngstown	148	10
Denver	147	11
Los Angeles–Long Beach	145.5	12
Detroit	143	13
Baltimore	141	14.5
Birmingham	141	14.5
Kansas City	140	16.5
York	140	16.5
New York–Jersey City–Newark–Passaic–Patterson–Clifton	135	18
Akron	134	20
Boston	134	20
Cleveland	134	20

From *Air Quality Criteria for Particulate Matter,* AP-49, National Air Pollution Control Administration, 1969.
[a] Geometric mean of center city station.

Although refuse incineration accounted for only about 9 percent of the 11.5 million tons discharged in 1966, local conditions varied widely. Because incinerators are often centrally placed in a city to minimize haul costs, urban atmospheres are likely to feel their effects more keenly. Space heating provided another 9 percent of the total emissions. While mobile sources—buses, trucks, and cars—accounted for only about 4 percent of the total quantity, in many areas particulates from vehicle movement do constitute a large share of the total weight of particle emissions. Data have shown the weight fraction of particles which arises from motor vehicle operation to be as high as 38 percent in Los Angeles and 15 percent in Washington, D.C., and New York City. The particles in the exhaust are mainly metal salts (lead salts principally), carbon, and droplets of liquid hydrocarbons. On the basis of pounds of particles emitted per gallon of fuel utilized, vehicles with diesel engines produce nearly 10 times the weight of particles that gasoline-powered vehicles do.

The remaining 6 million tons of the total arose from industrial processes: iron and steel manufacture, cement plants, coke manufacture, smelting, etc. Of the 1.1 million tons of particles from industrial fuel combustion, coal was responsible for about 80 percent of the total.

The results of a sootfall in Boston.

EFFECTS

ON PLANTS

Controversy surrounds the effects of particles on plants. While some investigators stress the potential of dust to clog the pores of leaves, others dismiss the resulting effect on gas exchange as insignificant. Claims are also made for a reduction in photosynthesis on extensive deposition of particles, but conclusive evidence is wanting. Injury to vegetation by acidic particles is, however, well demonstrated (see **Sulfur Oxides**).

ON MATERIALS

Materials such as metals, painted surfaces, and textiles are subject to damage by particles from the air. If the settled particles do not absorb moisture or if the surface is resistant, the effect may simply be one of soiling. This condition requires cleaning or painting. When the particles absorb moisture (as some chloride and sulfate salts do), the points of deposition may become the focus of corrosion or deterioration. The solution of certain gases such as sulfur dioxide in the water collected by the particle creates an acid condition. It is well documented that metallic corrosion rates are much higher in industrial-urban areas than in rural areas, but the accelerated corrosion is not due merely to particles. It is due to the presence of such acidic gases as sulfur dioxide in the urban atmosphere. The cost of painting steel structures to remedy the effects of polluted air has been placed at about $100 million yearly.

ON WEATHER, VISIBILITY, AND CLIMATE

Particles resulting from man's activities have an effect on weather. By providing additional nuclei upon which water vapor may condense to water droplets and ice, particles may alter the rainfall and snow patterns from their natural condition. Consider the city of La Porte, Indiana, where precipitation has been found to follow the level of industrial activity in nearby Chicago. Violent hailstorms occur in this city with greater frequency than in surrounding areas which are not downwind from Chicago. Tulsa, Oklahoma, has seen an increase in average annual rainfall as the city has increased in size since its founding over 50 years ago. The rainfall increase is thought to be linked to the atmospheric particles arising from human activities there. Elsewhere, it has even been shown that average Sunday rainfall is less than on weekdays presumably because diminished traffic lowered particle levels.

The scattering of radiation by particles is known to reduce the amount of sunlight received by cities. The effect is observed most dramatically in winter when particulates arising as products of combustion are at a maximum. The average rate at which radiation reaches a unit of the earth's surface may be decreased by 15 to 20 percent in cities as compared to rural areas. Leningrad is reputed to have received 70 percent less sunlight one winter than nearby rural areas.

Although visibility may be cut down by natural hazes such as the fog which forms on salt particles from the sea, it is also affected by atmospheric particles arising from man's activities. Visibility is important to airplane and automobile operation. Decreased visibility also diminishes aesthetic values; witness the pall which fills the valley of Los Angeles.

Finally, climate may be affected by particle concentrations. A massive volcanic eruption in the 1880's on the island of Java was followed by two extremely severe winters worldwide. The average worldwide air temperature has been decreasing since 1940, and this decline may be related to the increased reflection by particles of electromagnetic radiation away from the earth. Some observers speculate that increased particle concentrations are countering the warming trend predicted by others due to increased levels of **Carbon Dioxide** in the atmosphere.

ON MAN'S HEALTH

The effects of particulates on man are determined by the chemical composition of the particles and by the organs which the contaminants reach. Of course, the primary site at which the effects of particles are noticed is the lungs, but other organs may be impaired. For instance, some of the particles which penetrate the respiratory system may be captured by the upward flow of mucus that bathes a portion of the system. Ultimately, the mucus is swallowed and the particles enter the stomach and intestinal tract. Not all particles are cleared from the lungs; some remain and are surrounded by tissue. Thus, not only may substances be absorbed into the bloodstream at the level of the lungs, but they also have the opportunity to be absorbed in the stomach or intestine. A high incidence of stomach cancer has been noticed in regions of high air pollution, but direct causal links have not been established.

The class of compounds known as polynuclear aromatic hydrocarbons deserves special mention. On the basis of experiments with mice and statistical evidence, one compound in that group, benzo(a)pyrene, is suspected of being a human carcinogen. Benzo(a)pyrene is found free in the air, and it may be associated with soot. Nonurban concentrations may be lower than 2 micrograms per 1,000 cubic meters of air, but concentrations in urban areas above 20 micrograms per 1,000 cubic meters are quite common. Lung cancer in urban environments is more common than in rural environments, and the polynuclear aromatic hydrocarbons are prime suspects as causative agents.

Statistical studies of the effects of atmospheric particulates on man's health may focus either on deaths from various causes as related

TABLE P.2
PARTICLE CONTROL EQUIPMENT

Device	Operating Principle	Application
Settling chamber	Decreased gas velocity allows particles to settle	Kiln or furnace
Cyclone collectors	Whirling gas throws particles to rim of device	Control dusts from grains, grinding operations, wood operations, etc.
Wet collectors (scrubbers)	Particles are wetted by water mist to improve collection efficiency	Metal and chemical industries, fertilizer industry, ore mining and processing
Mist eliminators	Filters woven of fiber or wire mesh, beds packed with inert materials, baffle, or vane collectors are used	Manufacture of acids (sulfuric, nitric, phosphoric)
Electrostatic precipitators	Particles are charged by an arc of electric current and drawn to collecting surface	Fly ash removal in coal-fired power plants; chemical, iron, steel, and cement industries; manufacture of sulfuric acid
Fabric filters	Air is passed through fabric bags hung in baghouse; straining removes particles	Grain processing, cement operations, asbestos manufacture
Afterburners	Combustible particles are oxidized to water vapor and carbon dioxide	Control combustible particles

to pollution levels or on illnesses or work absences as related to pollution levels. The studies concerned with relating deaths to particulate levels have almost all observed that deaths from stomach cancer increase with increasing particle concentrations. Results concerning deaths from bronchitis as related to particles have been mixed, but several studies have concluded that there is a correlation. Results are mixed also on the relationship of deaths from lung cancer and particulate levels; only one study seems to have found an association.

It has been shown that respiratory illnesses such as bronchitis are more common in industrialized areas than rural areas. Industrial areas are likely to exhibit both high particle concentrations and high sulfur dioxide concentrations. One study indicated that bronchitis illnesses are linked more closely to sulfur dioxide than to particles, but another was unable to relate total respiratory illnesses to either sulfur dioxide or particles. The latter study did, however, find an association between total illnesses and sulfur dioxide and particulate matter. Extensive studies on large numbers of children have found that lower respiratory tract infections increase with pollution levels greater than 130 micrograms per cubic meter of sulfur dioxide and 130 micrograms per cubic meter of particles. From all these analyses it can be concluded that health is affected by particulate air pollution; exact relations are unknown and may remain so, but an overall association is becoming clear.

CONTROL

Many steps can be taken to diminish particulate levels. Hydroelectric power plants and **Nuclear Power Plants** may supplant **Fossil Fuel Power Plants**; oil or gas may be substituted for coal. For each unit of energy derived, oil combustion produces about one-tenth of the particles that coal combustion does, even when 90 percent of the particles from coal combustion are removed. As a second alternative, pricing may be utilized to discourage increased power demands. An obvious way to reduce local particle concentration in urban areas would be to place power sources at some distance from cities. Nuclear power plants are already coming into existence to meet growing needs for electricity. New hydropower starts, however, are few in number. **Lead** is being removed from gasoline, thus reducing particles from that source. Table P.2 lists the control equipment which has been developed for removing particles from the gas streams of stationary combustion sources. Thorough descriptions of the devices are given in *Control Techniques for Particulate Air Pollutants* [2].

REFERENCES

1. *Air Quality Criteria for Particulate Matter*, AP-49, National Air Pollution Control Administration, 1969.
2. *Control Techniques for Particulate Air Pollutants*, AP-51, National Air Pollution Control Administration, 1969.

The National Air Pollution Control Administration is now the Office of Air Programs of the U.S. Environmental Protection Agency.

PCBs

PCB is an abbreviation for the group of compounds known as polychlorinated biphenyls. Their structure is similar to that of DDT. They consist of two phenyl groups joined together; chlorine atoms may be attached to the phenyl groups in a number of places (see Figure P.1).

There are 210 different possible PCBs, depending on the number of chlorine atoms (0 to 10) attached to the basic structure and where they are located. Commercial products are not pure compounds but mixtures of the various possible PCBs. Monsanto's Arochlor 1254 contains 18 different polychlorinated biphenyls. In addition, industrial preparations can contain polychlorinated *ter*phenyls as well as the *bi*phenyl derivatives. This lack of homogeneity leads to problems in detecting the chemicals in the environment, since one is looking not for a single contaminant but for a group of contaminants.

FIGURE P.1
STRUCTURE OF PCB AND DDT
(1) Dotted circles indicate positions where chlorine atoms may be attached. (2) Solid circles indicate positions where chlorine atoms are attached.

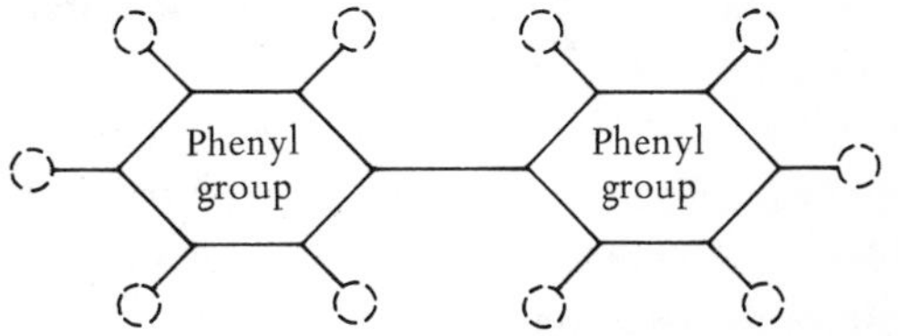

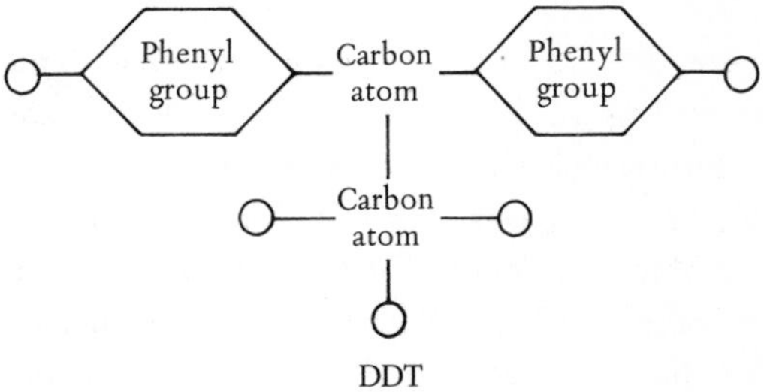

USES

PCBs are found as ingredients in a wide variety of products: lubricants, hydraulic fluids, waxes, adhesives, capacitor and transformer fluids, and asphalt. They are used to give plastic properties to such products as nitrocellulose lacquers, polystyrene, and crepe rubber. Their use as heat exchange fluids and as ingredients in inks has lead to contamination of foodstuffs. They can serve as insecticides against a few insects (e.g., mosquito larvae), but, more importantly, they influence the effectiveness of such pesticides as chlordane, aldrin, dieldrin, lindane, DDT, and parathion.

TOXICITY

PCBs are poisonous compounds, although their toxicity varies with the species being tested. The chemicals are perhaps half as toxic to rats as DDT is and about one-fourth to one-fifth as toxic as DDT to chickens. Less than fatal doses to mammals have been shown to produce liver defects; such doses to birds have resulted in kidney, spleen, and heart troubles. Reproduction in mammals, birds, and fish is adversely affected by PCBs.

A number of investigators feel that certain chemicals which contaminate the PCBs (the dibenzofurans) may be responsible for some of the toxic effects noted. Purification of the PCB mixtures before testing will be necessary to resolve the question.

Incidents of human poisoning by PCBs have been recorded. The most serious incident took place in Japan in 1968, when over 1,000 people were affected and five deaths occurred. The symptoms of PCB poisoning in humans are primarily eye discharges, a general feeling of weakness, an acne-like skin condition, and a dark-brown coloration of the skin. The disease is called yusho, and there is at present no specific treatment for it. The 1968 incident was traced to consumption of rice oil contaminated with Kanechlor 400, a PCB formulation produced by a Japanese chemical company. The Kanechlor, which was used as a heat-exchange fluid, apparently leaked from old sections of pipe directly into the rice oil during a production step.

Humans were not the only species affected, however. A by-product of the rice oil manufacturing process is dark oil, sold as an ingredient for poultry feeds. Six months before the epidemic of yusho was recognized by Japanese public health authorities, a mysterious, large-scale epizootic struck more than 2 million chickens in western Japan. Over 400,000 birds died. After the cause of the human epidemic was discovered, scientists were able to identify the use of contaminated dark rice oil in feeds as the cause of the carnage among chickens.

Human victims are estimated to have consumed an average of 2 grams of Kanechlor. About 0.2 gram was the minimum dose ingested by anyone showing symptoms. Thirteen women were pregnant when they ate contaminated rice oil, and most of their babies were born with the characteristic yusho pigmentation and eye discharges. Further studies of the children will be necessary to determine whether more serious defects also occurred. About half the people affected had not improved three years after they were poisoned.

The possible effects of long-term exposure to low levels of PCBs are not known. The compounds are fat-soluble and do accumulate in fatty tissues as DDT derivatives do.

CONTAMINATION OF THE ENVIRONMENT

PCBs were identified in birds' feathers as long ago as 1944. Many workers have since reported PCB levels in wildlife in Canada, Germany, Great Britain, the Netherlands, Sweden, and the United States. Polar bears and fish in the Arctic tundra lakes have PCB residues, as do birds living in Antarctic waters. Humans contain PCB residues in their fatty tissues. Some 40 to 45 percent of the United States population appear to have one part per million or more of PCBs in their adipose tissues (according to data obtained in the Environmental Protection Agency Human Monitoring Survey).

Residues found in fish may reach as high as 3 parts per million (wet weight) in Tokyo sea bass and Long Island silversides and as high as 5 parts per million in Coho Salmon from the Great Lakes. Even lake trout from the Arctic tundra lakes, however, contain 0.04 to 0.6 part per million PCBs. These lakes are not exposed to any contamination except from atmospheric fallout. Birds such as ospreys, which feed on fish, can have concentrations of 300 to 700 parts per million PCBs. These levels may be responsible for observed hatching failures among ospreys in industrialized areas of the United States.

In general, higher concentrations of PCBs are noted in organisms which occupy relatively high positions in the food chains. For instance, mussels taken from the Baltic were found to average 4.3 parts per million PCBs in their fat. Herring averaged 6.8 parts per million, seals 34 parts per million, and guillemot eggs 250 parts per million. There are exceptions, however, and it may be that differing fat solubilities or other factors affect the magnification of PCBs by organisms.

ENVIRONMENTAL SOURCES

In contrast to the case of DDT, in which the contaminating chemical was introduced directly and intentionally to the environment (see **Pesticides**), the environmental sources of PCB contamination are indirect and to some extent unidentified. The American manufacturer of PCBs is the Monsanto Company. The chemicals are also produced in France, Japan, Italy, Germany, Great Britain, and Russia. Monsanto sold about 500,000 tons of PCBs between 1930 and 1970. It has been estimated that about 100,000 tons per year were produced worldwide around 1970. This is an amount comparable to the world production of DDT at that time.

PCBs can reach the environment in a number of ways—for instance, via leaks from transformers, heat exchangers, or hydraulic systems. The manufacturing process may involve spills, as is indicated by the fact that catfish found near Anniston, Alabama, one of the two United States sites of PCB manufacture, contain high levels of PCBs. The disposal of PCBs or products containing them into water is another potential source of water contamination. PCBs can vaporize from products in which they are ingredients. The burning, in dumps and incinerators, of plastics and resins, products which may contain PCBs, probably contributes larger amounts to the atmosphere. One estimate [6] is that 1,500 to 2,000 tons of PCBs are lost to the air and that 4,500 tons are added to the water each year. Another 18,000 tons are presumed to be buried in dumps and landfills annually.

FOOD CONTAMINATION

Although PCBs are subject to degradation by photochemical processes and bacterial action, large amounts persist in the environment and can be absorbed into man's food supply. Food may be contaminated directly, as was the rice oil in Japan. Animal feeds may also be contaminated, with the result that meat, milk, and eggs may contain PCBs.

A case in point involved contamination of milk with PCBs. Arochlor 1254 had been used to coat concrete silo walls. Because it was absorbed by the silage, which was then fed to dairy cows, the Arochlor appeared in milk. In another instance, a large poultry producer,

Holly Farms in North Carolina, suffered a hatching failure. An investigation uncovered the presence of PCB residues in a minor component of the poultry foods, fish meal, at levels of 14 to 30 parts per million. It was determined that the meal had been contaminated during pasteurization in an apparatus using PCBs as the heat-exchange fluid. Although some 16,000 tons of fish meal, used in poultry and fish feeds, had been contaminated, only about 1,500 tons were returned to the plant. The FDA did seize other shipments of the meal, and a program to monitor poultry, eggs, and fish was instituted by the Department of Agriculture. No poultry or fish were reported to contain PCB residues in excess of the FDA-allowed limits, but a large number of egg samples were found to have excessive PCB levels.

There have been several recent seizures of poultry contaminated with PCBs. Fifty thousand turkeys were found to be contaminated in Minnesota in August 1971. In New York state 146,000 chickens were destroyed because of PCB tainting in 1970; another 88,000 were destroyed in North Carolina in July 1971.

In Market Basket Studies carried out by the Food and Drug Administration in 1970 and 1971, PCBs were found in 22 out of 720 "baskets." The sources were meat, fish, poultry, dairy products, and grain or cereal products. Table P.3 utilizes data obtained in the raw agricultural commodities surveillance program of the FDA and indicates levels of PCBs in some foods.

TABLE P.3

PCBS IN SELECTED FOOD COMMODITIES, JULY 1, 1970–SEPTEMBER 30, 1971

Food	Positive Samples (%)	Average Level (ppm)	High Level (ppm)
Fish	54	1.87	35.29
Cheese	6	0.25	1.0
Milk	7	2.25	27.8
Eggs	29	0.55	63.74

From A. Kolbye, "Food Exposures to Polychlorinated Biphenyls," *Environmental Health Perspectives*, no. 1, NIH 72-218, U.S. Dept. of Health, Education and Welfare, April 1972.

PCBs have not yet been found in fresh fruits or vegetables. Present guidelines, which the FDA is following, call for seizure of meat, fish, or poultry having more than 5 parts per million PCBs. Milk is subject to seizure if it has over 0.2 part per million (5 parts per million in the milk fat), and eggs or animal feeds may be seized if they contain more than 0.5 part per million. These are, however, limits above which the commodity is seized, rather than levels considered safe for consumption. Albert Kolbye, Deputy Director of the FDA's Bureau of Foods, notes that if the prevalence of PCBs is found to be increasing, these limits will be lowered.

Two events, in particular, illustrate the hazards involved in the widespread use of such potentially hazardous chemicals as PCBs. Both are related to the contamination of man's food. In August 1969, the FDA was informed that PCBs were found in milk in Baltimore. Investigation uncovered the fact that PCBs were used in electrical transformers in the West Virginia area where the milk had been produced. It was eventually learned that the electric company had allowed the use of transformer oil as a base for defoliant sprays employed in maintaining right of ways. Apparently enough spray had fallen on adjacent pastureland to contaminate milk from resident cows.

One fairly widespread use of PCBs has been in "carbonless" carbon papers. These copying papers usually consist of three sheets of paper. Writing on the upper sheet is transferred to the bottom sheet. PCBs are part of the ink solution involved in the transfer. Two investigators,

Masanori Kuratsune and Yoshito Masuda, have warned against the use of the PCBs in such papers. Their studies reported that 11 to 50 micrograms of PCB could stick to one's fingers from handling but 32 sets of the copying paper sheets. In Japan, the use of PCBs in carbonless copying paper was discontinued by the manufacturers in April 1971; in the United States, use was discontinued in June 1971.

In late 1970 the presence of PCBs was detected in dried fruit cartons. The paperboard manufacturers who made the cartons examined all the materials involved in carton manufacture: the printing ink, varnishes, adhesives, etc. They determined, however, that the PCBs were components of one of the paper stocks used in cardboard manufacture, specifically the high-quality white and colored paper stocks. These in turn were found to contain carbonless copying paper. Control of the source of paper stocks has effected dramatic reductions in the PCB concentrations of some cardboards; one example cited a reduction from 100 to 150 parts per million in August 1970, to less than 2 parts per million in January 1971. In fact, however, even virgin stock may contain PCBs, apparently as the result of general environmental contamination.

Studies by the Boxboard Research and Development Association at the Institute of Paper Chemistry indicate that recycled fiber cardboards currently contain an average of 6.4 parts per million PCBs. Something less than one-third of the PCB present in the packaging material will migrate to high-fat-content food if no barrier exists between the food and the carton. The FDA packaging material study in 1971 included the findings shown in Table P.4.

Thus, PCBs have found their way, via devious routes, into man's food supply. Contamination has resulted from misuse of PCB-containing products, as in the case of the transformer oil used as a defoliant spray base; from apparently unrelated uses, as in carbonless copying paper; and from general contamination of air and water, exemplified by the contamination of fish in Arctic lakes.

RESTRICTIONS ON USE

In September 1970, the Monsanto Company voluntarily began restricting sales of PCBs to uses in confined systems (mainly electrical uses) with the expectation that by 1972 all the PCBs then sold would be used in closed-system applications. In addition, the company has set up a disposal operation involving high-temperature incineration, which it will make available to its customers. This system is believed to result in no release of PCBs to the atmosphere. Quantities of PCBs from previous sales—and from other countries—can, of course, still be expected to reach the environment in the future. The effect of the voluntary restriction is reflected in data (see Table P.5.) obtained in a

TABLE P.4

MEAN PCB LEVELS IN PACKAGING MATERIALS AND FOOD CONTENTS (IN PARTS PER MILLION)

Material or Food	Pretzels and Potato Chips	Dry Infant Cereal	Macaroni and Noodles	Dry Breakfast Cereals	Cookies
Box or carton	13.5	23.5	19.0	9.5	7.0
Inner wrapper	13.0	—	1.5	2.0	0.5
Dividers	18.0	2.5	—	—	0.25
Food contents	0.5	0.5	Not detectable	0.5	0.5

TABLE P.5

PCB LEVELS OF RIVERS FLOWING INTO GREEN BAY, WISCONSIN (IN PARTS PER BILLION)

River	December 1970	May 1971	July 1971	August 1971
Peshtigo[a]	0.31	0.38	b	b
Oconto[c]	0.45	0.16	b	b
Pensaukee and Big Suamico[d]	b	b	b	b
Fox[e]	0.18	0.26	0.16	0.15

From G. Veith, "Recent Fluctuations of Chlorobiphenyls (PCBs) in the Green Bay, Wisconsin, Region," *Environmental Health Perspectives*, no. 1, NIH 72-218, U.S. Dept. of Health, Education and Welfare, April 1972.

[a] Receives wastes from city of Peshtigo.

[b] Less than 0.01.

[c] Receives wastes from Oconto Falls and Oconto.

[d] Drain agricultural and forest lands.

[e] Receives wastes from Green Bay and Marinette.

monitoring study of the water inputs to Green Bay, Wisconsin; the investigation began shortly after Monsanto's action.

In March 1972, the FDA issued notice of proposed rules to prevent excessive contamination of foods with PCBs. In view of the lack of information on the possible harmful effects of long-term exposure to low levels of PCBs, the FDA recommends that human exposure be reduced as much as possible. Provisions to decrease direct accidental contamination of animal feeds and human foods and to exclude PCB-containing pulp from food packaging materials are advocated. A temporary tolerance of 5 parts per million in food packaging materials is proposed, with an eventual decrease to zero. The time scale would allow for an "orderly withdrawal" of contaminated packaging materials. Temporary food tolerances are suggested, but the goal is stated to be: no PCB residues at the earliest possible time.

REFERENCES

1. D. B. Peakall and J. L. Lincer, "Polychlorinated Biphenyls—Another Longlife Widespread Chemical in the Environment," *Bioscience* 20, no. 17 (1970): 958.
2. R. D. Lyons, *New York Times*, September 30, 1971.
3. *Federal Register* 37, no. 54 (March 18, 1972): 5705.

References 4 through 12 all appear in *Environmental Health Perspectives*, no. 1, NIH 72-218, U.S. Department of Health, Education and Welfare, April 1972.

4. J. W. Cook, "Some Chemical Aspects of Polychlorinated Biphenyls (PCBs)."
5. I. Nisbet and A. Sarofim, "Rates and Routes of Transport of PCBs in the Environment."
6. R. Risebrough and B. deLappe, "Accumulation of Polychlorinated Biphenyls in Ecosystems."
7. G. Veith, "Recent Fluctuations of Chlorobiphenyls (PCBs) in the Green Bay, Wisconsin, Region."
8. M. Kuratsune and Y. Masuda, "Polychlorinated Biphenyls in Non-Carbon Copy Paper."
9. P. Trout, "PCB and the Paper Industry—A Progress Report."
10. H. Price and R. Welch, "Occurrence of Polychlorinated Biphenyls in Humans."
11. A. Kolbye, "Food Exposures to Polychlorinated Biphenyls."
12. M. Kuratsune, T. Yoshimura, J. Matsuzaka, and A. Yamaguchi, "Epidemiologic Study on Yusho, a Poisoning Caused by Ingestion of Rice Oil Contaminated with a Commercial Brand of Polychlorinated Biphenyls."

Pesticides

Pesticides are chemical compounds which are employed to kill pests. The definition of pest, however, is a subjective one. In general, any living thing which successfully competes with man for food or living space is considered a pest.

Food competitors include insects, fungi, bacteria, rodents, birds, mollusks (e.g., snails), mites, and nematodes. To deal with these creatures, insecticides, fungicides, bactericides, rodenticides, avicides, molluskicides, acaricides and nematocides have been developed. Living-space competitors, in addition to insects, birds, and rodents, include weeds and undesirable fish. The last two are controlled by **Herbicides** and piscicides.

Many types of chemical compounds have been employed in the control of pests. The major chemical groups are **Chlorinated Hydrocarbons,** an example of which is DDT, **Organophosphorus Compounds** such as parathion, and **Carbamates** such as carbaryl. There are also a variety of inorganic pesticides containing lead, arsenic, and mercury. Unfortunately, most such poisons do not discriminate between pest and nonpest or, for that matter, the pesticide user. Residues of chlorinated hydrocarbons have been found in air, water, soil, and animal tissues, including human tissues. Numerous cases of poisoning have occurred among pesticide users and small children.

More detailed descriptions of the major chemical groups of pesticides are found under their separate headings. The discussion here investigates the environmental effects of pesticides and examines the philosophy underlying the use of these compounds.

INTRODUCTION

Two groups of creatures living on the earth represent high points in evolutionary progress and have shown a marvelous ability to adapt to any situation—humans and insects. It is perhaps inevitable that their interests should conflict.

The battle between people and insects for food and fiber (and often, it seems, for our very flesh) was in the nature of guerrilla warfare for several centuries. During this time humankind discovered certain substances which could deter insects. For instance, the flowers of the Dalmatian daisy were ground up and sprinkled on plants.[1] Arsenic found similar use. The materials for modern warfare on insects were not at hand until about 1945. In that year, DDT, an inexpensive and broadly toxic insecticide, was synthesized. Many other compounds were soon devised.

Insects have fought back, with a combination of biochemical adaptability and sheer numbers. When a particular species is subjected to a lethal chemical, its survival may depend on the fact that out of the millions or billions of offspring born, some will have a difference in their chemical makeup, a mutation of their genes, that allows them to convert the chemical to some harmless material. The mutant can then repopulate the poisoned areas. An example of biochemical adaptability is found in the enzyme system called NADPH-dependent oxidase. This enzyme allows insects to convert insecticides to harmless substances. It is found in both susceptible and resistant insects but in higher concentrations in the latter. Synthesis of the enzyme is believed to be accelerated when

[1] Dalmatian daisy powder contains the insecticide pyrethrin, still in wide use.

the insect comes in contact with a toxic pesticide. Thus, insecticides themselves stimulate the insects to detoxify insecticides. Houseflies seem to have won a number of skirmishes in this manner. They have become resistant not only to chlorinated hydrocarbons such as DDT, but also to **Organophosphorus Compounds** which were developed next. Finally, they have become resistant to the most recently developed major group of pesticidal chemicals, the **Carbamates.**

The catalogue of modern pesticides is large and includes the following chemical groups:

1. *Inorganic pesticides.* These are usually compounds of copper, arsenic, lead, or mercury. Elemental sulfur is also in this group, as are halogen derivatives like sodium fluoride. Many of the compounds are characterized by extreme toxicity and persistence. Tobacco plants grown where arsenic was used even decades before have been shown to contain arsenic.
2. *Botanicals.* Compounds which are extracts of plants fall into this group. Besides the pyrethrins, examples are: sabadilla, which is used as louse powder by South American natives, and red squill, derived from the Mediterranean Sea onion and used against rats. A number of synthetic pyrethrins have also been produced. The group is generally characterized by low toxicity to humans.
3. **Chlorinated Hydrocarbons.**
4. **Organophosphorus Compounds.**
5. **Carbamates.**
6. *Oils.* Petroleum oils and coal tar oils are used in sprays to control pests of trees and bushes. Their use is decreasing as synthetic pesticides replace oil sprays.
7. *Organic mercury compounds* (see also **Mercury**). These compounds, such as Panogen, have been used as seed dressings to prevent fungus growths. Since mercury is a persistent environmental contaminant, the wisdom of adding it to the soil in this manner has been questioned. In 1972 the U.S. Environmental Protection Agency canceled registration for the use of mercury compounds as seed dressings. Both Great Britain and Sweden have outlawed their use because of the large number of birds poisoned from eating the contaminated seed.
8. *Antibiotics.* Chemicals such as streptomycin, produced by microorganisms, have found some use in preventing bacterial and fungal diseases of plants.
9. *Phenol derivatives.* These compounds, which often contain nitrogen, chlorine, bromine, or fluorine atoms, include insecticides (dinitrocresol), fungicides (Dinocap), and herbicides (pentachlorophenol).
10. *Organic acids.* Derivatives of acids, such as phenylacetic, benzoic, and phthalic, form an important group of herbicides. For more information, see **Herbicides.**
11. *Anticoagulants.* Compounds in this group are used in rodent control. Chemicals such as warfarin prevent blood coagulation and so cause death by internal hemorrhaging. Low-dose formulations are mixed with food in such a way that one accidental dose is not likely to kill. Their effect, rather, depends on repeated doses. Hence they are *relatively* safe for use around humans and pets. Rodents must receive a dose four to five times in the course of several days for the pesticide to be effective.

There are many other chemicals. The interested reader is referred to Melnikov's excellent work on the chemistry of pesticides for further information.

Finally, mention should be made of repellants and attractants—two types of compounds which are not intended to kill pests. Repellants repel pests rather than destroy them. They tend to be of lower toxicity than pesticides in general. Included are insect repellants like "6-12" (ethylhexanediol) and "Off" (N,N-diethyl-m-toluamide).

Attractants are used as lures in traps for insects or to get early warning of an infestation by a particular pest. Included are sex attractants like gyplure, first isolated from the glands of virgin female gypsy moths, and feeding attractants. Good control of Mediterranean fruit flies has been obtained with medlure, which attracts only the male. Most attractants are specific for only one species and often for only one sex.

One billion pounds of pesticides were used in the United States in 1970. When toxic chemicals are added to the environment in this quantity, it becomes essential to examine their impact.

SOIL CONTAMINATION

A large portion of pesticide usage is concentrated in agriculture. In fact, some 5 percent of the land area of the continental United States was treated with pesticides in 1968. The soil, then, is often the first repository for pesticides.

Contamination of soil with a pesticide becomes a problem when the chemical is persistent. If a pesticide is not degraded by bacteria, sunlight, or chemical processes within a growing season, concentrations of the chemical can build up in the soil, it may be absorbed by crops planted in later seasons, and it may slowly leach into water supplies. Table P.6 illustrates how persistence is defined in relation to the various pesticidal chemicals. The problem of accumulation, then, centers on the chlorinated hydrocarbons, the heavy metal compounds (lead, mercury), and the arsenicals, because these are the compounds which soil microorganisms and other natural processes are unable to use or degrade. Lead, mercury, and arsenic compounds remain undegraded in the soil almost permanently. In some orchard areas of the United States, pesticides containing arsenic have been used for decades. Such use has caused a buildup of the toxic chemical which affects the life-span of trees in the orchards; the residual pesticide may also act as an herbicide preventing use of former orchard land for forage crops.

Chlorinated hydrocarbons also build up in the soil. Residues in woodlands regularly sprayed with these substances were found to have increased from 0.169 kilogram per hectare in 1958 to 0.628 kilogram per hectare in 1961.

Various factors such as soil type, acidity, organic matter content, and mineral content affect persistence. Chemical stability, volatility, and water solubility also affect the length of time a chemical will remain in the soil. Volatile chemicals will be more easily carried off by air, and water-soluble compounds by water. Residues which are not cultivated into the soil disappear more readily, as do those distributed on fields not planted with a cover crop. These latter effects are probably due to evaporation of

TABLE P.6

PERSISTENCE OF PESTICIDES

Designation	Duration	Group
Non-persistent	1–12 weeks	Organophosphates, carbamates
Moderately persistent	1–18 months	2,4-D, Atrazine, most others
Persistent	2–5 years	Chlorinated hydrocarbons
Permanent	Virtually not degraded	Mercury, arsenic, and lead compounds

From Report of the Secretary's Commission on Pesticides and Their Relationship to Environmental Health, U.S. Dept. of Health, Education and Welfare, December 1969, p. 104.

the pesticides, in which case they can become an air pollution problem.

Soil residues of pesticides can cause the following problems.

1. Crops grown on contaminated soils may absorb the residues, making the crops unfit for human or animal food. For instance, sweet corn grown in fields previously planted with alfalfa which was treated with lindane at half a pound per acre will absorb enough of the chemical to exceed legal tolerances for lindane in corn. Plants do not, however, appear to *concentrate* pesticides from the soil.

2. The microcosm of soil organisms may be destroyed. The microorganisms in soil are essential in the recycling of nutrients. They degrade dead plants and animals to chemicals that may be used over again. Special concern has been expressed about the valuable microorganisms which convert atmospheric nitrogen to nitrogen usable by plants.

Most common insecticides are not toxic to these soil microorganisms in recommended quantities. Soil fumigants and some herbicides, however, are. Studies have shown that small areas sterilized in this way are quickly repopulated from adjacent areas. It is not known whether these results can be extrapolated to large areas. There are also no guarantees that the new population contains the same proportions of the various types of microorganisms. It is possible that an undesirable mixture will result. On the positive side, some pesticides can stimulate microbial growth if the pesticides are usable (degradable) ones. Nutrients may be contributed to the soil by such pesticides, thus enhancing plant growth.

3. Animals living in the soil may concentrate pesticidal chemicals. Invertebrates, such as earthworms, which eat large quantities of soil and extract nutrients from it, may extract and concentrate pesticide residues as well.

4. Some pesticidal chemicals, used for purposes other than plant control, are also herbicidal. They prevent plant growth and may make land areas unusable until they disappear. The example of arsenic was given above. Overuse of some chlorinated hydrocarbons or other insecticides may also have this effect.

5. Pesticides may be leached from the soil into water or may evaporate into the air.

WATER CONTAMINATION

Pesticides reach water in several ways. Spraying to control such pests as mosquitoes, in their larval stage, or accidental spraying, could result in the direct introduction of pesticides. Sewage effluents from populated areas and from pesticide manufacture may introduce pesticides to the water. Rain may carry down pesticides from the air. Finally, soil particles on which pesticides are adsorbed may be incorporated in stream sediments or lake bottom muds. Aerial spraying and industrial effluents are probably the main sources of pesticides in water.

Pesticides applied directly to waters are rapidly concentrated into wildlife and plants or adsorbed to bottom muds. Concentrations in the water may thus amount to only a fraction of the total amount of pesticidal contamination. They are therefore a less valuable indicator of contamination than pesticide residue levels in aquatic life or bottom sediments.

Since chlorinated hydrocarbons are relatively insoluble in water, it is unlikely that when applied to the land they will leach into groundwater or be dissolved and carried off in drainage. However, these compounds do attach themselves very tightly to soil particles and are thus carried into streams, lakes, and rivers. Here the particles tend to settle out. The pesticides may then either redissolve slowly into the water or be resuspended during fall overturns[2] or flood times. Monitoring studies of United States

---

[2] See **Reservoirs.**

waters have reported the presence of traces of all of the chlorinated hydrocarbons. As of 1969, DDT was found in the greatest quantities; dieldrin and endrin were next. Levels found in several studies have been generally no higher than 100 to 200 parts per trillion except for areas below chlorinated hydrocarbon manufacturing plants or where agricultural and industrial runoff enters estuaries. In these areas values can reach 100 parts per billion or even 11,000 parts per billion. In Great Britain only DDT, BHC (benzene hexachloride), and dieldrin have been detected. Some northern British rivers (about 2 percent) were found to contain enough insecticide to be toxic to trout, and about 30 percent had one-tenth of the lethal dose. These levels are attributed to industrial contamination.

There is little experimental evidence relating usage to pesticide levels in seawater. But it seems likely that quantities of pesticides are added to the ocean yearly. It is estimated that the San Joaquin River probably carries some 4,000 pounds of pesticides into San Francisco Bay per year and the Mississippi some 20,000 pounds into the Gulf of Mexico. The ocean itself may absorb large quantities of chemicals before harmful levels are reached. However, the pesticides in river waters are entering the ocean via estuaries. These bodies of water are the fragile breeding grounds for a great many fish and shellfish. Effects of pesticide pollution on estuarine species are discussed in a later section.

A case cited in the 1969 Report of the Secretary's Commission on Pesticides is illustrative of the hazards involved in direct applications of pesticides to waters. For several years DDD[3] was applied at Clear Lake, California, to control nuisance insects. It had been determined that the levels of 14 to 20 parts per billion, which were achieved in 1949, 1954, and 1957, would not be toxic to fish and other aquatic organisms. In 1954, 1955, and 1957, however, grebes were found dead around the lake. These birds were shown to contain up to 1,600 parts per million of DDD in their tissues. Further study also disclosed high concentrations in fish. Apparently aquatic organisms had concentrated DDD from the lake waters. Fish, feeding on these organisms, magnified the original water concentrations of pesticide. The fish-eating grebes then absorbed and concentrated DDD into their fatty tissues. Thus, predictions of the effects of pesticide application are complicated by the phenomenon of biological magnification. As of 1972, no official standards were set by the U.S. Public Health Service for pesticide levels in public drinking water. Suggested guidelines are given in Table P.7.

[3] A pesticide very similar in structure to DDT.

AIR CONTAMINATION

In 1963, 80 percent of pesticides used commercially were applied by air. To the amount intentionally disseminated by air can be added the amount that evaporates from the soil after

TABLE P.7

SUGGESTED ALLOWABLE LEVELS OF PESTICIDES IN PUBLIC DRINKING WATER

Chemical	Parts per Billion
Aldrin	17
Chlordane	3
DDT	42
Dieldrin	17
Endrin	1
Heptachlor	18
Heptachlor epoxide	18
Lindane	56
Methoxychlor	35
Organophosphates	100
Carbamates	100
Toxaphene	5
2,4-D	100
2,4,5-T; 2,4,5-T-P	100

From H. S. Nicholson, "Occurrence and Significance of Pesticide Residues in Water," *Proceedings of the Washington Academy of Science* 59, no. 4–5 (1969): 77.

Dusting grape vines with sulfur to retard mildew, near Fresno, California.

other forms of application. A third source of airborne pesticides is dust which has adsorbed pesticides and is picked up by winds and blown around. In 1965, for example, there was an unusually severe dust storm in Texas. Dust monitoring stations as far east as Cincinnati, Ohio, detected residues of DDT, chlordane, DDE, Ronnel, heptachlor epoxide, 2,4,5-T, and dieldrin, carried on dust particles by the storm. Probably the greater portion of pesticides in the air is carried on dust particles which are periodically washed down by rain. One author has estimated that in the 1960's 40 tons of DDT per year were dropped on England by rain. Thus, there is a pesticide fallout problem similar to the problem of radioactive fallout, which also travels large distances and is then washed down by rain.

Nine air sampling stations located across the United States,[4] to monitor air concentrations of pesticides, reported the following facts in 1968. Only DDT was found in all localities, and it was highest in southern agricultural areas. Values found for DDT ranged from 0.1 milligram per cubic meter to 1,560 milligrams per cubic meter. Toxaphene reached concentrations of 2,520 milligrams per cubic meter and parathion 465 milligrams per cubic meter. The highest levels were monitored when spraying was reported in the area before sampling.

The extent of contamination due to spraying depends mainly on weather conditions. If the air temperature decreases with height, the sprayed particles will tend to move vertically and there is little drift. If temperature increases with height,[5] movement is lateral and drift is likely. In one study, 45 percent of the chemical sprayed from an airplane landed on the ground 10 to 10,000 yards downwind rather than beneath the plane. The same study relates a case in which the spraying of a half-acre plot of deciduous trees resulted in a 70 percent waste of the spray since it drifted out of the plot. Turbulence also contributes to drift from spraying, as does particle size. Larger particles tend not to drift badly although smaller particles give better crop coverage. Large particles also may not penetrate the thick, leafy cover of a forest area.

The hazard of inhalation of pesticides is probably greatest for industrial workers, for farm laborers, and for pilots who apply pesticides. Nevertheless, measurable excretion of the breakdown product of parathion is found in the urine of people whose only exposure to parathion is that they live near orchards.

In 1964 and 1965 a study attempted to determine whether DDT had reached isolated areas such as Antarctica. No DDT was found in

[4] Baltimore, Maryland; Dothan, Alabama; Fresno, California; Iowa City, Iowa; Orlando, Florida; Riverside, California; Salt Lake City, Utah; and Stoneville, Mississippi.

[5] This phenomenon is referred to as an inversion and is described more fully in **Air Pollution, meteorological aspects.**

water or snow samples (the limit of detection was 0.005 part per million), but some residues were detected in the animals. Seals were found with residue levels of 0.042 to 0.12 part per million, penguins with 0.015 to 0.18 part per million, and skuas with concentrations up to 2.8 parts per million. One group of fish contained 0.44 part per million of DDT. The study was not conclusive, since the creatures with the highest DDT concentrations, the skuas, were also the most far-ranging animals. It did, however, indicate that traces of pesticides are being transported all over the globe.

A good deal of this type of contamination of soil, water, and air is apparently unnecessary. The following statement indicates professional opinion from the Report of the Secretary's Commission on Pesticides:

> Much contamination and damage results from the indiscriminate, uncontrolled, unmonitored and excessive use of pesticides. . . . The careful application of many of the pesticides and the use of techniques presently available and being developed can be expected to reduce contamination of the environment to a small fraction of the current level without reducing effective control of the target organisms [1, p. 101].

EFFECTS ON ORGANISMS OTHER THAN MAN

The addition of pesticides to the environment can reasonably be expected to affect plants and animals, including man. Exposure can occur through eating food with residues or breathing contaminated air or drinking contaminated water.

AQUATIC ORGANISMS

The aquatic environment is host to an integrated community of creatures, which comprise what is known as a *food chain.*[6] Tiny single-celled organisms, plankton, are food for larger creatures, small fish or shellfish, which in turn are eaten by larger fish and birds. Phytoplankton not only serve as dinner for some species of organisms but may also be responsible for the production, through photosynthesis, of most of the oxygen we breathe.[7]

Experiments have shown that in high enough concentrations pesticides interfere with the process of photosynthesis. In one set of experiments the productivity of a phytoplankton culture was reduced by 70 to 94 percent by an exposure of four hours' duration to 1 part per million of DDT, dieldrin, heptachlor, aldrin, methoxychlor, or toxaphene. Endrin, lindane, or Mirex can produce a 28 to 40 percent decrease under the same conditions. When the alga chlorella was examined with respect to the effect of DDT, a decreased rate of photosynthesis and changes in shape of the organism were detected after three days at only 0.3 part per billion [1]. The photosynthesis of marine algae may decrease 10 to 40 percent at DDT concentrations of 100 parts per billion. These values should be compared to the levels of pesticides found in natural waters.

Shrimp and crabs, which are close biological relations of insects, are very sensitive to many insecticides. At a DDT level of 0.5 part per million in the water 50 percent of the exposed blue crabs are killed within 24 hours. A concentration of 0.3 to 0.4 part per billion heptachlor, endrin, or lindane kills 50 percent of the shrimp exposed within 24 hours. Larval stages of these creatures may be particularly sensitive to many pesticides. Reduced growth and failure to grow into the adult form are related to pesticide concentration. Adult oysters and other shellfish are vulnerable because they are often attached to a surface and thus cannot

[6] For further information on food chains, see [26].

[7] *Phytoplankton* is derived from the Greek and means "wandering plants," an apt description, for the plants float in the water, instead of being rooted in soil.

move to escape noxious concentrations of chemicals. Shellfish are very good indicator organisms for pesticide pollution since they concentrate pesticides many times over the water concentrations; oysters have been known to concentrate DDT up to 700,000 times.

Many pesticides have been isolated from fish. The most common residues are DDT and dieldrin. Endrin, toxaphene, and heptachlor epoxide are found in fish from United States waters but not in the fish of Great Britain, a fact which reflects the patterns of use in the two countries. Some BHC and aldrin have been found in fish in both countries. These chlorinated hydrocarbons are concentrated in the fatty tissses of the fish. Freshwater fish, reflecting a greater exposure, generally contain higher residue concentrations than salt-water fish, although exceptions are known. Levels of DDT found in British fish commonly range from 0.01 part per million to 0.71 part per million (salmon) but values up to 23.3 parts per million have been reported. In the United States, reported DDT levels have ranged from 0.06 part per million to 136 parts per million, and in Canada from 0.01 to 30 parts per million.[8]

The effects of sublethal pesticide concentrations on fish include lowered resistance to disease, reproductive failure, and feeding aberrations. Some development of resistance to pesticides has been seen. However, the resistant fish then carry larger than normal amounts of pesticides in their tissues and can be a greater hazard to birds and humans who consume them. Baby lake trout are nourished by a yolk sac during late stages of their development. If the yolk sac contains 5 parts per million or more DDT, the trout will not survive. In 1955 a hatchery on Lake George experienced 100 percent mortality of its 350,000 eggs because of such a high DDT concentration. Previous to this time, 10,000 pounds of DDT had been applied, per year, to the Lake George watershed for control of the gypsy moth.

Large enough concentrations of pesticides will, of course, kill fish outright. A number of spectacular pesticide-caused fish kills have occurred in recent years. According to one report, a million fish were killed in Florida on account of sand fly control efforts. Five million fish were killed in the lower Mississippi in another instance.

The toxicity of pesticides (and other chemicals) to aquatic organisms has been shown to vary with physical characteristics of the water such as pH and water hardness. For instance, the molluskicide Bayluscide appears to be much more toxic to both snails and fish in soft water than in hard water. Another molluskicide, Frescon, is not stable in waters with a pH below 7.5 and thus is not effective in acidic waters. Studies with fish have shown that most pesticides become more toxic as temperatures rise. A number of exceptions are noted, however. As an example, DDT appears to be more toxic to bluegills and rainbow trout at 13°C than at 23°C [25].

A major problem which arises from pesticide contamination of water is that the concentrations of pesticides found in organisms living on the water can increase phenomenally as residues move up the food chains, i.e., as one species consumes another for food. This is the phenomenon called *biological magnification.* If a substance is not excreted as rapidly as it is ingested, it tends to build up in the tissues. Such concentrations become greater and greater at the upper levels of the food chain. Each organism passes on the concentrated dose to its predator, which in turn concentrates the substance even more (see Figure P.2).

According to recent data, Lake Michigan

[8] Residue data collected in Clive Edwards, "Persistent Pesticides in the Environment," *Critical Reviews in Environmental Control* 1, no. 1, Chemical Rubber Publishing Co. (February 1970).

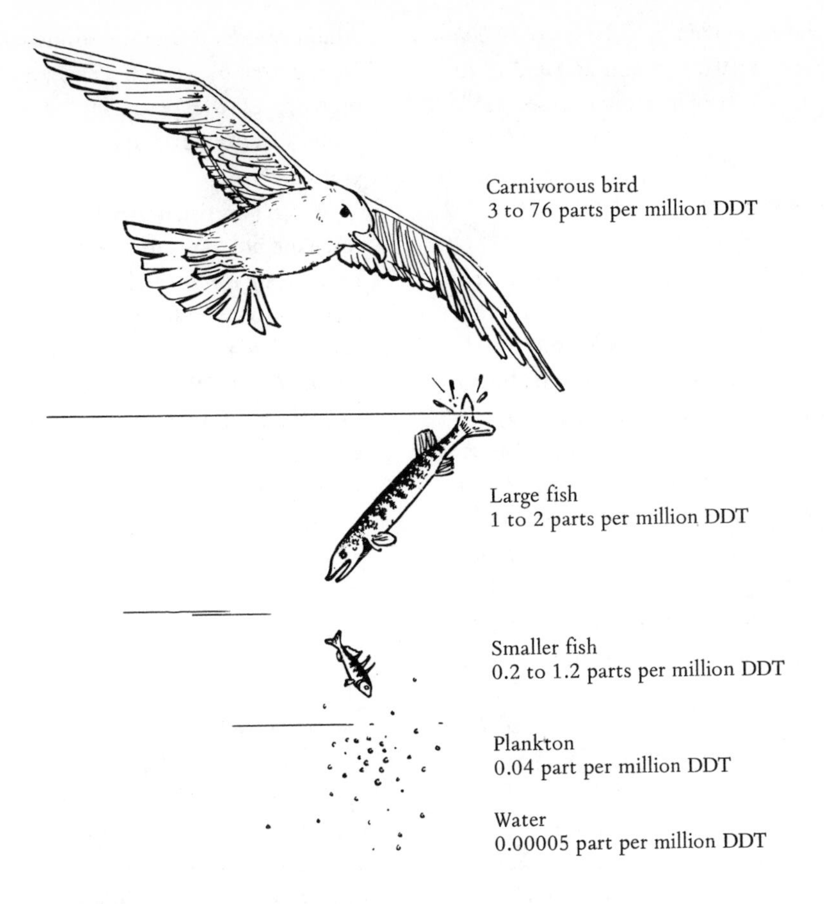

FIGURE P.2

BIOLOGICAL MAGNIFICATION

Carnivorous birds in a Long Island salt marsh estuary are found to contain about a million times more DDT than is present in the water. (DDT values given are the sum of the DDT, DDE, and DDD residues and have been rounded.)

From data in G. M. Woodwell and C. F. Wurster, Jr., "DDT Residues in an East Coast Estuary: A Case of Biological Concentration of a Persistent Insecticide," *Science* (May 12, 1967), p. 821.

contains 0.014 part per million of DDT in its bottom soil; 0.41 part per million DDT is found in its invertebrates, 3 to 8 parts per million in the fish, and nearly 2,500 parts per million in the fatty tissues of herring gulls. Plankton can concentrate DDT 17,000 times; oysters concentrate DDT 15,000 to 700,000 times, dieldrin 3,000 times. Clams concentrate DDT 1.5 to 70,000 times. Invertebrates in general may concentrate toxaphene 2,000 times. Fish can concentrate DDT up to 40,000 times, heptachlor up to 1,000 times. Aldrin and dieldrin may be magnified by fish up to 19,000 times, toxaphene to 19,000 times, and endrin to 70,000 times. For further information, see [25].

LAND DWELLERS

Insects

Insects not killed by pesticides may be affected by them in several ways. Many species can carry high concentrations of pesticides in their bodies. They are thus a hazard to the insects, birds, and fish that eat them. Insect fertility may be reduced or in some cases enhanced. As an example of induced fertility changes, 2,4-D appears to stimulate the reproduction of certain aphids. Growth of some species may be slowed or accelerated by pesticide applications. For instance, 2,4-D slows the growth of the predacious coccinellid beetle larvae which eat mites, aphids, and scale insects. On the other hand, growth of the rice stem borer is accelerated by 2,4-D. Finally, insects may become resistant to pesticides. This last phenomenon is one of the real difficulties in chemical pest control.

The development of insect resistance to pesticides was noted even before the DDT era. The problem has become truly serious, however, because of the very efficiency of modern pesticides. The development of insect resistance has so far been found to depend on several conditions. One necessary condition is that, in a large population of insects, a few individuals have a mutation which makes them resistant to the pesticide used. In view of the huge numbers in which insects occur, this is not an unlikely possibility. Another necessary condition is that those individuals be given a chance to multiply. Such an opportunity is readily presented by modern pesticides. If competing insects in an area are killed off, the food supply of the resistant species may be ensured. In addition, predators and parasites which inhibit increases in the species' population may be destroyed. As mentioned previously, at least one mechanism which confers resistance on insects may actually be induced by certain pesticides. That is, the presence of the pesticide causes the insect to produce enzymes capable of degrading the pesticide.[9]

When a resistant strain develops, it may be resistant not only to one pesticide or chemical group but to several. A strain of houseflies resistant to carbamates was also found to be resistant to a chemical sterilant. Another strain, resistant to insecticides, showed great resistance to an insect hormone which was being investigated as a control method.

By 1969, over 224 pests were known to be resistant to one or more pesticides. In this group were 97 public health or veterinary pests and 127 pests which attack crops or other vegetation. Salt marsh sand flies developed resistance to DDT after only three applications of 1 pound per acre. The resulting strain also had increased resistance to heptachlor, chlordane, and BHC. Anopheles mosquitoes are a particular target for chemical pest control since they carry malaria. Thirty-five anopheline types are now resistant to pesticides, 34 to the chlordane-heptachlor-endrin group and 12 to DDT; some are resistant to both. These mosquitoes all started to develop resistance during malaria eradication programs.

Resistance is a serious problem in cotton production in the United States. At least 18 cotton pests are resistant to chlorinated hydrocarbons and 8 are also resistant to the organophosphates. The cotton bollworm and tobacco budworm are resistant to four classes of synthetic pesticides. The potential course of events following extensive use of chemical pesticides is exemplified in the situation which occurred in the cotton-producing Rio Grande Valley in Texas. Before 1945 the major cotton pest in the area was the boll weevil and the major pesti-

[9] The NADPH-dependent oxidase system is believed to act this way.

cides used were arsenic compounds. In the words of F. W. Plapp, Jr. [15]:

> Shortly after World War II, the introduction of the synthetic chlorinated insecticides dramatically changed the picture. Control of insect pests, primarily the boll weevil, greatly increased the yield of cotton. Further, the high degree of control encouraged growers to use fertilizer and irrigation as crop management techniques. The net effect of the new program was a spectacular increase in yield and profit.
>
> In the late 1950's, the situation deteriorated. The boll weevil became resistant to the chlorinated insecticides which had proven effective for a dozen years. Growers switched to organophosphates, primarily to . . . methyl parathion. At doses which gave good control of the boll weevil, methyl parathion was ineffective against two other major cotton insect pests. These insects, the bollworm (*Heliothis zea* Boddie) and the tobacco budworm (*H. virescens* Fabricius), were now major pests on Texas cotton. Their numbers had increased because of their release from natural control as a consequence of the destruction of their natural enemies by the potent insecticides in use. Control was obtained by adding DDT to the methyl parathion used for boll weevil control.
>
> By 1962 the chlorinated insecticides no longer controlled Heliothis on cotton. Control was now obtained by using high doses of methyl parathion. By 1968 resistance to methyl parathion was present in the bollworm. High levels of methyl parathion were used in attempts to control the resistant populations. Doses of as much as 2 pounds per acre were applied 10 to 15 times per season. Even so, control was not satisfactory.

Today, cotton growing in the valley has become so expensive that cotton produced there cannot compete on the market.

This story illustrates the possible economic consequences of pesticide use. It also shows how the phenomenon of resistance can increase the total number of species considered to be pests. Many insects are kept under control by a predator. If the predator is killed off by pesticides and the insect it was eating is not, that insect may become a pest. Most of us are familiar with the fact that the praying mantis eats undesirable insects and thus should be protected rather than exterminated. Spiders, too, are felt by some to be beneficial creatures. Yet predacious insects and spiders can be almost wiped out in certain spraying programs. In one study parathion reduced plant eaters 8 percent, but parasitic and predacious forms 95 percent. This type of situation allows a rapid overgrowth of the plant pests, and extensive crop damage can result.

Honeybees, necessary for pollination of many crops, are especially sensitive to carbaryl. Ladybugs, which eat red mites, are sensitive to DDT. Orchards sprayed with DDT may be subject to red mite damage since the mite is resistant to DDT and is ordinarily kept under control by ladybugs.

Soil Inhabitants

Earthworms burrow through the ground by swallowing earth and then excreting it. As the earth passes through the worm, nutrients are extracted, but pesticides are also absorbed by the worms' tissues and may be concentrated. DDT, for instance, can be concentrated up to 74 times over soil levels. Other soil inhabitants such as beetles, lice, slugs, and cutworms may also concentrate pesticides. Birds feeding on these organisms can thus be poisoned. Robins have been poisoned in this manner after DDT spray programs to control Dutch elm disease. Spray and contaminated leaves falling on the soil contributed to high DDT concentrations in earthworms. Birds which ate the earthworms then received a toxic dose of DDT.

Birds

A great deal of controversy surrounds the issue of the effects of pesticides on birds. Much of the data has been in the form of population studies, in which attempts are made to relate declining populations to pesticide concentrations in the tissues of the remaining birds. Demographic data, however, are frequently open to many interpretations. Nevertheless, when such data are combined with experimental laboratory data, a picture emerges which indicates serious effects on several species of birds, including the European sparrow hawk, Scottish golden eagle, English kestrel, sharp-shinned hawk, Cooper's hawk, osprey, peregrine falcon, and bald eagle.

Population declines seem to be due to reproductive failure in the birds. Effects are believed to be both hormonal and behavioral and to result from chronic pesticide ingestion. As an example of one type of effect, William Ruckelshaus, Administrator of the U.S. Environmental Protection Agency, in his July 1972 order reaffirming the cancellation of registration of DDT, stated [16]:

> I am persuaded that a preponderance of the evidence shows that DDE[10] causes thinning of eggshells in certain bird species. . . . Birds in the laboratory, when fed DDT, produced abnormally thin eggshells. In addition, researchers have also correlated thinning of shells by comparing the thickness of eggs found in nature with that of eggs taken from museums. The museum eggs show little thinning, whereas eggs taken from the wild after DDT use had become extensive reveal reduced thickness.

Birds which have been acutely poisoned show an inability to fly, suffer convulsions, and die in a stiff position with their legs extended. In Great Britain large numbers of birds were sometimes found dead near fields, and examination showed that they contained pesticide residues sufficient to have killed them. As a result of the poisonings, the use of seed dressings containing chlorinated hydrocarbons has been banned in Great Britain, except in the autumn. Heptachlor is no longer allowed as a seed dressing at all. Mortalities of the type described have decreased dramatically since these measures were taken.

Birds feeding on fish and other aquatic creatures may be poisoned when pesticide concentrations in water become magnified in the birds' prey (see Figure P.2).

Residues found in wild birds depend more on the habitat of the species than the country in which they live. DDT residue levels vary from 0.12 part per million in Irish pigeons to 194 parts per million in United States pelicans! Dieldrin and aldrin residues vary from 0.03 part per million in migrating birds to 13.5 parts per million in meadowlarks.[11]

Mammals

Data on mammals are much more scarce than data on birds. Mass mortalities of mammals have been reported after the use of aldrin, endrin, and dieldrin. Residue values found in mammals are generally lower than those found in birds although the relative values for the various pesticides are similar. DDT and dieldrin are seen most often and in the greatest quantities.

DDT appears to slow the growth of white-tailed deer. It also slows their reproductive rate and lowers the survival rate of fawns. Mice show some reproductive failure at 200 to 300 parts per million DDT in their tissues and deer are affected at 25 parts per million. Some resistance has been noted; e.g., pine mice in Virginia orchards are becoming resistant to endrin.

EFFECTS ON MAN

The effects of pesticides on people are of two general types. Large doses produce acutely toxic effects while smaller doses result in chronic effects.

[10] DDE is a product derived from the biological breakdown of DDT.

[11] Values obtained from 1963 to 1969.

Acute effects are most likely to be experienced by workers in pesticide manufacturing plants, by agricultural workers who apply pesticides, by crop dusters, and by workers who load the pesticide into crop-dusting aircraft. Also included in this group are children who are accidentally poisoned by swallowing pesticides left within their reach. In 1961, 8.5 percent of all accidents and poisonings occurring among children under five involved pesticides.

All groups of pesticides are hazardous when ingested or inhaled. Many organophosphorus compounds are also easily absorbed through the skin and thus present a further hazard to those working with them. Dermatitis, or skin irritation, is a common and often unrecognized ailment associated with all types of pesticide use. Eye irritation and respiratory symptoms are not unusual. A discussion of the symptoms of pesticide poisoning can be found in **Chlorinated Hydrocarbons, Organophosphorus Compounds, Carbamates,** and **Mercury.** Examples of doses sufficient to cause visible symptoms are shown in Table P.8. A 150-pound person could be expected to show symptoms of poisoning after ingestion of about 0.5 gram of parathion. There are, however, variations in susceptibility among different people.

A major concern about the use of pesticides is not the acute or chronic effects on occupationally exposed persons but exposure of the general population to much lower levels of pesticides. It does not seem possible for the average person to escape daily exposure to small amounts of pesticides. Although air contamination is a possible route of exposure, the main source is believed to be food. Both plant and animal tissues introduce pesticides to the human diet. While food and feed crops may be contaminated with excessive pesticide residues from pesticides applied to them intentionally, laws governing allowable residues (discussed later) should in general guard against this hazard. Rather, in a process of secondary contamination, general pollution of the environment with persistent pesticides appears to have led to contamination of the food supply.

There is evidence that plants concentrate pesticides from water. In extreme cases, concentration factors of up to 100,000 times have been found. Plants grown with their roots underwater, like rice and cranberries, absorb more chlorinated hydrocarbons than do plants grown only in soil. Some studies have also noted that plants can concentrate pesticides from the air.

As a rule, plants do not concentrate pesticides from the soil. Lindane, however, can concentrate in the roots of peas and carrots, and high-fat seeds like peanuts will concentrate fat-soluble pesticides.

Residues in feed or on pasturelands[12] may be ingested and concentrated by animals which are eventually used by man for food. Thus, general environmental contamination can lead to contamination of foods even when there is no deliberate use of persistent pesticides on the foodstuffs themselves.

Organophosphates and carbamates are not concentrated in human or animal fatty tissues, but chlorinated hydrocarbons are. Table P.9, adapted from a review by Edwards [2], illus-

TABLE P.8

DOSES OF PESTICIDE THAT PRODUCE VISIBLE SYMPTOMS IN MAN (IN MILLIGRAMS PER KILOGRAM OF BODY WEIGHT)

Pesticide	Amount
Organophosphates	
Parathion	7.5
Systox	6.75
Methyl parathion	11–19
Chlorinated hydrocarbon	
DDT	6–10

[12] Such residues can, for example, result from accidental drift during spraying operations or from pesticide fallout.

TABLE P.9

PESTICIDE RESIDUES IN HUMAN TISSUE IN VARIOUS COUNTRIES (IN PARTS PER MILLION)

Country	DDT	Lindane	Aldrin/Dieldrin	Heptachlor and Heptachlor Epoxide
Great Britain	3–6	0–0.3	0.2–0.3	Trace
Holland	2–7	0–0.1	0.2	0.009
France	5	—	—	—
Italy	8	0.06	0.5	0.2
Israel	5–8	—	—	—
India	12–28	1–2	0.03–0.06	—
New Zealand	5	Trace	0.3	—
Australia	2	—	0.05	—
United States	5–10	0.2–0.5	0.1–0.3	0.1–0.2
Canada	4	0.06	0.2	0.07

From C. Edwards, "Persistent Pesticides in the Environment," *Critical Reviews in Environmental Control* 1, no. 1, Chemical Rubber Publishing Co. (February 1970). Residue values were published from 1963 to 1969.

TABLE P.10

PESTICIDE LEVELS IN UNITED STATES DIET (IN MICROGRAMS PER KILOGRAM OF BODY WEIGHT)

Compound	FAO/WHO Acceptable Daily Intake	Dietary Intake, 1968
Aldrin/dieldrin	0.1	0.06
DDT[a]	10.0	0.7
Lindane	12.5	0.04
Heptachlor epoxide	0.5	0.031
Malathion	20.0	0.04
Parathion	5.0	Less than 0.001

From R. E. Duggan and G. Q. Lipscomb, "Dietary Intake of Pesticide Chemicals in the United States (2), June 1966–April 1968," *Pesticides Monitoring Journal* 2, no. 4 (1969): 153.

[a] Values given are the combined levels of DDT, DDE, and TDE.

trates the chlorinated hydrocarbon residues found in human tissues in various countries. It has been pointed out that since the DDT residue limit on meat recommended by the World Health Organization is 7 parts per million, most people would be inedible according to WHO/FAO standards.

It is interesting to compare the World Health Organization's recommended daily intake levels with actual diet levels of pesticides in the United States. The data on the dietary intake of pesticidal chemicals in Table P.10 were calculated by Duggan and Lipscomb [27] for 16- to 19-year-old males from five regions of the United States. It was reasoned that this group eats a greater quantity of food than any other does. Most pesticide levels were at least 10 times less than their recommended limits except for aldrin and dieldrin, which were approximately equal to their recommended limit. For some dietary patterns the aldrin/dieldrin level probably exceeds the recommended daily intake. Duggan further points out [28, p. 2]:

It is noteworthy that the combination of meat, fish, poultry and dairy products accounts for over half the intake of chlorinated organic pesticides. There are few registered uses of pesticide chemicals known to result in residues in meat and poultry. No registrations have been granted which are calculated to result in residues in meat and poultry. Therefore, there are environmental factors contributing 'unavoidable' residues in these commodities.

In fact, other studies have shown that the body fat of people who do not eat meat contains approximately one-half of the DDT of people who do eat meat. Duggan and Lipscomb did not find a pattern of increase for pesticide residues in their diet studies over the years 1965 to 1968.

Pesticide residue levels in foods could be decreased if insecticide levels in animal feed were decreased, if animals were not slaughtered until some time after pesticide treatment for parasites, and if proper time intervals were observed between spraying and crop harvest. The consumer can decrease consumption of pesticide residues by washing foods thoroughly (with soap) and peeling foods, when possible.

Human milk, since it contains human fat, has detectable concentrations of pesticides. In the United States, DDT residue levels range from 0.05 to 0.37 part per million with an average around 0.1 to 0.2. Dieldrin and BHC are also present but in smaller quantities. It has been calculated that the breast-fed baby takes in 0.02 milligram of DDT per day.

One other type of food contamination should be mentioned, that occasioned where food is accidentally contaminated with large amounts of spilled pesticides during transportation. Several instances of flour contamination are recorded. Over 500 people were poisoned and 16 died in Tijuana, Mexico, in 1967, when bread ingredients were contaminated with parathion. Also in 1967, in Qatar and Saudi Arabia a similar incident involved endrin contamination of flour; the flour had been stored along with the pesticide in the holds of two ships. Twenty-six people died and 1,874 were hospitalized after eating bread made from the contaminated flour. The restriction of vehicles used for transport of pesticides to only that function would prevent such accidents.

The storage of chlorinated hydrocarbons in fatty tissues is not an unlimited process. The amount stored is related to the daily dose. After a certain period of time at a certain dose rate, tissue levels reach a maximum concentration. The higher the intake, the higher will be this leveling off point, as is shown both in animal studies and in studies of workers exposed occupationally to much higher concentrations than the general public. When exposure ceases, the residue in human tissue decreases at a rate which depends on the compound and on the particular tissue.

Although the available data indicate that food is the only major route for dieldrin absorption, this may not be the case for DDT. DDT is not evenly distributed in the population, people in warmer states having higher residues (9.21 parts per million average) than those in cooler states (4.85 parts per million). The same thing holds in Europe, levels being higher in the south and east. Middle Eastern and Asian values also tend to be high. The differences seem too large to be explained by food preferences. Pesticide levels in dust and dirt in the home may have something to do with the differences. Sex and race appear also to affect DDT storage. Since these factors were not taken into account in most residue studies, according to the Public Health Service Report, no conclusions can be drawn at the present time about whether pesticide levels are increasing or decreasing or staying the same in the general population.

Although the presence of chlorinated hydrocarbons in human fatty tissue is an unquestioned fact, what effects those residues may have is very much in dispute. A good deal of research has dealt with the industrial exposure of workers in pesticides-related occupations. In large groups of generally exposed workers one commonly finds blood and enzyme changes, gastrointestinal problems, hormonal disturbances, and kidney or liver dysfunctions. However, differences are slight, mostly within normal ranges, and may represent adaptive rather than pathological changes.

In one study of men working in a dieldrin, aldrin, and endrin factory, the pesticide levels in the tissues, even of those working in the office, were nine times that in the general population. No correlation was found between residue levels and the amount of sick leave taken. In another study of men working in a DDT factory for 9 to 11 years, the average daily intake was determined to be 18 milligrams per day. Residues were from 38 to 647 parts per million DDT. No unusual illnesses were reported. Although one study did show more cardiovascular disease than would have been expected, most investigators report similar results; i.e., no adverse effects are shown despite high residue levels.

A different type of study among Florida structural pest control workers, however, indicated some disturbing trends. In this study, which covered the years 1948 to 1965 and involved 20,000 person-years, mortality data for the workers were compared with data on the general population. In the first place, a rising death rate was noted for the workers while the rate for the whole population was fairly stable (data were age-adjusted). In the second place, comparison of workers in the 35 to 44-year-old age group with the same age group in the population revealed that the workers had a greater death rate. In the 55- to 64-year-old age group the death rate was the same as for the general population. It would have been expected to be lower, because people in any occupational group, since they are well enough to work, should be healthier than the general population. No unusual causes of death were found for the workers. Thus, if pesticides are not causing unusual diseases but rather increasing the susceptibility to normal ones, effects may well have escaped the scrutiny of previous studies.

One of the diseases to which pesticides might increase susceptibility is cancer, and a number of pesticides have been tested for carcinogenic (cancer-producing) effects in animals such as mice, rats, and hamsters. There are always problems and uncertainties involved in extrapolating animal data to humans. First, the life-span of these animals is so much shorter than that of humans that comparable long-term exposure is impossible. Second, doses are generally made very large to find effects quickly and unequivocally, but this procedure gives no more than an idea of what low-dose, long-term effects might be. Third, animals do vary in their susceptibility to different chemicals. Thus animal effects are not perfect indicators of human effects. On the other hand, warm-blooded animals are more similar than they are different from man in their drug responses. Chemicals which are highly toxic to rats are, with few exceptions, highly toxic to man.

A number of pesticides have been designated as carcinogenic by the United States Public Health Service 1969 Secretary's Commission on Pesticides. They cause a significant number of tumors in one or more species of warm-blooded animals when compared with a control group of nonexposed animals. The carcinogenic compounds are aldrin, aramite, chlorobenzilate, DDT, dieldrin, Mirex, Strobane, heptachlor epoxide, Amitrole, diallate, bis (2-chloroethyl)-ether, N-(2-OH ethyl)-hydrazine, and PCNB. On the same basis, a few compounds were judged noncarcinogenic. These include chlorpropham, rotenone, and carbaryl (sevin). A large number of other compounds fell into the possible-positive and possible-negative category because of insufficient evidence.

Studies which have tried to relate cancer deaths and pesticide levels found in the tissues of victims have been inconclusive since the disease itself could have caused altered storage of pesticides. Also, since cancer is a wasting disease, the loss of fatty tissue might cause increased levels in remaining tissue. One such study did find twofold to threefold increases in pesticide residues in people who had died of lung, stomach, rectal, pancreatic, prostate, and bladder cancer over those in a control group who had died of other causes. The Report of the Secretary's Commission on Pesticides summarized the uncertainty in specifying limits on the concentrations of carcinogens:

> Since the effects of carcinogens on target tissues leading to tumor formation appear irreversible, with accumulation of effects over extended periods of exposure, the reduction of exposure to carcinogenic substances to the lowest practicable levels may be one of the most effective measures toward cancer prevention. Many different factors may influence dose-response in carcinogenesis in man and animals. Their complexity is such that no assuredly safe level for carcinogens in human food can be determined from experimental findings at the present time.

A second major concern about long-term effects of pesticides involves their possible mutagenic effects. A mutation is a permanent change in the genetic makeup of an individual and is passed on to his offspring. Mutations can be caused by many agents. X-rays and certain chemicals are examples of mutagenic agents. Mutations may be harmful or beneficial, but if one considers the extremely complex makeup of the human body, with its interrelationships and interdependencies, it is easy to understand why most mutations are harmful. Mutant genes may not be discovered until several generations later, and their effects are usually mild. That is, they will not kill the individual who possesses the mutation but rather weaken the population in general.

Mutagenic effects, like carcinogenic effects, are difficult to detect. Animal experiments and experiments with isolated human cells cultured in test tubes have implicated certain pesticides [1].

The third concern about long-term pesticide exposure involves possible teratogenic effects—the effects pesticides may have on the development of a fetus before birth. The thalidomide disaster in 1963 is an example of a teratogenic chemical's affecting the population. Researchers have determined that DDT and DDE do pass across the placenta; fetal blood levels have been found to be approximately half the maternal blood levels. However, studies have indicated no adverse effects even when the mothers had significant exposure during the first three months of pregnancy. On the other hand, pesticides containing alkyl mercury compounds have been shown to be teratogens (see **Mercury**), as have Captan, PCNB, and Phaltan. A discussion of the possible teratogenic effects of the herbicide 2,4,5-T is found in **Herbicides.**

The results on teratogenic effects were obtained by experiments on at least two mammalian species. The test animals were treated in the same way that humans would be expected to encounter the pesticide (orally, by skin contact, etc.) during the period of fetal organ formation.

Perhaps the best summary of the effects of pesticides on humans is given in the 1969 Secretary's Report on Pesticides:

> The field of pesticide toxicology exemplifies the absurdity of a situation in which 200 million Americans are undergoing life-long exposure; yet our knowledge of what is happening to them is at best fragmentary and for the most part indirect and inferential. While there is little ground for forebodings of disaster, there is even less for complacency. The proper study of mankind is man. It is to this study that we should address ourselves without delay.

STANDARDS, LAWS, AND MONITORING PROGRAMS

In the United States, pesticide regulation is the function of the Environmental Protection Agency (EPA). The laws which relate to pesticides are the Federal Insecticide, Fungicide and Rodenticide Act (passed in 1947), the Pesticide Amendment to the Federal Food, Drug and Cosmetic Act, 1958, (sometimes called the Delaney amendment), and the Federal Environmental Pesticide Control Act of 1972.[13]

Although the 1947 law provided for registration of pesticides which were to be transported in interstate commerce, the new law requires registration of all pesticides regardless of where they are manufactured and used. Manufacturers desiring registration for a pesticide must present the EPA data to show that the pesticide, when used as directed, is effective against the pest species listed on the label and will not harm nontarget animals, people, or the general environment.

Once given, registration can be revoked if further experience shows the use of a pesticide is more hazardous than was realized. If registration is *canceled,* the manufacturer may appeal the cancellation and continue to ship his product until avenues of appeal have been exhausted. If there is evidence of imminent hazard to the public health and welfare, however, registration will be *suspended.* In this case the manufacturer must halt shipment immediately, even if he appeals the suspension.

Pesticides to be used on food or feed must, in addition to meeting normal registration requirements, be shown to be safe for humans and animals under proposed conditions of use. In some cases tolerance limits are set for residues. These are acceptable levels below which pesticides may be found in or on food or feed crops. Tolerances are set by EPA and enforced by the Food and Drug Administration in the Department of Health, Education and Welfare. Occasionally government scientists refuse to set a tolerance; i.e., no residue is considered safe. Such was the case when the Shell Company was forced to change the label on its No Pest Strip, to say that it must not be used where food is eaten or prepared. Tests had shown that detectable residues of DDVP, the insecticide in the strip, were found on food left in the same room with a strip. The government did not feel that any residue of DDVP in food would be safe.

In 1961, 385 tons of agricultural commodities were seized because they contained pesticide residues in excess of legal limits. Reasons for contamination of produce were determined in one Texas study to include: drift from spraying; earlier maturing of crops than expected, resulting in too short a time between spraying and harvest; absorption of endrin from the soil by carrots; deliberate misuse, and inadvertent misuse (one worker mistook BHC for fertilizer).

Another provision of the 1972 act involves classification of pesticides into either general-use or restricted-use categories. Pesticides in the latter category may be used only under the supervision of certified applicators. Each state is expected to present a plan, for certification of applicators, to the EPA for approval. The EPA Administrator has stated that most farmers should be able to qualify for certification for the use of restricted pesticides on their own lands.

The EPA is given broader enforcement powers in the new act, since penalties are provided for misuses of pesticides by applicators, and for various other violations by registrants and distributors.

The Delaney amendment to the Food, Drug and Cosmetic Act states that no carcinogenic substances may be present as residues in foods. Since the time the amendment was written, analytical techniques have improved substan-

[13] The EPA has until October 1976 to put into effect all the provisions of the 1972 law.

tially; thus, if the amendment were enforced as written, most food of animal origin including all meat, butter, milk, ice cream, cheese, eggs, fowl, and fish could not be transported between states. The 1969 Secretary's Commission on Pesticides recommended repeal of the amendment.

Other scientists feel that the amendment should be kept in force, for otherwise it would become necessary for the EPA to set tolerance levels for known carcinogenic substances. These scientists argue that insufficient data are available to set such tolerances. Furthermore, they assert that if the setting of tolerances were attempted, the application of reasonable safety factors would result in standards so close to zero that the Delaney clause might just as well have been used.

A number of governmental pesticide surveillance programs are at present in operation. A Community Studies program is being carried out by EPA to attempt to measure the exposure of the population to pesticides from all sources. Since 1965, permanent programs have measured occupational exposures, as well as the exposure of members of the general population from food, soil, water, and air sources. The program is operating in communities in 14 states. Another 15 states are aided in smaller projects. Additionally, EPA is conducting a national study of pesticide residue levels and their trends in the population. Pesticide levels in the air are also monitored in selected locations. The FDA directs a continuing Market Basket Study aimed at determining the exposure of the population to pesticides through foods.

PHILOSOPHY OF PESTICIDE USAGE

The discussion of pesticide usage so far has been rather one-sided in that only effects, and primarily damaging effects, have been catalogued. Were only this information available, one might conclude that the use of pesticides was completely unjustified. Of course, this is not the case. Pesticides are generally employed in an attempt to improve the quality of life. It is the complexity of living systems which foils the would-be benefactor and confuses the results.

An example of beneficial use of pesticides is the control of carriers of human diseases. The list of miseries which pesticides help control is impressive. Malaria, viral encephalitis, louse-borne typhus, amoebic dysentery, bacillary dysentery, cholera, Rocky Mountain spotted fever, tularemia, filariasis, dengue, yellow fever, leishmaniasis, bartonellosis, onchocerciasis (river blindness), sand-fly fever, trypanosomiasis, yaws, infectious conjunctivitis, Chagas' disease, scrub typhus, scabies, rickettsialpox, and tick-borne relapsing fever can all be controlled, at least to some extent. The pesticides are employed to control the insects or animals which transmit the disease, or which act as reservoirs from which humans are infected.

Malaria can be attacked in two ways: (1) by use of DDT to kill the mosquitoes which transmit the disease or (2) by engineering methods, primarily draining and filling of the swamps and ponds where the mosquitoes breed. The latter methods have been quite successful in the southern United States. The extensive irrigation systems in the western part of this country, however, obviate such methods. Engineering methods are really successful only in populated areas where mosquito habitats are easily wiped out by the need for housing and commercial building sites. In a large portion of the world, swamps exist in rural or sparsely populated areas. Draining and filling operations would be astronomical in cost. Consequently, reliance is placed on chemical pesticides.

DDT is the pesticide of choice and the mainstay of the World Health Organization's campaign to eradicate malaria. DDT is sprayed on the inside walls of houses where it effectively kills mosquitoes on contact for months. Most experts agree that there is no pest which cannot be killed by some pesticide other than DDT. However, DDT is the only pesticide currently available which is cheap enough to be used by developing countries in large-scale malaria control programs. Experts in charge of these efforts feel that most programs would collapse if DDT were unavailable for use.

An example of the effects of such a collapse is provided by experiences in Ceylon. In the years 1934–1935, 1.5 million cases of malaria resulted in 80,000 deaths. DDT was introduced in 1945, and by 1963 there were only 17 cases of malaria. Vigilance was relaxed and a tremendous increase in cases has been seen; 600,000 cases were reported during 1968 and the first quarter of 1969 with a projected total of 2 million cases.

Barry Commoner, however, interprets the success of DDT differently. He claims that malaria control programs have been effective only in relatively dry areas like Greece or islands such as Ceylon. He reports that programs in Central America, using DDT and other insecticides, have failed. The incidence of malaria remains as high as it was 12 years ago because of the development of insect resistance to insecticides.

In the United States the use of DDT was banned, except for public health measures, as of December 31, 1972, by the Environmental Protection Agency. In the order rescinding registration for DDT, EPA head William Ruckelshaus lists as the reasons for cancellation [16, p. 13373]:

(1) that DDT and its metabolites are toxicants which persist in the soil and the aquasphere, (2) that once unleashed DDT is an uncontrollable chemical which can be transported by leaching, erosion, runoff and volatilization, (3) that DDT is not water soluble and collects in fatty tissue, (4) that organisms tend to collect and concentrate DDT, (5) that these qualities result in accumulations of DDT in wild life and humans, (6) that once stored or consumed DDT can be toxic to both animals and humans and in the case of fish and wildlife inhibit regeneration of species, and (7) that the benefits accruing from DDT usage are marginal, given the availability of alternative insecticides and pest management programs, and also the fact that crops produced with DDT are in ample supply.

DDT use has been banned in a number of countries, including Sweden, Canada, Denmark, and Hungary. There are currently no United States restrictions on the manufacture of DDT for export.[14] Clearly some moral dilemma is involved in manufacturing and exporting a chemical considered too dangerous to be used in the country of manufacture. Proponents of DDT exportation insist that we have no moral right to make the choice of use or nonuse for other countries, especially those with a serious malaria control problem. The registration of other persistent **Chlorinated Hydrocarbons** is under review by the EPA.

As noted in the 1969 Secretary's Report, reliance is placed on chemical companies to develop new pesticides. This is not realistic. The cost of development for a new chemical pesticide has been estimated to be about $2.1 million. Unless the company can be sure of a fairly large market for a pesticide, it will have no incentive to spend such a sum to develop it. The most desirable types of pesticides, therefore, those with a very specific target, are exactly those which are poor economic risks. A system of either government incentives to industry or government support of noncommercially linked research is necessary to discover or develop more suitable pesticides.

Besides protecting people from disease, pesticides contribute to their well-being by increasing agricultural yields, an effect which lowers food costs while increasing available supplies. Studies have been insufficient to show

[14] Some 30 million pounds of DDT are exported annually.

unarguable figures for improved agricultural yields due to pesticide usage. Nevertheless, some of the figures which have been quoted include: 25 to 1,000 percent increase in wheat, oats, and barley in 1951–1956 due to parathion treatment; 35 percent increase in sugar cane production after use of endrin to control the sugarcane borer; 96 to 118 percent increase in corn yields with various pesticide treatments for corn rootworms; 90 to 100 percent control of cattle grubs, which damage hides and cause weight loss, with systemic insecticides. The use of toxaphene, methoxychlor, and malathion has been claimed to add 30 to 70 pounds per steer by controlling horn flies. It has been stated that apple and peach yields might decrease by 25 percent because of coddling moths or by 100 percent because of oriental fruit moths if pesticides were not used. Herbicides for weed control can increase yields of wheat, corn, and soybeans up to 10 bushels per acre, while rice yields can be doubled in some cases. Pests which destroy stored food can also be controlled, further increasing the food available.

Wise choices of pesticides are possible for these uses. Nonpersistent pesticides can take the place of the chlorinated hydrocarbons, or arsenic, lead, or mercury compounds. It has been calculated that in this country the cost of pesticides used is between one and 1.3 percent of the retail cost to the consumer of food or clothing. Thus we can well afford more costly pesticides if they are less likely to damage the environment.

Growers themselves could decrease pesticide usage. They could calculate the amount of chemical needed to bring residues already present in the soil up to desired levels rather than using the same dose each time. They could use seed dressings or apply small amounts of the pesticides close to seeds or plants. Roots of transplants could be dipped into pesticides and whole soil treatments avoided.

Although such measures begin to confront the problem, some environmentalists feel that pesticide usage can be even further diminished. Barry Commoner has stated that if herbicides like 2,4-D and 2,4,5-T were not used in growing wheat only about 4 percent of the total production would be lost. If these herbicides were not used on 62 million acres of crops now being treated, he estimates the additional production cost to be $290 million or 6.6 percent of the value of herbicide-treated crops. In his opinion, moreover, using land in retirement programs accompanied by a 70 to 80 percent decrease in pesticide usage would result in no decrease in agricultural production. The Agency for International Development personnel feel that a gradual withdrawal of persistent pesticides might be accomplished. If the withdrawal were combined with education in the use of nonpersistent pesticides and alternative controls, it would not be expected to have a very harsh effect on food supplies in other parts of the world.

In fact, there are many alternatives to chemical control of pests. The following is a summary of some of them:

1. *Prevention of the introduction and spread of pests.* This involves inspection and quarantine procedures. Of the several hundred pests in the United States, 123 are known to have been introduced from some other country.
2. *Cultural methods of control.* A wide variety of farming procedures decrease crop damage. Examples include:
   a. Destruction of crop refuse.
   b. Rotation of crops to prevent the buildup of a specific pest.
   c. Shifting of crops to areas of the country where the crop pest cannot survive. For instance, cotton might be shifted from the South to the West, where conditions

Contour strip cropping. The planting of mixtures of crops can be part of a program to reduce or avoid the use of pesticides.

are not so favorable for the boll weevil.

d. Timing planting to avoid pests. This is done in southern Texas by establishing deadlines for the planting of cotton and destruction of the stalks after harvest. By the timing procedure, periods when large numbers of insects hatch do not correspond to periods when food is available to them. Wheat can be planted to avoid high points in the Hessian fly population.
e. Strip harvesting. This affords the natural enemies of pests a reservoir. Alfalfa fields harvested in strips were shown to have 56 of the alfalfa aphid predators per square foot while clear-cut fields had 14.
f. Planting of mixtures of crops rather than monocultures. This again provides better habitats for predators and parasites.
g. Thorough cultivation of the soil. This will kill some pests which overwinter in the soil. About 98 percent of corn earworm pupae are killed by thorough soil preparation.
h. Removal of diseased plants to prevent spread to healthy ones. Dutch elm disease is 90 percent controllable by such methods. Only another 2 to 5 percent control is effected by the use of chemicals against the beetles involved in the spread of Dutch elm disease.

3. *Introduction of natural predators or parasites.* This is an especially good control method if a pest has been imported without its natural predator. Care must be taken, of course, that the introduced species will not itself become a pest. One successful introduction was the vedalia beetle, which was imported from Australia in 1888 and controls cottony cushion scale on citrus. Small wasps were im-

ported in 1964 to control olive scale. Japanese beetles are controllable by milky spore disease, marketed under the name Doom. Parasites can be extremely specific in their requirements for host species and thus can present little danger to nontarget organisms. The only case we have found recorded in which a human was infected with a plant parasite is that of James Thurber's uncle, who died of Dutch elm disease. Parasites and predators are mainly useful for pests of relatively stable perennial crops such as citrus. About 650 species of predators, parasites, and pathogens have been introduced into the United States. Of these, about 100 have become established and are controlling 20 pests. Occasionally mistakes are made. For instance, the mongoose was introduced into Puerto Rico and Jamaica to control rats. It did destroy the population of ground-nesting Norway rats but by so doing removed competition for the tree-nesting rats, which became a further nuisance. Besides, the mongoose eats poultry and wild birds and has become a reservoir for rabies.

4. *Sterile insects.* Some success has been achieved by releasing male insects which are sterile. Females that mate with them do not produce young. The screwworm in the southeastern United States has been controlled very well in this manner, as has the melon fly on the island of Rota. Indications are that coddling moth, boll weevil, and fruit fly control may be possible.
5. *Traps.* Some chemicals have been isolated or developed which are attractants for insects and which may be used as lures in traps. The Mediterranean fruit fly has been controlled in this way in Florida. Attractants are specific for one species of insect, and often for only one sex.
6. *Resistant varieties of plants.* Many varieties of plants which are resistant to major pests have been developed. Examples include wheat resistant to Hessian fly, alfalfa resistant to the spotted alfalfa aphid, cereals resistant to stem rust.

The 1969 Secretary's Report on Pesticides notes that 80 percent of all pesticides used are intended to control less than 100 species of pests. Thus the search for alternatives to chemical control of pests does not involve huge numbers of possible target species. Even if a great deal of research is needed to find alternative control methods for this small target group of species, the expense might be justified by the consequent large reduction in pesticide usage.

The use of pesticides in food production is partly related to consumer preference. One of the pressures on farmers to use pesticides is provided by consumer insistence, translated in many cases into law, on blemish-free foods with no insects in the shipment. Perhaps consumers should reconsider these demands and be willing to accept a less perfect product in return for a more perfect environment in which to live.

An area of pesticide exposure wholly within the individual's power to control is household use. Yet the 1969 Secretary's Report on Pesticides states that household use "and misapplication have reached the point where contamination by household pesticides may constitute a significant proportion of the total population exposure."

Several studies of household use have been carried out, notably by the Environmental Protection Agency's Division of Community Studies. An EPA study in Salt Lake County, Utah, a county of 440,000 people, disclosed that 200,811 pounds of pesticides were applied from July 1967 to July 1968. Of this amount 102,490 pounds, or about half, were for domestic uses and the other half for agricultural uses. In Arizona, on the other hand, during the same period, about 9.5 million pounds were used in agriculture and about 63,000 pounds were applied to a total of 475,362 households. For some reason domestic pesticide usage on a per-person basis was much higher in Salt Lake County than in Arizona.

Another study, contrasting heavy pesticide users (once a week or more) with light users (less than once a week), found that those in the light-use category performed better in respiratory tests (forced vital capacity and forced expiratory volume) than those in the heavy-use category. Also, those in the heavy-use group had more respiratory problems such as asthma, chronic sinusitis, and nasal allergy.

A Mississippi Community Studies report pointed out that 25 out of 41 commercial dry cleaning establishments used moth-proofing chemicals, generally during the spring and summer months. All products used contained DDT, and clothing was treated as a general practice. Customers were not asked whether they wanted it or not.

COMMENTARY

Considering both the complexity and the toxicity of pesticides, it is amazing that their use has been left solely up to the individual farmer or householder. If people are not allowed to prescribe medical drugs unless they have a medical school education, it seems only reasonable that choice of the proper pesticide or other control measure would be best put into the hands of someone especially trained for such decisions. It has also been pointed out that a person prescribing medical drugs for himself stands to hurt only himself, while a person misusing pesticides can harm whole populations.

One proposal, put forth by a number of experts, would entail the education and state licensing of accredited pest control experts in much the same fashion that doctors are trained.[15] Farmers or householders would consult with the pest control expert, who would then recommend a course of action which might also entail prescribing a pesticide. Such a plan would remedy the present situation, in which the main source of information available to the pesticide user is the company salesman. Agricultural extension services are an alternate source of information, but it is apparently not reaching the consumer, as shown by the following story related by the insect ecologist, R. Van den Bosch [12].

Azodrin, a pesticide produced by the Shell Chemical Company, was tested from 1964 to 1967 at Shell's request by the Agricultural Experiment Station at Berkeley. It was not recommended by the experimenters, for use on cotton, since they felt that it killed too many predators and parasites of cotton pests and would therefore leave the cotton open to attack by a number of pests. Shell promoted the use of the chemical through its salesmen anyway and succeeded in convincing growers to apply the new chemical to almost a million acres of cotton in 1967. In addition, Shell recommended application on a scheduled basis without regard for the presence or nonpresence of cotton pests. This is recognized as a bad practice in pesticide usage since it contributes to the development of insect resistance, and causes unnecessary environmental contamination. No obvious benefit resulted from the use of the insecticide, and harm may have been done. In 1967 the cotton crop was the smallest in the past 10 years. In addition, wildlife kills were extensive in treated fields.

The pest control expert, beside serving as a safeguard against overzealous sales practices by the chemical industries, would also be well

[15] Although state certification of pesticide applicators, as required by the 1972 pesticides law, is a step in this direction, the pest control expert referred to here is envisioned as being a much more highly trained person.

versed in nonchemical control methods. It is hoped that he would be able to advise farmers and householders of the best combination of methods for solution of a particular problem. Such integrated control techniques seem to represent our best hope for maintaining a high standard of agricultural production and public health while safeguarding the environment.

REFERENCES

1. Report of the Secretary's Commission on Pesticides and Their Relationship to Environmental Health, U.S. Dept. of Health, Education and Welfare, December, 1969.
2. C. Edwards, "Persistent Pesticides in the Environment," *Critical Reviews in Environmental Control* 1, no. 1, Chemical Rubber Publishing Co. (February 1970).
3. J. C. Headley and J. N. Lewis, *The Pesticide Problem: An Economic Approach to Public Policy*, Johns Hopkins Press, Baltimore, 1967.
4. Dahlsten et al., eds., *Pesticides: A Scientists Institute for Public Information Workbook*, S.I.P.I., New York, 1970.
5. N. N. Melnikov, "Chemistry of Pesticides," *Residue Reviews* 36 (1971).
6. "The Wilson Committee Report," *Pesticide Articles and News Summaries* 16, no. 2 (1970); 259.
7. L. E. Rozeboom, "DDT, the Life Saving Poison," *The Johns Hopkins Magazine*, Spring 1971, p. 29.
8. "Shell's No-Pest Strip," *Consumer Reports*, November, 1970, p. 701.
9. A. Bevenue and Y. Kawano, "Pesticides and the Law," *Residue Reviews* 35 (1971): 103.
10. J. J. Hickey, O. B. Cope, J. C. George, and D. E. H. Frear, "Pesticides in the Environment and Their Effects on Wildlife," *Journal of Applied Ecology* 3, suppl. (1966).
11. D. B. Peakall, "p,p'-DDT: Effect on Calcium Metabolism and Concentration of Estradiol in the Blood," *Science*, May 1, 1970, p. 592.
12. R. Van den Bosch, "The Toxicity Problem —Comments by an Applied Insect Ecologist," in *Chemical Fallout*, ed. M. W. Miller and C. G. Berg, Charles G. Thomas, Springfield, Ill., 1969.
13. D. Pimentel, *Ecological Effects of Pesticides on Non-Target Organisms,* Executive Office of the President, Office of Science and Technology, June 1971.
14. George M. Woodwell, Paul P. Craig, and Horton A. Johnson, "DDT in the Biosphere: Where Does It Go," *Science*, December 1971, p. 1101.
15. Frederick W. Plapp, Jr., "Where the Present Strategy of Pest Control Has Failed," Paper presented to the AAAS Conference on Environmental Sciences and International Development, Philadelphia, December 27, 1971.
16. *Federal Register* 37, no. 131 (July 7, 1972).
17. Samuel W. Simmons, "Living Labs That Study How Pesticides Affect Man," Publication 308, Pesticides Program, National Communicable Disease Center, Atlanta, Ga.
18. "Pesticides and Community Studies, FS P-1," *Environmental News*, U.S. Environmental Protection Agency, December 7, 1971.
19. "Pesticides and the Law, FS P-2," *Environmental News*, U.S. Environmental Protection Agency, December 7, 1971.
20. Barry Commoner, *New York Times*, December 11, 1971.
21. Nicholas Wade, "Delaney Anti-Cancer Clause: Scientists Debate on Article of Faith," *Science*, August 18, 1972, p. 588.
22. Robert Gillette, "DDT: Its Days Are Numbered," *Science,* June 23, 1972, p. 1313.

23. A. V. Holden, "The Effects of Pesticides on Life in Fresh Waters," *Proceedings* of the Royal Society of London, Series B 180 (1972): 383.
24. G. W. Cox, ed., *Readings in Conservation Ecology*, Appleton-Century-Crofts, New York, 1969. See especially pp. 142–217, 325–82.
25. R. C. Muirhead-Thompson, *Pesticides and Freshwater Fauna*, Academic Press, New York, 1971.
26. E. P. Odum, *Fundamentals of Ecology*, 3rd ed., Saunders, Philadelphia, 1971.
27. R. E. Duggan and G. Q. Lipscomb, "Dietary Intake of Pesticide Chemicals in the United States (2), June 1966–April 1968," *Pesticides Monitoring Journal* 2, no. 4 (1969).
28. R. E. Duggan, "Residues in Food and Feed," *Pesticides Monitoring Journal* 2, no. 1 (1968).

A bibliography of a large part of the available literature through 1970 can be found in:
29. W. R. Benson and X. Jones, "The Literature of Pesticide Chemistry," part 1, *Journal of the AOAC* 50 (1967): 22.
30. W. R. Benson and C. R. Blalock, "The Literature of Pesticide Chemistry," part 2, *Journal of the AOAC* 54 (1971): 192.

Photochemical Air Pollution

Photochemical air pollutants are substances that are *not* directly emitted to the atmosphere; they are created by reactions stimulated to proceed by the presence of light. Pollutants that are emitted directly to the atmosphere and contribute to the creation of photochemical air pollutants are hydrocarbons and nitrogen oxides.

HYDROCARBONS

While hydrocarbons may be considered air pollutants, their *direct* effects are quite small. However, the products of the atmospheric reactions of hydrocarbons have substantial direct and harmful effects.

Hydrocarbons are composed of only carbon and hydrogen. The two classifications of greatest interest are saturated and unsaturated hydrocarbons. Saturated hydrocarbons are the most abundant hydrocarbons in urban air, but they are essentially unreactive in the atmosphere. It is the unsaturated hydrocarbons which have the potential to enter into those atmospheric reactions which produce photochemical air pollutants. Thus control measures aimed at reducing hydrocarbons ought to be directed first toward those compounds which contribute to photochemical air pollution: the unsaturated hydrocarbons.

SOURCES

Methane, the simplest saturated hydrocarbon and also the most abundant, has a relatively high natural background level, between 0.7 and 1.1 milligrams per cubic meter; the gas arises from gas and petroleum fields, swampy areas, coal fields, fires, and other natural sources. It does not, however, enter into photochemical reactions. Most of the hydrocarbons which react in the atmosphere to give photochemical smog are by-products of man's activities.

Transportation activities account for about 50 percent of the total nationwide emissions of hydrocarbons, although on an individual city basis their contribution has been observed to be as low as 37 percent. Nationwide, industrial operations account for about 14 percent and solvent uses for another 10 percent of the hydrocarbon emissions. Solid waste disposal including incineration and burning in landfills may account for 5 percent and gasoline marketing for another 4 percent. Other sources are less important. Local situations may be markedly different from the national picture; for instance, fuel combustion or solid waste incineration may exhibit large effects in certain areas.

Hydrocarbons arise from transportation activities as a result of the inefficient combustion of fuels. The transportation mode which must bear the greatest guilt in this regard is the motor vehicle (see also **The Automobile and Pollution**). The conventional automotive engine burns gasoline with insufficient air. As a consequence, in the cars of the 1960's about 60 percent of the hydrocarbon pollutants emitted by the auto were found in the exhaust stream. Another 20 percent of the total appeared in the crankcase blow-by (gases which blow by the pistons during compression). Evaporation of about equal portions from carburetor and fuel tank accounted for the remainder. These figures are for cars without the special emission control devices which appear on 1971 and later model cars. The pollutants from the auto include hundreds of different hydrocarbons as well as organic compounds. Fuel evaporation and hydrocarbon losses in crankcase blow-by are much less for the diesel engine. In addition, more efficient combustion in diesel engines leads to lower hydrocarbon emissions in the exhaust. However, the exhaust stream is rich in formaldehyde, which may account for the characteristic diesel odor.

Industrial hydrocarbon emissions arise from most operations of the petroleum industry and from certain segments of the chemical industry. Organic solvents, some of them hydrocarbons and others closely related, are released from the many commercial and domestic activities involving paints, lacquers, and varnishes. They also derive from rubber and plastic manufacture as well as the restoration of metal parts by degreasing, and from dry cleaning.

All these sources produce a wide range of hydrocarbons. If one recalls that methane is a significant background contaminant and additionally unreactive, it becomes clear that any measurement of atmospheric hydrocarbons should discriminate between methane and other hydrocarbons. An extensive sampling in Los Angeles in 1965 showed methane at an average of 3.22 milligrams per cubic meter. The same survey put total hydrocarbons at 3.79 milligrams per cubic meter and nonmethane hydrocarbons at 0.57 milligram per cubic meter (equivalent to 570 micrograms per cubic meter). As expected, hydrocarbon levels follow the level of traffic; they peak in the morning rush hours and again during the return hour and exhibit a trough at night. The decrease in concentrations between peak hours is accounted for by dispersion of the substances by wind and by reactions which produce new nonhydrocarbon substances (such as PAN—peroxyacyl nitrate—compounds, aldehydes, and ketones).

EFFECTS

As stated earlier, we are concerned with hydrocarbons largely because of the compounds to which they give rise, but ethylene, a hydrocarbon found in automotive exhaust, both participates in reactions in the atmosphere and has a toxic effect on plants. It has been reported in urban air at concentrations ranging from 25 to 150 micrograms per cubic meter. Although green plants produce ethylene as a growth regulator, in excess, the compound suppresses growth or may lead to a yellowing of leaves or even kill plant tissue. A cotton crop was decimated near a polyethylene plastic plant in Texas in 1957. Besides ethylene's effects, though, little in the way of direct harmful effects of hydrocarbons may be noted.

NITROGEN OXIDES

Nitrogen oxides are of major importance because of their contribution to photochemical air pollution. In contrast to hydrocarbons, however, they do seem to have physical and physiological effects, independent of their role in photochemical air pollution. Nevertheless, that role is fundamental. The action of sunlight on nitrogen dioxide appears to be the beginning step in a long chain of interwoven reactions whose end products are the photochemical air pollutants.

NITROGEN OXIDE COMPOUNDS

A single atom of nitrogen bound to a single atom of oxygen is known as nitric oxide, a colorless, odorless gas. A single atom of nitrogen combined with two atoms of oxygen is known as nitrogen dioxide. The gas is reddish brown (have you noticed the color of the haze over some urban areas?) and offensive to smell. One frequently sees a pseudochemical formula for these two oxides of nitrogen in the form NO_X, which represents the presence of both compounds without distinguishing the relative amounts of each. The other oxides of nitrogen are of little importance to photochemical air pollution.

Nitric oxide will react with oxygen to produce the more noxious nitrogen dioxide. The exhaust gases from an automobile are rich in nitric oxide and poor in nitrogen dioxide. Usually less than 0.5 percent of the total quantity of nitrogen oxides is present as nitrogen dioxide. Perhaps 10 percent of the nitric oxide is converted to nitrogen dioxide during the initial mixing of exhaust with the air. However, once the nitric oxide has been diluted in the air, the rate of conversion by direct oxidation becomes negligible.

Following the concentrations of these oxides through the day, one observes that while nitric oxide levels diminish in the period just after the morning rush hour the nitrogen dioxide concentrations increase. The nitrogen dioxide production must proceed by some other mechanism than oxidation once the nitric oxide is dispersed in the atmosphere. Whatever the mechanism, nitrogen dioxide becomes an important atmospheric pollutant even though it enters the air in relatively small quantities. It is significant because the energy of sunlight can convert the dioxide to nitric oxide and atomic oxygen.

Atomic oxygen's reactions with certain hydrocarbons provide the very reactive molecules called free radicals. These in turn react with nitrogen dioxide to yield some of the principal photochemical air pollutants. Little importance to air pollution is attributed to the reaction of nitrogen dioxide with water vapor.

SOURCES

Nitrogen oxides stem from natural sources, giving rise to background concentrations of about 10 micrograms per cubic meter. These background levels are presumably due to production of NO_X by bacteria. Although natural sources annually produce about 10 times more NO_X than technological sources, urban concentrations are 10 to 100 times those in rural areas. The dominant source of NO_X in urban areas must be technological.

During the high-temperature combustion of fuels such as coal, gasoline, oil, and natural gas, nitrogen from the air and evidently nitrogen from the fuel will react with oxygen from the air. Nitric oxide is the predominant product of the reaction. Technological activities produced nearly 21 million tons of nitrogen oxides in the United States in 1968, and nearly 90 percent of this output derived from the burning of fossil fuels. With respect to the amount of nitrogen oxides produced by various fuels, the burning of natural gas for heat and power accounted for 23 percent of the total output; coal, for 18 percent; and fuel oil, for 5 percent. The combustion of gasoline produced 32 percent of the total output and diesel fuel another 3 percent. Interestingly, natural gas, an innocent with regard to **Sulfur Oxides**, is a villain in nitrogen oxides emissions. An important technological source in addition to motor vehicles was the generation of **Electric Power;** about 4 million tons of the total arose from fuel combustion at power plants.

The daily pattern of nitrogen oxide concentrations in cities provides conclusive evidence of the impact of man's activities on NO_X levels. Peaks in NO_X concentrations are observed to coincide with increased traffic volumes, although morning peaks are generally higher than evening peaks because of lower wind speeds and generally more stagnant conditions. Peak concentrations as high as about 2,400 micrograms per cubic meter of nitric oxide and 2,400 micrograms per cubic meter of nitrogen dioxide have been observed in Los Angeles, although these are extreme values. However, peaks exceeding about 1,200 micrograms per cubic meter of nitric oxide and about 400 micrograms per cubic meter of nitrogen dioxide are not uncommon in Philadelphia, Washington, Chicago, and Los Angeles. The peaks in NO_X levels tend to be higher in fall and winter when fuel consumption is high and when inversions (see **Air Pollution, meteorological aspects**) are more frequent.

EFFECTS

Nitrogen oxides, especially nitrogen dioxide, are causative agents in the fading of dyes (particularly blue dyes) of acetate rayon. Although new and more costly dyes were developed in the 1950's to deal with the problem, fading still occurs to a certain extent because of some continued use of lower-cost dyes in inexpensive articles. Fading of colors on nylon and viscose rayon has been noted when garments are placed in gas dryers, and the production of NO_X during the combustion of the gas seems to be responsible for it. White fabrics of nylon and other synthetics may yellow, apparently because of nitrogen oxides, causing economic losses to sellers and buyers.

One well-publicized incident involved the effect of NO_X on the fabric itself. In 1964, in an area of New York City, women reported excessive running of nylon stockings. Nitrogen oxides produced during dynamite blasting operations were pinned down as the culprit in the affair. Nitrate particles in the air have been implicated in the relatively rapid corrosion of relay springs and crossbar switches in telephone equipment.

Few studies can be pointed to which indicate epidemiological effects of NO_X. However, a study of an area in Chattanooga in 1968–1969 did provide some interesting results. The area was unusual in that the concentration of nitrogen dioxide was elevated over surrounding regions because of the presence of a TNT plant, and particulates and sulfur oxides were low. Areas of low nitrogen dioxide levels were studied simultaneously. Analysis of the data detected a 19 percent higher rate of respiratory illness among families living in the area with elevated nitrogen dioxide levels. Since these levels are actually quite common in large cities, nitrogen dioxide and/or derived substances may, in fact, be exerting a substantial effect on health.

PHOTOCHEMICAL REACTIONS, PRODUCTS, AND EFFECTS

As the working day begins in the typical urban area, hundreds of thousands of motorists sip their last ounces of coffee and slip behind the wheel of their mobile polluters ready to create a day of running, itchy eyes, irritated throats, poor visibility, and generally obnoxious atmospheric conditions. Their vehicles exhaust hydrocarbons and oxides of nitrogen into the air, and their contribution is matched by further emissions from industrial and domestic sources. The confluence of these emissions and the consequences of the confluence are the subject of this section.

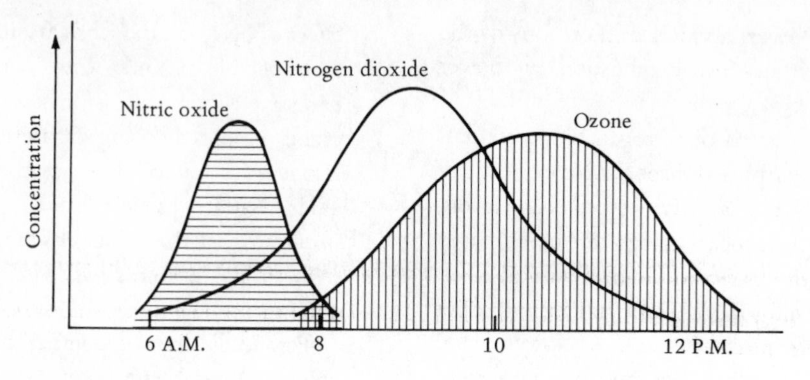

FIGURE P.3
PEAK ATMOSPHERIC CONCENTRATIONS OF NITROGEN OXIDES AND OZONE

The day begins with significant emissions of nitric oxide and hydrocarbons and some nitrogen dioxide. Gradually the levels of these compounds build up. In the hours before dawn, few reactions of significance occur, but with daylight, nitrogen dioxide, originally present in only small concentrations, increases. Nitric oxide continues to rise with continuing emissions but reaches a peak between 6 and 8 A.M. and then declines. The same sequence is seen for nitrogen dioxide; it too reaches a peak and then falls, although the peak is later, between 8 and 10 A.M. Following the rise of nitrogen dioxide, ozone begins to increase, reaching its peak after 10 A.M. and then decreasing. The events are indicated schematically in Figure P.3. It looks as though nitric oxide is being oxidized to nitrogen dioxide by the oxygen of the air, but this reaction takes place to only a negligible extent under conditions in the atmosphere. In fact, sunlight is breaking up the dioxide molecule into nitric oxide and atomic oxygen.

Enter the unsaturated hydrocarbons. The atoms of oxygen produced by the action of sunlight on nitrogen dioxide react with hydrocarbons at an enormous rate to form the extremely reactive *free radical hydrocarbons.* These substances, in turn, apparently react with the original nitric oxide to give more nitrogen dioxide. More, in fact, is produced than was consumed in the original dissociation reaction which provided the atomic oxygen. Nitrogen dioxide levels rise, and nitric oxide levels fall. When most of the nitric oxide is gone, the free oxygen atoms react with molecular oxygen to give ozone, which now begins to build up. Nitrogen dioxide levels are diminished by reactions with free radical hydrocarbons to give peroxyacyl nitrate (PAN) compounds. Ozone may be depleted by reactions with hydrocarbons to give aldehydes. Formaldehyde, acetaldehyde, and acrolein are typical products. Ozone may also be blown away and diluted by winds. The initial emissions of nitric oxide and hydrocarbons coupled with this pattern of reactions explain the sequential rise and fall of nitric oxide, nitrogen dioxide, and ozone.

Photochemical pollutants are reported as oxidant levels. By *oxidant,* we mean a compound capable of oxidizing substances which atmospheric oxygen could not oxidize. The most important oxidants in urban air are ozone, nitrogen dioxide, PAN compounds, and aldehydes. A chemical technique measures the total oxidant in the air without regard to which of these substances are present.

Oxidant concentrations are commonly reported as either the maximum hourly average over the 24 hours of a day or as the instantaneous peak concentration for a day. In the period from 1964 to 1967, measurements of oxidants in various cities showed Pasadena and Los Angeles way out in front, so to speak. In both Pasadena and Los Angeles a maximum hourly oxidant concentration exceeding about 300 micrograms per cubic meter occurred on over 30 percent of the days. No other city, in a list including Washington, Chicago, Philadelphia, St. Louis, and Denver, was over the 300 level more than 6 percent of the time. More typically, concentrations of oxidants were less than 100 micrograms per cubic meter nearly 50 percent of the time.

Within a given day, oxidant levels would characteristically reach a maximum between 10 A.M. and noon. Maximums may be expected on workdays with high light intensities and low wind speeds. Remarkably, the yearly average concentration in all cities ranged only from 37 to 82 micrograms per cubic meter, which is very close to normal background levels.[16] This is explained by noting that the time span within a day when oxidant levels are elevated is very short, being less than six hours.

What are the effects of photochemical pollutants on plants and animals? In the middle 1940's in Los Angeles, investigators first encountered an unusual form of plant damage, tiny dark spots on leaves, which were called stipples. Stipples are now known to be caused by ozone in the atmosphere. Since that time, damage to plants by photochemical air pollutants has become widespread, having been observed across most of the United States. Ozone has been implicated by laboratory studies in injury to onions, spinach, peanuts, tomatoes, radishes, tobacco, grapes, and soybeans. Citrus crops have been damaged in California by ozone; the size of fruit is diminished, fruits drop early, and leaf fall occurs.

Peroxyacyl nitrate (PAN) compounds, in contrast to ozone, produce a characteristic glazing or bronzing of the undersurface of plants' leaves. The effect of PAN compounds has been observed on spinach, celery, romaine lettuce, and pepper plants, to name a few favorite foods. It has also impaired tobacco, endive, beets, Swiss chard, and alfalfa, and some species of flowers.

Ozone is known to decrease the breaking strength of cotton, nylon, and polyester fabrics. Certain blue and red dyes, the anthraquinones, are susceptible to fading due to ozone attack. The appearance of dyes faded by ozone is pale and washed out as opposed to dyes faded by nitrogen oxides, which are reddened.

What are the effects of photochemical pollutants on man? A number of the photochemical air pollutants are known to be eye irritants. Those implicated thus far are the aldehydes (formaldehyde and acrolein), peroxyacetyl nitrate, and peroxybenzoyl nitrate. Acrolein and formaldehyde are also known to irritate the respiratory tract. So far, ozone has escaped blame as an eye irritant, but it is known to irritate the nose and throat above a certain concentration. The research results are far from complete, and other substances, yet unknown, may be involved.

Most studies have been unable to show a relation between levels of oxidants (photochemical air pollutants) and daily death rates even when the studies focused on older ages or the chronically ill. When ambient temperatures are elevated, the daily death rate is seen to increase. This phenomenon makes the interpretation of data difficult since elevated temperatures are indicative of sunshine and hence of the potential for photochemical air pollution.

Nor have hospital admissions been firmly

[16] The background is due in large part to ozone from the upper atmosphere.

linked to oxidant levels, although some studies have sought to show an association. On the other hand, asthma attacks and oxidants do appear to be related above certain oxidant levels. Additionally, the aggravation of asthma may be especially severe for some individuals. Studies of people with chronic respiratory ailments such as emphysema and bronchitis have shown mixed results. Even though the frequency of automobile accidents has been seen to increase on days of high photochemical air pollution, the effect cannot be solely attributed to oxidant levels, because simultaneous high concentrations of **Carbon Monoxide** and **Particulate Matter** are likely.

In a survey of 3,500 families in California, three-quarters of the Los Angeles study population said they were "bothered" by air pollution. Further, 75 percent of the physicians surveyed in a different Los Angeles study were of the opinion that the health status of their patients was related to air pollution. One-third of the doctors surveyed had considered moving from the area on account of air pollution.

CONTROL OF HYDROCARBONS

There are two principal ways to limit hydrocarbon and solvent emission from stationary commercial sources. First, process changes may be made which capture and recycle formerly wasted materials. Second, where vapor concentrations are low and recovery costs high, the hydrocarbon vapors may be burned in an afterburner to carbon dioxide and water. This combustion of vapors, while diminishing hydrocarbon emissions, raises the possibility of increased nitrogen oxide emissions.

Hydrocarbon emissions from automobiles are currently being controlled in two ways. Fuel losses from the crankcase are being prevented by positive crankcase ventilation; the system recycles to the combustion chambers the gases which blow by the pistons. Emissions from the exhaust stream are currently being reduced by improved engine systems. A central feature of all the systems in use is a larger ratio of air to fuel and better fuel metering. Another method for controlling exhaust emissions at present is air injection. In this technique, used mainly on manual transmission vehicles, air mixes with the hot exhaust gases prior to ejection, further combusting the remaining hydrocarbons and carbon monoxide.

Evaporative control systems are on all 1971 and later model cars. Fuel lost from the carburetor or fuel tank after the engine is shut off is stored temporarily either in the crankcase or in a canister filled with activated carbon. On start-up, the vapors from the crankcase are returned to the combustion chamber by the positive crankcase ventilation system. Those vapors adsorbed on carbon are freed by the blowing of air through the canister to the combustion chamber.

In prospect for the future is a catalytic converter, which will consist of two catalysts in series. The first will reduce NO_X to nitrogen; the second will oxidize hydrocarbons and carbon monoxide to carbon dioxide and water. A flow of secondary air to supply more oxygen for combustion will enter just upstream from the second catalyst. The catalyst is expected to be "poisoned" by **Lead;** hence the need for use of lead-free gasoline. (See also **The Automobile and Pollution**).

CONTROL OF NITROGEN OXIDES

Nitrogen oxides in urban air originate nearly entirely from combustion processes. One way to avoid nitrogen oxides production is to utilize a different fuel with less nitrogen; for instance, gas or oil could replace coal in the generation of **Electric Power.** If the fuel is used for power generation, alternative sources of power such as hydropower and nuclear power may be weighed. At the present time, there are apparently no promising techniques for removing NO_x from the flue gas of power plants. The limestone injection process, currently being demonstrated for the removal of **Sulfur Oxides** at several large power stations, does appear, however, to accomplish about a 20 percent reduction in the nitrogen oxides in the flue gas.

On the other hand, changes in the combustion process itself present practical alternatives for control. For instance, a decrease in flame temperature, achieved by injecting steam or water into the combustion zone, or an increase in the distance between burners, or a recirculation of flue gas, reduces nitrogen oxide emissions. Recirculation of exhaust gases may hold promise for the internal combustion engine and power plant boilers.

The technique of diminishing excess air has been in use on oil-fired boilers since the early 1960's to reduce emissions, and its use on gas-fired boilers is thought to be possible. Another combustion modification is two-stage combustion, which first admits less air than needed, then admits the additional air required for complete combustion to the combustion zone. This technique has been used to effect on oil- and gas-fired boilers.

Controlling nitrogen oxide emissions from the 29 million domestic heating units (1967 figures) appears difficult, especially since virtually all units have already been converted from coal to gas or oil. Switching to electric heat decreases dwelling emissions but shifts the burden of pollution control to the power plant (see **Electric Power**).

No system to control oxides of nitrogen from motor vehicles has yet been installed on production vehicles. However, the technique of choice to meet 1976 nitrogen oxide emission requirements appears to be the use of the catalytic converter, which would reduce nitrogen oxides to nitrogen (see also **The Automobile and Pollution**).

REFERENCES

1. J. Pitts and R. Metcalf, eds., *Advances in Environmental Science*, Wiley, New York, 1969.

Six recent and authoritative books from the National Air Pollution Control Administration, now the Office of Air Programs of the U.S. Environmental Protection Agency, are:

2. *Air Quality Criteria for Hydrocarbons*, AP-64, 1970.
3. *Air Quality Criteria for Nitrogen Oxides*, AP-84, 1971.
4. *Air Quality Criteria for Photochemical Oxidants*, AP-63, 1970.
5. *Control Techniques for Carbon Monoxide, Nitrogen Oxide and Hydrocarbon Emissions from Mobile Sources*, AP-66, 1970.
6. *Control Techniques for Hydrocarbon and Organic Solvent Emissions from Stationary Sources*, AP-68, 1970.
7. *Control Techniques for Nitrogen Oxide Emissions from Stationary Sources*, AP-67, 1970.

Phthalates

INTRODUCTION

Past experience with chemicals such as **Mercury** and DDT should lead us to be wary of the harm which might be done by any chemical in a position to become widely distributed in the environment. We have witnessed the concentration, by aquatic organisms, of mercury believed to be safely sunk to the bottoms of lakes.

Science has documented the concentration in our own fatty tissues of insecticides sprayed on crops. It should not be necessary to experience a disaster such as the poisoning of people in Japan by **PCBs** in order to galvanize studies on phthalates, chemicals which have not yet caused identifiable harm but which are becoming widespread environmental contaminants.

Phthalates are chemicals[17] which have the ability to "plasticize" the polymer polyvinyl chloride (PVC). That is, they transform PVC from a hard, glasslike solid into a flexible and elastic material which can be fabricated into products such as plastic storage bags, food wraps, and flexible tubing. However, the phthalate compounds are not integral parts of the plastic but are interfused at levels up to 40 percent with the molecules of PVC. Under the proper conditions phthalates can migrate out of the plastic material and into the substance it touches. Migration is more likely to occur into fatty substances than into aqueous ones, for phthalates are more soluble in fats than in water. High temperatures also facilitate migration.

Polymers plasticized with phthalates are found in nearly every category of products, as can be seen in Table P.11. Major use categories include construction materials, home furnishings, and automotive equipment.[18] Some 1 billion pounds of phthalates were produced in 1972, the major portion of which was used as plasticizers. A lesser amount (50 million pounds) was used in pesticides, cosmetics, fragrances, munitions, and industrial oils.

TABLE P.11

PRODUCTS MANUFACTURED FROM PHTHALATE-CONTAINING PLASTICS

Product	Approximate Percentage of Phthalate Production
Construction products	42
Wire and cable	
Flooring	
Swimming pool liners	
Weather stripping	
Automotive products	12
Upholstery and seatcovers	
Tops	
Mats	
Home furnishings	22
Upholstery	
Wall coverings	
Housewares	
Hose	
Appliances	
Apparel	8
Footwear	
Outerwear	
Baby pants	
Packaging	2
Food wrap film	
Closures	
Medical products	2
Tubing	
Blood storage bags	

From data in P. R. Graham, "Phthalate Ester Plasticizers: Why and How They Are Used," *Environmental Health Perspectives*, experimental issue no. 3, U.S. Dept. of Health, Education and Welfare, Public Health Service, National Institutes of Health, January 1973.

ENVIRONMENTAL CONTAMINATION

Phthalates used as plasticizers can contaminate the environment when plastics are burned in incinerators. Levels in the air near a municipal

[17] Phthalates (pronounced "thal lates") are esters of phthalic acid and various alcohols. Commonly used phthalates include: di-2-ethylhexyl phthalate (DEHP or DOP), dibutyl phthalate (DBP), diethyl phthalate, and butylglycol butyl phthalate.

[18] A study of the atmosphere inside a new car found organic compounds at about 12 milligrams per cubic meter. Depending on temperature, phthalate levels ranged from 0 to 6 percent, the higher concentration occurring at higher temperatures. Based on these data, phthalate concentrations could have reached 0.72 milligram per cubic meter.

TABLE P.12

PHTHALATE LEVELS IN NATURAL WATERS (IN MICROGRAMS PER LITER)

LOCATION	DI-N-BUTYL-PHTHALATE	DI-2-ETHYLHEXYL PHTHALATE	UNSPECIFIED PHTHALATE ESTERS
Charles River, Mass.	—	—	1–2
Hammond Bay, Lake Huron, Mich.	0.04	—	—
Lake Huron, Mich.	—	5.0	—
Missouri River, McBaine, Mo.	0.09	4.9	—
Black Bay, Lake Superior, Ont.	—	300	—

From R. A. Hites, "Phthalates in the Charles and the Merrimack Rivers," *Environmental Health Perspectives*, experimental issue no. 3, U.S. Dept. of Health, Education and Welfare, Public Health Service, National Institutes of Health, January 1973; F. L. Mayer et al., "Phthalate Esters as Environmental Contaminants," *Nature* 238 (1972).

incinerator in Hamilton, Ontario, reached 1.75 micrograms per cubic meter. Phthalates can also leach out of wastes buried in landfills. Industrial effluents which undergo at least secondary sewage treatment (see **Water Treatment and Water Pollution Control**) may not be a major source of environmental pollution since the activated sludge procedure is believed to degrade 90 to 99 percent of added phthalates. Levels of phthalate reported in water are given in Table P.12. Generally higher levels have been found in the more populated or industrialized areas.

Phthalate esters have also been found in soil combined with fulvic acid. Fulvic acid is a humic substance, one of the decomposition products of plant materials. Phthalates combined with humic acid become much more water soluble.

There is not a large body of data on the occurrence of phthalates in organisms. Values of up to 3.2 parts per million in Channel catfish from Mississippi and Arkansas and 100 parts per million in the deep-sea jellyfish *Atolla* have been noted. Fish and tadpoles from a hatchery in Iowa were found to have 0.6 to 0.8 part per million phthalates, probably because of phthalate contamination of fish food. Invertebrates have been reported to concentrate phthalates several hundred times over the water levels, and fish have been found capable of concentrating phthalates 7,000 to 13,000 times over water concentrations.

TOXICITY

It is generally agreed that the acute toxicity of phthalate compounds is low compared to that of chemicals such as DDT. An oral dose of 30 grams of di-2-ethylhexyl phthalate (DEHP) per kilogram of body weight is necessary to kill one-half of a test group of rats. It has been estimated that humans could survive a daily dose of 2.4 to 480 milligrams of phthalate, depending on the particular compound.

Some data on the chronic effects of phthalate exposure can be obtained from a study of Russian workers in artificial leather and polyvinyl chloride film industries. Levels of plasticizers (mainly phthalates) in the air were found to range from 1.7 to 66 milligrams per cubic meter. Polyneuritis—consisting of pain in the extremities, numbness, and spasms—was a common complaint. Of workers with 6 to 10 years of exposure, 57 percent reported symptoms. Among workers with more than 10 years of exposure, the percentage rose to 82.

However, most concern about the effects of phthalates has been focused on two areas. Teratogenic effects (on the developing fetus) and mutagenic effects (on genetic material) constitute one area of concern. The second is the effects of phthalates on recipients of blood transfusions. Blood is routinely stored or "banked" in plastic bags. Recent investigations have disclosed that blood stored for 21 days can absorb up to 11.5 milligrams of phthalates per 100 grams of blood from the plastic storage bags. Patients receiving massive transfusions can then receive large doses of phthalates. A case is reported of a patient with a gunshot wound who, it is calculated, received about 600 milligrams of DEHP in a 24-hour period. Hemophiliacs receiving emergency or maintenance transfusions may receive similarly high doses. No effects on the quality or performance of the blood itself have been documented, although there are indications that phthalates may cause aggregation of the platelets in blood. Such aggregations could conceivably interfere with circulation at the capillary level.

Several investigators have begun to report on the subtle effects low doses of phthalate compounds have on living systems. DEHP, at levels comparable to those in blood which has been stored for 4 days in plastic bags, is lethal to 97 percent of beating chick embryo heart cells in tissue culture. Phthalate compounds injected into the peritoneum in rats caused increased incidence of skeletal and gross abnormalities (such as lack of eyes) in offspring. In addition, the number of fetuses which implanted was reduced, and greater maternal death rates at birth were noted. Dillingham and Autian [5] suggest that phthalate esters may be deleterious to dividing cells. Thus, although phthalates have low toxicity for the adult animal, they may be capable of affecting the developing embryo or fetus.

In humans who have received doses of phthalates via blood transfusions or other medical treatments, the compound may localize in the spleen, liver, and lungs. Oral doses of DEHP given to rats were, however, excreted rapidly with no concentration in the tissues. DEHP has been found in particular fractions (mitochondria) of the heart muscle of beef cattle, rats, rabbits, and dogs. Tissues other than the cardiovascular system contained no phthalates.

Preliminary evidence suggests that aquatic organisms are sensitive to low levels of phthalates. Reproduction in the water flea is impaired at 3 micrograms per liter, a level already present in some natural waters.

In summary, phthalates have not been shown to be acutely toxic to most species, but there are a number of subtle effects of phthalates which give cause for concern. It is not possible at this time to know whether evidence from studies involving, for instance, intraperitoneal injections in rats are indicative of effects which might be expected among humans receiving phthalates primarily in food or by transfusion. While such questions are being resolved, it would seem wise to decrease unnecessary uses of phthalates. A simple switch to other plasticizers is not indicated, for these compounds too are toxicological unknowns. If the use of flexible plastics were decreased, however, sources of environmental contamination such as incineration of plastics could be reduced.

REFERENCES

1. F. L. Mayer et al., "Phthalate Esters as Environmental Contaminants," *Nature* 238 (1972): 411.
2. R. Morris, "Phthalic Acid in the Deep Sea Jellyfish, *Atolla*," *Nature* 227 (1970): 1264.

3. G. Ogner and M. Schnitzer, "Humic Substances: Fulvic Acid-Dialkyl Phthalate Complexes and Their Role in Pollution," *Science*, October 16, 1970, p. 317.

The following references are from *Environmental Health Perspectives*, experimental issue no. 3, U.S. Dept. of Health, Education and Welfare, Public Health Service, National Institutes of Health, January 1973.

4. E. F. Corcoran, "Gas-Chromatographic Detection of Phthalic Acid Esters."
5. E. O. Dillingham and J. Autian, "Teratogenicity, Mutagenicity and Cellular Toxicity of Phthalate Esters."
6. R. M. Gesler, "Toxicology of Di-2-ethylhexyl Phthalate and Other Phthalic Acid Ester Plasticizers."
7. P. R. Graham, "Phthalate Ester Plasticizers: Why and How They Are Used."
8. R. A. Hites, "Phthalates in the Charles and the Merrimack Rivers."
9. R. J. Jaeger and R. J. Rubin, "Phthalate Ester Metabolism in the Isolated, Perfused Rat Liver System."
10. R. J. Jaeger and R. J. Rubin, "Extraction, Localization and Metabolism of Di-2-ethylhexyl Phthalate from PVC Plastic Medical Devices."
11. L. G. Krauskopf, "Studies on the Toxicity of Phthalates via Ingestion."
12. Y. L. Marcel, "Determination of Di-2-ethylhexyl Phthalate Levels in Human Blood Plasma and Cryoprecipitates."
13. T. Maxwell et al., "Plasticizers in Blood: Real or Artifactual."
14. F. L. Mayer, Jr., and H. O. Sanders, "Toxicology of Phthalic Acid Esters in Aquatic Organisms."
15. L. E. Milkov et al., "Health Status of Workers Exposed to Phthalate Plasticizers in the Manufacture of Artificial Leather and Films Based on PVC Resins."
16. D. J. Nazir et al., "Di-2-ethylhexyl Phthalate in Bovine Heart Muscle Mitochondria: Its Detection, Characterization and Specific Localization."
17. J. W. Peters and R. M. Cook, "Effects of Phthalate Esters on Reproduction of Rats."
18. R. J. Rubin and R. J. Jaeger, "Some Pharmacologic and Toxicologic Effects of Di-2-ethylhexyl Phthalate (DEHP) and Other Plasticizers."
19. C. O. Schulz and R. J. Rubin, "Distribution Metabolism and Excretion of Di-2-ethylhexyl Phthalate in the Rat."
20. S. I. Shibko and H. Blumenthal, "Toxicology of Phthalic Acid Esters Used in Food-Packaging Material."
21. D. L. Stalling et al., "Phthalate Ester Residues—Their Metabolism and Analysis in Fish."
22. M. S. Stein et al., "Some Aspects of DEHP and Its Action on Lipid Metabolism."
23. G. H. Thomas, "Quantitative Determination and Confirmation of Identity of Trace Amounts of Dialkyl Phthalates in Environmental Samples."
24. C. R. Valeri et al., "Accumulation of Di-2-ethylhexyl Phthalate (DEHP) in Whole Blood Platelet Concentrates, and Platelet-Poor Plasma."

R

Radiation Standards

See also Nuclear Power Plants.

Numerous factors are involved in the setting of radiation standards, and an understanding of these factors is central to understanding the standards themselves. By what mechanism do the radiations from radioactive elements arise? How does radiation affect man, and on what philosophy should the setting of allowable environmental exposures be based? What is the role of background radiation? How do food chains influence environmental exposure? Consideration of these factors leads to an explanation of how environmental radiation standards have been set. Controversy surrounded the standards in the late 1960's, and changes may occur.

IONIZING RADIATIONS

There are three common forms of ionizing[1] radiations from the naturally occurring radioactive elements: alpha particles, beta particles, and gamma rays. An alpha particle is simply the nucleus of a helium atom and consists of two protons and two neutrons. A beta particle is merely an electron. Gamma rays are not particles but a form of electromagnetic radiation; they are energy, not mass. Unstable elements, such as uranium-238, naturally give off these particles and rays. Typically, either an alpha particle or a beta particle is emitted, rarely both.

[1] "Ionizing" because they frequently collide with neutral atoms and break them into positive and negative components (called ions).

TABLE R.1
IONIZING RADIATIONS

RADIATION	IDENTITY	PENETRATING ABILITY[a]	SOURCE
Alpha particle	Helium nucleus	Several sheets of paper or a few centimeters of air	Decay of radioactive elements
Beta particle	Electron	Several millimeters or more of aluminum	Decay of radioactive elements
Gamma rays	Electromagnetic radiation	Several centimeters *or more* of lead	Decay of radioactive elements
Cosmic rays	Primary particles are mostly high-energy protons and helium nuclei; secondary rays[b] are pions, muons, electromagnetic radiation, positrons, and electrons	Soft component of secondary rays attenuated by 10 centimeters of lead; hard component of secondary rays goes through 10 centimeters of lead without much attenuation	Primary particles are mostly from the sun and the Milky Way galaxy
X-rays	Electromagnetic radiation	Several centimeters or more of lead	Medical and dental radiographic procedures

[a] Material and thickness required to stop radiation.
[b] Produced by reactions of primary particles with nuclei in the atmosphere.

At the same time as the alpha or beta particles are emitted in a particular decay, gamma rays may also be given off.

In addition to the radiations from naturally occurring radioactive elements, two other classes of radiation are of interest: X-rays, a man-made radiation; and cosmic rays, which bombard the earth from space. Table R.1 summarizes the properties of these radiations.

RADIOACTIVE DECAY

The three radiation types arising from radioactive elements are by-products of changes in the nuclear structure of atoms. These spontaneous changes in nuclear structure are referred to as decay—decay in the sense that, as time goes by, less and less of the original substance is present. Decay of a substance implies instability, and this instability is due to the relative number of protons and neutrons in the nucleus. When an unstable nuclear arrangement exists, spontaneous decay takes place; that is, a change in the ratio of protons to neutrons is made. This change is accompanied by the emission of radiation, in the form of particles or of electromagnetic radiation or both.

Many familiar elements have forms characterized by unstable nuclear arrangements. There is an isotope[2] of carbon called carbon-14 (where 14 is the sum of protons and neutrons) which is unstable and decays. Potassium-40 (19 protons and 21 neutrons) is a radioactive isotope of potassium in relative abundance in the earth's crust. Some isotopes are stable (they are not subject to decay), but carbon-14, potassium-40, and uranium-238, among many others, are

[2] An element is characterized by the number of protons in the nucleus. Isotopes of a particular element have the same number of protons but different numbers of neutrons in the nucleus.

radioactive isotopes. We will use carbon-14 and uranium-238 to illustrate radioactive decay reactions:

$$\text{U-238} \xrightarrow{\text{decays to}} \text{Th-234} + \text{alpha particle (helium nucleus, atomic mass} = 4)$$

$$\text{C-14} \xrightarrow{\text{decays to}} \text{N-14} + \text{beta particle (an electron with negligible atomic mass)}$$

Frequently the decay to a "daughter" element is accompanied by the emission of excess energy in the form of a gamma ray.

The rate at which an element decays (that is, the rate at which atoms undergo nuclear rearrangement) is characteristic of the element and is proportional to the number of atoms of the element present. The greater the number of atoms, the greater the rate of decay. Since radiation is a result of decay, it follows that the rate at which radiation is emitted falls as the quantity of the element decreases because of decay. Thus, the radiation from an element will diminish as rapidly as the quantity of the element diminishes.

In specifying the maximum permissible concentrations of a radioactive element in air and water, it is common to avoid mass units and instead refer to a surrogate for mass, the permissible rate at which radiation is emitted from the substance. By rate of radiation emitted is meant the number of disintegrations per second. The fundamental unit is the curie,[3] the quantity of the radioactive element giving 37 billion disintegrations per second. The millicurie is the quantity giving 37 million disintegrations per second. The microcurie is the quantity giving 37,000 disintegrations per second. The maximum permissible concentration for strontium-90 in water is 0.0001 microcurie per liter. This means that no more strontium-90 may be present in water than the quantity which yields this rate of disintegration: about 70 trillionths of a gram of strontium-90 per 1,000 liters of water.

[3] Named for the famous atomic scientists, Pierre and Marie Curie.

EFFECTS OF RADIATION ON MAN AND ANIMALS

In a discussion of the effects of radioactivity exposure, it is useful to divide exposures into two types, those which result in acute symptoms arising within days or weeks, and those low-level exposures which yield no effects for months, years, and possibly generations. The latter are referred to as long-term effects. We shall focus on the long-term effects, for such radiation levels are likely to be the only ones to which the general population will ever be exposed, except in the instance of nuclear war. The acute effects are described in Eisenbud's *Environmental Radioactivity* [1].

Long-term effects become an increasingly important topic as the number of exposure points in the environment increases. Our special concern is with **Nuclear Power Plants,** whose low-level wastes enter water and air.

Although a more precise discussion is given shortly, a portion of the 1960 Memorandum of the Federal Radiation Council (FRC) is used here to introduce the long-term "hazards of ionizing radiation":

> 4. The delayed effects from radiation are in general indistinguishable from familiar pathological conditions usually present in the population.
>
> 5. Delayed effects include genetic effects (effects transmitted to succeeding generations), increased incidence of tumors, life-span shortening, and growth and development changes.
>
> 6. The child, the infant, and the unborn infant

appear to be more sensitive to radiation than the adult.

7. The various organs of the body differ in their sensitivity to radiation.

8. Although ionizing radiation can induce genetic and somatic effects (effects on the individual during his lifetime other than genetic effects), the evidence at the present time is insufficient to justify precise conclusions on the nature of the dose-effect relationship at low doses and dose rates. Moreover, the evidence is insufficient to prove either the hypothesis of a "damage threshold" (a point below which no damage occurs) or the hypothesis of "no threshold" in man at low doses.

9. If one assumes a direct linear relation between biological effect and the amount of dose, it then becomes possible to relate very low dose to an assumed biological effect even though it is not detectable. It is generally agreed that the effect that may actually occur will not exceed the amount predicted by this assumption.

Look again at statements 8 and 9 above. As indicated at the beginning of the discussion, the standards were recently the subject of controversy, and these two statements were at the heart of the disagreements.

There are two types of long-term effects—somatic effects, which are cellular changes deleterious to the individual; and genetic effects, which may be noted only in future generations. Somatic effects of radiation are manifested by such diseases as leukemia, bone cancer, thyroid cancer, and lung cancer. Life-span shortening is also counted as a somatic effect, even though the specific causes of death are not separable from those in the general population. It should be emphasized that no one has ever been able to link these effects in man with exposure levels which were near the FRC-recommended standards for the general population. Massive numbers of individuals would need to be considered in such a study, and the exposures to them would be unwarranted, for although the standards are thought of as maximum levels, the FRC warns that "every effort should be made to encourage . . . doses as far below this guide as practicable."

However, somatic effects have been observed in individuals who were exposed to much higher levels of radiation. Individuals who developed the diseases received in excess of 100 rems and generally far above this value. According to the FRC, the dose for the general population should not exceed 0.17 rem per person per year. Thus, in listing these somatic effects, we are careful to note that, while they were observed at high dose levels, one could not predict their occurrence at low levels with any degree of confidence. In the absence of firm conclusions, atomic scientists concerned with health effects have utilized the "linear hypothesis" mentioned in the FRC memorandum. The FRC felt the linear hypothesis was cautious, and in some sense it is, for it says that every dose of radiation has an effect, no matter how small. One is led, therefore, to be more conservative about standards than if one believed in a damage threshold below which no effect occurred. The remainder of the linear hypothesis discussion is postponed to the last section.

Much of the evidence for increases in leukemia with exposure levels stems from studies of survivors of the atomic bomb, from studies of radiologists (who were exposed over many years), and from studies of children who as infants were irradiated for enlargement of the thymus. Increased bone cancer, which is closely tied to the deposit of radium,[4] strontium, and plutonium[5] in the bone has also been observed. Excess lung cancer was noted among the Czech miners who worked in pitchblende mines. An

[4] This was observed among the group of women who painted luminous radium dials on watches and in the course of their work "tipped" their brushes with their mouths.

[5] Bone cancer due to strontium and plutonium has been found in *animal* experiments.

elevated incidence of thyroid cancer has been seen in the irradiated children mentioned above, and life-span shortening has been observed, primarily in rodents. The shortening was about 2 to 4 percent per 100 rems per year where doses exceed 200 rems.

Genetic effects are changes in the genes, the hereditary material. Animals born with one or more genes which do not correspond to any genes of the parents are mutants; the new genes are called mutations. Fortunately, most new genes are recessive; that is, unless they are complemented in the individual by an exactly similar gene from a mate, the character (e.g., albino coloration) is not expressed. Thus, the chances of the mutation's being expressed as a characteristic of an individual is much diminished. Moreover, mutations are generally not beneficial, and the possibility of the live birth of an individual in which the character is actually expressed may be low. With mutations, also, the linear relation between effect and dose has been postulated, but again, this is a hypothesis, probably not susceptible of proof.

PERMISSIBLE DOSES

Because radiation exposure is such a complex subject, we will avoid the factors, definitions, and units which bear primarily on occupational exposures. Permissible doses for those in occupations involving periodic or regular exposure to small amounts of radiation are set with an entirely different philosophy from that applied to the general population. First, those involved in radiation work are said to be willing to accept some on-the-job risk just as workers in other occupations accept associated risks. Second, since those involved in radiation work constitute so small a portion of the total population, the maintenance of the race should not be endangered by the few who may develop faulty hereditary material. These are two of the reasons cited for allowing the permissible dose of radiation workers to be 10 times the permissible dose for individuals in the general population. Of course, the converse of these statements is cited as a reason for the lower allowable exposure for the general population. In addition, it is thought that children and embryos may be more sensitive to radiation injury and, since these are in the general population, the whole of the general population must be treated in a way to protect children and embryos.

RADIATION BACKGROUND

There are numerous units and factors for radiation dose; some are obsolete or ignored in practice. The unit of interest here is the rem or millirem (one-thousandth of a rem); standards of exposure have been set in rems, and background exposures may be given in rems.

Cosmic rays are one source of the background radiation received by man. At higher altitudes and larger latitudes cosmic ray intensity increases. Naturally occurring radioactive elements are a second source of exposure. Radioactive elements in the earth's crust—potassium-40, radium, etc.—yield background radiations. Potassium-40 is also in the body (the body cannot distinguish it from stable potassium) and subjects the body to radiation internally. The total yearly background radiation in the United States ranges from 120 to 250 millirems [7]; background variations from place to place are due to differences in concentrations of radioactive elements in the crust and differences in cosmic ray intensity. One of the highest average backgrounds in the United States is in a portion of Colorado. The high cosmic ray background exposure in Leadville, Colorado (160 millirems), is due to its high elevation (about 10,000 feet) [13]. Fallout contributes to background exposures to a negligible extent. Exposures to X-rays in medical and dental procedures have been estimated at about 90 millirems per year, although the value is clearly variable from one individual to the next.

In several regions of the world natural background is exceedingly high. In a section of the state of Kerala, India, the thorium content of the soil is about 0.1 percent. Natives sleeping on bare ground may receive over 2,000 millirems per year of radiation dose. In the Brazilian states of Espirito Santo and Rio de Janeiro, there is a region in which background may reach 1,000 millirems per year. Here, too, the radiation is due to thorium in the soil. Besides these geographic areas there are localized concentrations of radioactivity. The spas such as Bad Gastein in Austria where individuals come to bathe in radioactive waters are an example.

MAXIMUM PERMISSIBLE DOSE AND ENVIRONMENTAL STANDARDS

The maximum permissible dose for those exposed *occupationally* should never cumulate over the working lifetime to an average of more than 5 rems per year for the years after age 18. Persons below age 18 are not to be exposed in occupational environments. The maximum permissible dose for individuals in the general population was set by the Federal Radiation Council (FRC) in 1960. From nonmedical sources (i.e., from environmental exposure by air, water, or food) it is 500 millirems per year, one-tenth of the allowable occupational exposure. However, this is to be the maximum allowable exposure.[6] In order to account for statistical variations in exposures above this level, the dose was divided by a factor of 3. The resulting standard is 170 millirems per year per person averaged over the population.

Such standards can be achieved only by setting environmental standards on the concentrations of radioactive elements. The task is to translate these maximum doses into a set of maximum permissible concentrations (MPCs) in air and water for the various radioactive elements. Air and water as well as food are the probable routes by which the population is exposed. The calculation involves knowledge of (1) the rate at which a substance decays to a stable, nonradioactive, and thus harmless state, (2) the rate at which the body "turns over" or eliminates the substance, and (3) the energy and effect of the radiations on the tissues, especially the effect on tissues in which the substance is likely to accumulate. The calculation presumes the characteristics of a "standard man" whose daily intake of air, water, and food is known, whose daily output of wastes is known, and whose physiological parameters and organ weights are also given.

For instance, strontium-90 has a slow rate of radioactive decay; only one-half the given initial quantity decays in 28 years. It is also eliminated by the body very slowly; for an initial level to decrease by a factor of one-half would require 50 years. Furthermore, it is a "bone-seeker," replacing calcium in bone structure; it affects the manufacture of red blood cells which takes place in the marrow of the bone. Consequently, the maximum permissible concentrations of strontium-90 in air and water are set very low.

Table R.2 lists the maximum permissible concentrations in water (MPCw) for certain important radioactive elements, which (with the exception of phosphorus) are common constituents of power reactor effluents. To see how to use these quantities, consider a water sample in which the concentrations of cobalt-60, iron-59, and zinc-65 are 5×10^{-6},

[6] It was called a Radiation Protection Guide in the 1960 Memorandum from the Federal Radiation Council. While the term seems to have fallen into disfavor, it did convey a message, for the guide "is defined as the radiation dose which should not be exceeded without careful consideration of the reasons for doing so; every effort should be made to encourage the maintenance of radiation doses as far below this level as practicable."

TABLE R.2

MAXIMUM PERMISSIBLE CONCENTRATIONS OF RADIOACTIVE ELEMENTS IN WATER (IN MICROCURIES PER MILLILITER)

ELEMENT	SOLUBLE	INSOLUBLE
Chromium-51	2×10^{-3}	2×10^{-3}
Cobalt-58	1×10^{-4}	9×10^{-5}
Cobalt-60	5×10^{-5}	3×10^{-5}
Copper-64	3×10^{-4}	2×10^{-4}
Fluorine-18	8×10^{-4}	5×10^{-4}
Iron-59	6×10^{-5}	5×10^{-5}
Manganese-56	1×10^{-4}	1×10^{-4}
Molybdenum-99	2×10^{-4}	4×10^{-5}
Phosphorus-32	2×10^{-5}	2×10^{-5}
Silver-110	3×10^{-5}	3×10^{-5}
Tungsten-187	7×10^{-5}	6×10^{-5}
Zinc-65	1×10^{-4}	2×10^{-4}

From Code of Federal Regulations, Title 10, Atomic Energy, Part 20, Standards for Protection Against Radiation, Appendix B.

NOTE: The number of microcuries indicates the allowable rate of radiation emission (number of disintegrations per second) from the particular element.

3×10^{-5}, and 2×10^{-5} microcuries per milliliter, respectively. No other radioactive elements are presumed to be present. The ratios of actual concentrations to maximum permissible concentrations are 1:10, 1:2, and 1:5, respectively. That is, over a year, the cobalt would deliver one-tenth of the permitted dose to the "standard man," the iron would provide one-half of the dose, and the zinc would give one-fifth of the dose. The total fraction of the dose provided by all three elements is eight-tenths, indicating that the dose would not be exceeded via the consumption of *water.* To be complete, the same manipulations would have to be performed for elements in air and these fractions would be included in building the total fractional dose.

In addition, when the identity of radioactive elements in water or air are unknown, and the medium does not contain certain specific hazardous substances, such as strontium-90, a single maximum permissible concentration in microcuries per milliliter may be applied to the unknown quantities of contaminants.

THE FOOD CHAIN

The radioactive elements in water may be concentrated by aquatic life. Radioactive elements in air may be deposited on grass and eaten by dairy animals. Can man receive additional exposures via the food he eats? There is that possibility. The MPCs in air and water may protect him from excessive exposure from either of these sources alone, and suitable calculations may be made to determine limiting concentrations in one medium, given the concentrations in the other medium. But the determination of exposures via foodstuffs is tricky and demands knowledge of the dietary habits of the population. Not only is it necessary to know dietary habits, but the concentrations in foods must be capable of estimation as well. This is the weak link in applying exposure standards. Eisenbud writes in *Environmental Radioactivity* [1]:

> Each radioelement has its own complex path through the food chain, and there is insufficient information about these pathways to permit even a qualitative understanding of how the elements move through the food chains. The quantitative relationships that govern the movement of radioelements are not understood sufficiently to permit one to forecast with any degree of precision the pathways that would be taken for a given radioelement in moving from the aquatic environment to man.

Honstead [5] observes three quite possible routes of radioelements to man: (1) eating fish which have been exposed to radioelements (phosphorus-32 is a particular element concentrated in fish), (2) eating oysters or clams grown in seawater which contains zinc-65, and

(3) drinking milk from cows which graze on pasture grass exposed to iodine-131. "There is no assurance," he notes, "that there are not others [other pathways] which have remained undetected."

Absorption of radioelements by algae and animals of the stream bottom which are eaten by fish have led to marked accumulation in fish. Particularly, phosphorus-32 has been noted to be concentrated in fish both in the Columbia River and in White Oak Lake, Tennessee. It is treacherous to cite numbers for a particular species because of seasonal variation, life cycle variation, and differences between tissue concentrations (e.g., bone and soft tissue of fish accumulate strontium-90 differently). However, having warned of these pitfalls, one can introduce some factors for concentration magnification. Such a factor would ideally represent the activity (the counted number of disintegrations per unit time) per gram of tissue of the organism divided by the activity in 1 gram of the water in which the organism lived.

Two biologically important radioelements are phosphorus-32 and zinc-65. Both have established roles in plant and animal metabolism. Phosphorus is a known component of the genetic material of plants and animals. Zinc-65, in particular, may be a component of the effluent of **Nuclear Power Plants.** Adult whitefish in the Columbia River downstream from the Hanford plant (an atomic energy research center) have concentrated phosphorus-32 (a minor component of the effluent) between 0 and 5,000 times, the higher values being reached in the summer and late fall. Zinc-65 was concentrated about 1,000 times, this value occurring in November.

Cobalt is also known to have a role in human metabolism (it is in vitamin B_{12}, cobalamin). As cobalt-60 and cobalt-58 are low-level components of reactor effluent, these elements, too, warrant study. Although copper is involved in metabolism, low-level copper-64 from reactor effluents should not pose a serious problem because of the radioelement's rapid decay (half of the quantity disappears every 13 hours). It should be pointed out that the data given thus far are conservative. Another study has noted a 20,000-to 30,000-fold magnification of phosphorus-32 in blue-gills and crappies in the Columbia River. Filamentous algae were reported to have concentrated iron-59 and phosphorus-32 by a factor of 100,000 in the Columbia River. Iron-59 is a common corrosion product in reactor effluents. Iron is a constituent of hemoglobin, the oxygen carrier in the blood, a participant in photosynthetic reactions, and involved in all aerobic biological oxidation schemes.

CONTROVERSY AND A NEW CONCEPT

In hearings before the Subcommittee on Air and Water Pollution of the Senate's Committee on Public Works in late 1969 and in hearings in 1969 and 1970 before the Joint Committee on Atomic Energy, Dr. John Gofman and Dr. Arthur Tamplin attacked the standards for radiation exposure. Their testimony was countered by statements from a number of scientists associated with organizations which recommended standards.

Gofman and Tamplin take issue with the apparent interpretation of the linear hypothesis made by the organizations recommending standards. These groups state the hypothesis as a basis on which to work. It implies that no amount of radiation is harmless. The scientists involved in setting standards state the hypothesis and do indeed set their standards on the basis of it. However, because they tend to disbelieve it, feeling instead that the threshold concept may be true, they see their recommendations as very conservative. See Figure R.1 for a comparison of the doses necessary, according

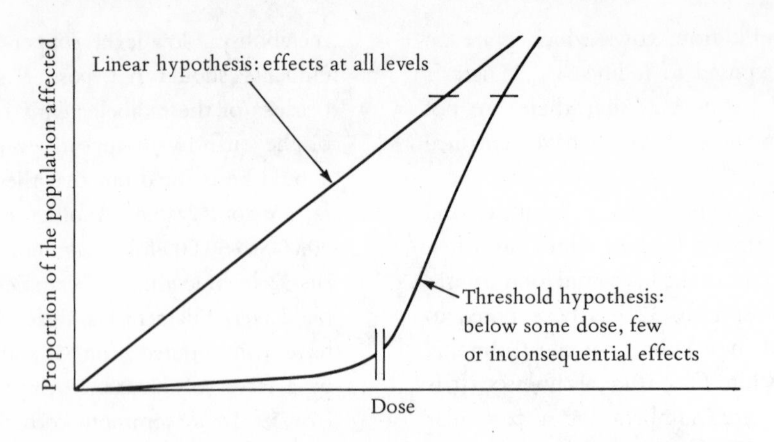

FIGURE R.1
COMPARISON OF LINEAR AND THRESHOLD CONCEPTS

to the two different hypotheses, to affect a given proportion of the population. Gofman and Tamplin, on the other hand, believe the linear hypothesis is nearly correct, and this conviction sets them apart from the scientific organizations. Gofman and Tamplin asked for a 10-fold reduction in standards from 170 millirems per year to 17 millirems per year. At the same time, Dr. Lauriston Taylor, Chairman of the National Council on Radiation Protection and Measurement (NCRP) said in the hearings that *if* today's information had been available when the current standards were first recommended in 1959, the standards might not have been so restrictive. Recently (1971), the NCRP released the report of a 10-year study; they found "no basis for any drastic reduction in the recommended exposure levels."

The hearings indicated a reluctance on the part of the AEC to adopt different environmental standards from those recommended by the FRC. It was pointed out that civilian **Nuclear Power Plants** actually produced but a small fraction of the allowable discharges (which had been derived from the environmental standards). Nevertheless, in June 1971 the AEC announced new rules *for consideration.* These proposed rules would generally limit the discharges from light water power plants (the BWR and PWR) to levels which would result in maximum environmental exposures to individuals of 5 millirems per year. This is a hundredfold reduction of the FRC's environmental standard for the maximum individual exposure. Further discussion of the proposed rules is found in **Nuclear Power Plants, causes for concern.** In early 1973, the rules had not yet been finalized or amended. Figure R.1, albeit highly simplified, is meant to illustrate the consequences if (1) the linear hypothesis is true and (2) the threshold concept is correct.

With the acceptance of Government Reorganization Plan No. 3, the Environmental Protection Agency inherited the functions of the Federal Radiation Council (October 1970). The Council, which recommended exposure standards to federal agencies, went out of existence, and that responsibility was transferred to the EPA. However, before the FRC was dissolved, it had in response to the criticism of Gofman and Tamplin commissioned a new study on exposure standards by the National Academy of Sciences. The report of the Committee on the Biological Effects of Ionizing

Radiation of the NAS was issued in late 1972 [9, 11]. While the panel called the calculations of Gofman and Tamplin "overestimates," it did concur with the conclusion of these scientists that the 170-millirem standard was too high.

One indication of EPA's thinking on radiation standards appears in an article in *Physics Today.* Dr. Joseph Lieberman introduces the concept of standards stated in man-rems per year. Man-rems are the sum of the exposures of all members of a given population. If all members of a population of 1 million were to receive 10 millirems (.01 rem) per year, the total annual man-rems of exposure to the population is $10^6 \times 10^{-2}$ or 10,000 man-rems. A standard which included not only a limitation on man-rems per year for a population but also the maximum exposure for any individual would be as complete as one which limited both average and maximum exposures. In addition, the man-rem concept, Lieberman argues, can help the public understand the risk which accompanies the standard. He cites the fact that risk can be stated as the number who may be affected ultimately (i.e., have disease induced) per man-rem of exposure. A uniformly applied standard in man-rems would also have the effect of lowering the average exposure to larger populations. The AEC in the proposed rules mentioned earlier has already proposed a man-rem standard. They suggest design objectives to limit the exposure of a population to 400 man-rems per year per 1,000 megawatts of installed nuclear electric capacity [8].

REFERENCES

1. M. Eisenbud, *Environmental Radioactivity*, McGraw-Hill, New York, 1963.
2. S. Glasstone, *Sourcebook on Atomic Energy*, 3rd ed. Van Nostrand, New York, 1967.
3. "The Environmental Effects of Producing Electric Power," *Hearings Before the Joint Committee on Atomic Energy*, vols. 1 and 2, 1969–1970.
4. J. Davis and R. Foster, "Bioaccumulation of Radioisotopes Through Aquatic Food Chains," *Ecology* 39 (1958): 530; reprinted in G. Cox, ed., *Readings in Conservation Ecology*, Appleton-Century-Crofts, New York, 1969.
5. J. Honstead, "A Survey of Environmental Dose Evaluations," *Nuclear Safety* 9, no. 5 (September–October 1968): 383.
6. J. Lieberman, "Ionizing-Radiation Standards for Population Exposure," *Physics Today*, November 1971, p. 32.
7. Press release of the U.S. Environmental Protection Agency, "Report Lists State-by-State Radiation Levels," November 28, 1971.
8. *Federal Register* 36, no. 111 (June 9, 1971): 11113.
9. *Science*, December 1, 1972, p. 966.
10. L. Sagan, "Human Costs of Nuclear Power," *Science*, August 11, 1972, p. 487.
11. "The Effects on Populations of Exposure to Low Levels of Ionizing Radiation," National Academy of Sciences, November 1972.
12. *Draft Environmental Statement Concerning Proposed Rule Making Action: Numerical Guides for Design Objectives and Limiting Conditions for Operation to Meet the Criterion "As Low as Practicable" for Radioactive Material in Light-Water-Cooled Nuclear Power Reactor Effluents.* U.S. Atomic Energy Commission, January 1973. This book refers to the impact of regulations proposed in [8].
13. A. Preston, "Artificial Radioactivity in Freshwater and Estuarine Systems," *Proceedings* of the Royal Society of London, Series B, 180 (1972): 421.
14. D. T. Oakley, "Natural Radiation Exposure in the United States," U.S. Environmental

Protection Agency, Office of Radiation Programs, 1972.
15. Code of Federal Regulations, Title 10, Atomic Energy (Revised as of January 1, 1972), Part 20, Standards for Protection Against Radiation, Appendix B. ("Revised as of" means that any changes made since the previous issuance of Title 10 were included. It does not imply that the numerical standards were modified during 1971. In fact, the standards quoted in Table R.2 have not changed in over 10 years.)

Rare and Endangered Animals

As mankind builds his cities and suburbs, roads and farms, he is increasing the habitat for one creature—man. It is commonly forgotten that at the same time the habitat for many creatures who once lived in the area is decreased. Some species adapt well and live happily close to humans. Sparrows, pigeons, and squirrels are familiar even to city dwellers. Most other species, however, cannot find suitable nesting sites or sufficient food or even enough living space to cohabit with us. In addition, some animals are considered food and hunted, and others are thought of as pests and exterminated. These conditions have led to the complete loss of 15 different animals and birds in the contiguous United States. Examples are the passenger pigeon, Carolina parakeet, eastern elk, giant sea mink, Badlands bighorn, San Gorgonio trout, and Ash Meadows killifish. The North American continent, of all the continents, has lost the greatest number of species in recent history because of its rapid transition from a wilderness to a highly industrialized status.

The islands bordering our continent have been even more severely affected. They tend to have simply organized animal communities, without the competition found on the mainland. Civilization brought with it diseases as well as several new species—rabbits, goats, cats, and mongooses. The introduced species preyed on the resident species or usurped their habitat, and, as a result, many native animals and birds are now extinct. Twenty-two birds were lost from the Hawaiian Islands alone.

Certain behavioral patterns make the extinction of a species more likely. Birds which flock in large groups are easily preyed upon. Passenger pigeons are a good example. They were once so numerous that they could be stoned as they flew overhead in great flocks, yet hunting eventually wiped them out. A certain species of bat hibernates in only four caves during the winter. Wanton destruction by thoughtless humans and disturbances by spelunkers threaten the bats with extinction. Birds which cannot fly are at a severe disadvantage. The flightless Laysan rail was an easy mark for introduced predators. Animals which must live in a very specific habitat can easily vanish if their habitat is destroyed. Prairie chickens, which require tall-grass prairie, are being slowly eliminated as their habitat is used for grazing or farming. Those species which eat only one type or a few types of food are likewise in danger. One bird species, which eats only a particular kind of snail, is dependent on a rapidly shrinking food source as the marshlands where the snails grow are drained and filled.

Nature's rule is unrelenting; creatures which cannot adapt to changing conditions are doomed. New species evolve to fill ecological niches as older ones fail to adapt. Man's rule, however, may be even crueler. Species are dying out faster than new ones are evolving. The satisfying diversity of creatures we enjoy is decreasing. The loss of species is probably not

Black-footed ferret. These large weasels make dens in holes dug by prairie dogs, their natural prey. The destruction of prairie dogs that results when grasslands are used for crops or grazing has brought the black-footed ferret to the verge of extinction.

Eastern timber wolf. Civilization has encroached upon the life of this fearsome and magnificent animal. Posted bounties have encouraged hunting and trapping, and much of the wilderness in which the wolf lives is being destroyed.

American alligator. This huge reptile is endangered by the poaching of commercial skin collectors and by the destruction of its habitat.

felt acutely by any except the devoted naturalist or animal lover. Yet almost everyone should appreciate the sadness attached to the thought that he can never see alive those creatures that once abundantly inhabited our land.

Of additional concern should be the fact that at least another 60 species are considered rare or endangered. These creatures may well become extinct in our lifetime unless measures are taken to protect the remaining individuals. If we allow them to be crowded out or hunted to extinction, we have in effect made an irrevocable choice for our descendants. Surely it is our responsibility to strike a balance between improving conditions for our own species and preserving some portion of the natural world as it was passed to us.

Of what value are these imperiled creatures to man? Henry Beston wrote in *The Outermost House*:

Remote from universal nature, and living by complicated artifice, man in civilization surveys the creature through the glass of his knowledge and sees thereby a feather magnified and the whole image in complete

California condor. North America's largest soaring land bird, the condor has a wingspread of from 9 to 9 1/2 feet. Because of poisoning, shooting, and loss of habitat, only 60 or 80 adult birds remain. It is estimated that each year no more than 3 young birds grow to flying size.

Humpback whale. Commercial whalers disregarding international conventions regulating whaling are the primary cause of this animal's endangered status. The humpback whale grows to a length of 40 or more feet. Females usually bear only one calf, in alternate years.

distortion. We patronize them for their incompleteness, for their tragic fate of having taken form so far below ourselves. And therein we err, and greatly err. For the animal shall not be measured by man. In a world older and more complete than ours they move finished and complete, gifted with extensions of the senses we have lost or never attained, living by voices we shall never hear. They are not brethren, they are not underlings, they are other nations, caught with ourselves in the net of life and time, fellow prisoners of the splendor and travail of the earth.[7]

Furthermore, despite the great amount of knowledge gathered by the science of ecology, we by no means understand all the interrelationships among the creatures who inhabit our planet. Who can be sure that the loss of some species might not have unhappy consequences for the human species? An example of a parallel situation was discovered during research on the ecological effects of pesticide usage. Orchards sprayed with DDT to control certain insects can be devastated by red spider mites because ladybugs, which ordinarily eat the spider mites, are killed by DDT while the mites are not. With their natural predator gone, the mites are free to reproduce and destroy the orchards. Ladybugs are, of course, in no danger of extinction, but the lesson should be clear. How unfortunate it would be to discover that a species was necessary after it had become rare or even extinct.

Animals, birds, and fish which are believed to be in some danger of extinction are placed in one of four categories: endangered, rare, periph-

[7] Henry Beston, *The Outermost House*, Viking Press, New York, 1962, p. 25.

TABLE R.3

RARE, ENDANGERED, AND EXTINCT AMERICAN ANIMALS

Animal	Where Found	Comments and Reasons for Problem
Extinct mammals		
Amargosa meadow vole	New York (1918)[a]	
Badlands bighorn	North and South Dakota (1910)	
Eastern cougar	East (1899)	Still some in Canada
Eastern elk	East of Great Plains (1880)	
Gull Island vole	New York (1898)	
Merriam elk	Arizona (1900)	
Plains wolf	Great Plains (1926)	
Sea mink	New England (1890)	
Steller's sea cow	North Pacific (1768)	
Extinct birds[b]		
Carolina parakeet	South Central (1920)	Hunted, infringed[c]
Great auk	North Atlantic (1844)	Hunted
Heath hen	East (1932)	Infringement, hunted
Labrador duck	Northeast (1875)	
Louisiana parakeet	South Central (1912)	Infringement, hunted
Passenger pigeon	North America (1914)	Infringed, hunted
Extinct fish		
San Gorgonio trout	California	Extinct about 1935
Pahranagat spinnedace	Nevada	Extinct between 1938 and 1959
Big Spring spinnedace	Nevada	Extinct between 1938 and 1959
Harelip sucker	Mississippi River, Tennessee River (1900)	
Leon Springs pupfish	Texas (1938)	
Ash Meadows killifish	Death Valley (1942)	
Endangered mammals[d]		
Black-footed ferret (32)	West	Infringement, hunted
Columbia white-tailed deer (300–400)	Southwest	Infringed, predation, hunted
Eastern timber wolf (300–400)	Mid-northern United States and Canada	Infringement, hunted
Florida manatee	Florida waters	Hunted, predation
Florida panther (100–300)	Florida	Infringement, hunted

From data in *Rare and Endangered Wildlife of the United States*, U.S. Dept. of the Interior, Committee on Rare and Endangered Wildlife Species, Bureau of Sport Fisheries and Wildlife, 1966 and 1968.

[a] Dates in parentheses show when the species was last seen.

[b] In addition, 26 species of birds have become extinct on the Hawaiian Islands in the last century. These include the Oahu oo, the Hawaii mamo, and the sandwich rail.

[c] *Hunted* means that although most endangered or rare animals are protected by law from hunters, poaching still threatens them. *Infringed* means that the normal habitat for the species was taken over or destroyed by some other creature, usually man.

[d] Numbers in parentheses are estimates of the numbers of animals left. For some species no reliable estimate can be made.

Animal	Where Found	Comments and Reasons for Problem
Endangered mammals (continued)		
Indiana bat (500,000)	Midwest, East	Wanton killing; pesticides; infringement; all congregate in winter in 4 caves
Key deer (300)	Florida	Infringement
San Joaquin kit fox (113 dens)	California	Infringement, rodenticides
Sonoran pronghorn (1,040)	Arizona, Mexico	Infringed, predation, hunted
Texas red wolf	Texas	Infringement, hunting
Whales, Atlantic right (several hundred)	Atlantic coast	Hunted
Whales, blue (1,500–2,000)	Pacific, Atlantic	Hunted
Whales, bowhead (1,000)	Northern oceans	Hunted
Whales, gray (8,000)	North Pacific	Hunted
Whales, humpback (5,000)	North Pacific	Hunted
Whales, Pacific right (few hundred)	Pacific coast	Hunted
Endangered birds[e]		
American ivory-billed woodpecker	South, Texas	Infringement, hunted
American peregrine falcon	Alaska to Mexico	Pesticides, hunted
Attwater's greater prairie chicken (1,000)	Texas	Infringement
Bachman's warbler	Virginia, South Carolina	Infringement (?)
California least tern	California coast	Infringement
California condor (60–80)	California	Infringement
Cape Sable sparrow (500–1,000)	Southwestern Florida	Infringement, natural causes
Dusky seaside sparrow (900–1,200)	Florida	Infringement: mosquito control for Cape Kennedy
Eskimo curlew	Atlantic coast	Hunted
Florida Everglades kite (47)	Florida	Infringement, hunted
Kirtland's warbler (15,000)	Texas	Infringement
Light-footed clapper rail	California, Mexico	Infringement
Masked bobwhite (400–1,000)	Mexico	Infringement
Mexican duck (500 U.S., 1,000 Mexico)	Southwest	Infringement, interbreeding
Northern red-cockaded woodpecker	Virginia to North Florida	Infringement
Southern bald eagle (235 nests)	East	Infringed, hunted, pesticides
Southern red-cockaded woodpecker	Florida	Infringement
Tule white-fronted goose	Far North to Mexico	Hunted—very unwary
Whooping crane (600)	Canada, Texas	Infringement, hunted
Yuma clapper rail (200–400)	Lower Colorado River	Infringement
Endangered amphibians		
Houston toad (few hundred)	Texas	Infringement, interbreeding
Santa Cruz long-toed salamander	California	Infringement, overcollected
Texas blind salamander	Texas	Infringement, overcollected
Endangered reptiles		
American alligator	South and West	Infringed, hunted, predation
Blunt-nosed leopard lizard	California	Infringement
San Francisco garter snake	San Francisco	Infringement

[e] In addition, 23 species of birds are in danger of extinction on the Hawaiian Islands. These include the Hawaiian coot, the Hawaiian dark-rumped petrel, the Kauai oo, and the Molokai creeper.

TABLE R.3 (CONTINUED)

Animal	Where Found	Comments and Reasons for Problem
Endangered fish[f]		
Arizona trout	Arizona	Infringement
Atlantic salmon	Maine	Pollution; reservoir barriers; endangered only in United States
Blue pike (very few)	Lake Erie, Lake Ontario	Commercial catch once 20 million pounds; pollution
Colorado River squawfish	Colorado River	Habitat change due to reservoirs
Gila topminnow	Arizona	Infringement, possible extinction in United States
Greenback cutthroat trout (more than 200)	Colorado	Infringement
Humpback chub	Green and Colorado rivers	Unknown
Lahontan cutthroat trout	California, Nevada	Infringement
Little Colorado spinnedace (less than 1,000)	Arizona	Infringement
Longjaw cisco	Lakes Michigan, Huron, Erie	Lamprey predation, commercial fishing, pollution in Erie; only 7 taken, 1962–1964
Montana westslope cutthroat trout	Montana	Infringement
Shortnose sturgeon	Hudson River	Pollution
Rare mammals		
Beach Meadow vole	Muskegat Island, Mass.	Infringement, predation
Block Island meadow vole	Rhode Island	Infringement, hurricanes
California bighorn (1,850)	Eastern Oregon, California, Canada	Infringement, disease
Glacier bear (500)	Alaska	Hunted
Grizzly bear (850, excluding those in Alaska)	Alaska, Northwest	Infringement, hunted
Kaibab squirrel (1,000)	Grand Canyon, Arizona	Disease, infringement
Peninsular bighorn (7,000)	Southern California, Northern Mexico	Hunted
Ribbon seal	Alaska, North	Has always been rare
Southern sea otter (500)	California coast	Hunted, persecuted by abalone fishermen
Spotted bat	West, Mexico	May be rarest U.S. mammal, reason not known
Tule (dwarf) elk (400)	California	Infringement, hunted
Rare birds[g]		
California clapper rail	California	Infringement; very secretive
Florida great white heron (2,000)	Florida	Hunted, hurricanes
Golden-cheeked warbler (15,000)	Texas	Infringement
Greater sandhill crane (6,000)	North Central, northwest, West, and South; South central and southwest Canada	Infringement

[f] In addition, the habitat and number of some species of endangered fish have always been small, according to some naturalists. These include the Big Bend gambusia, the Comanche Springs pupfish, the Devil's Hole pupfish, and the Maryland darter.

[g] In addition, two bird species are listed as rare in Hawaii—the large Kauai thrush and Newell's manx shearwater.

TABLE R.3 (CONTINUED)

Animal	Where Found	Comments and Reasons for Problem
Rare birds (continued)		
Florida sandhill crane (2,000–3,000)	Florida, Georgia	Infringement
Ipswich sparrow (4,000)	Nova Scotia to southern Georgia	Infringement, erosion of island home
Lesser prairie chicken (12,000–80,000)	West	Infringement
Northern greater prairie chicken	West	Dependent on a disappearing habitat
Prairie falcon	Western Canada, United States, Mexico	Possibly pesticides
Short-tailed hawk (200 in United States)	Florida and Mexico to Argentina	Infringement
Rare reptiles		
Bog turtle	Connecticut to North Carolina	Infringement, hunted
Rare amphibians		
Black toad (700–10,000)	California	Infringement
Limestone salamander (100–200)	California	Localized in commercially desirable habitat
Pine barrens tree frog	New Jersey, North Carolina	Infringement
Vegas Valley leopard frog (may be extinct)		Infringement, predation
Rare fish[h]		
Atlantic sturgeon	Atlantic coast, St. Lawrence to northern Gulf coast	Pollution and barriers in spawning streams
Arctic grayling	Montana, Michigan	Infringement
Blackfin cisco	Lakes Michigan, Huron	Last specimen seen in 1955; commercial fishing and lamprey predation
Blueback trout	Maine	Infringement
Deepwater cisco	Lakes Michigan, Huron	Last specimen seen in 1951; commercial fishing and lamprey predation
Lake sturgeon	Great Lakes	Overfishing

[h] In addition, a number of rare fish have very restricted habitats. Only a few specimens of others have ever been taken. These include the Olympic mudminnow, the Ozark cavefish, the Sunapee trout, and the Suwanee bass.

eral, or status undertermined. According to the Bureau of Sports Fisheries and Wildlife, a *rare species* is one which is not now in danger of extinction but is found in such small numbers that it might become so endangered. An *endangered species* is in jeopardy already. Unless measures are taken, it is very likely to die out completely and soon.

The third category, *peripheral species*, comprises those animals which are (1) at the limit of their range in the United States and (2) endangered in this country but not likely to become extinct in the world as a whole. Concern is felt that these species should be kept as a part of our country's fauna. *Status undetermined species* are those which may be in trouble but about which insufficient facts are known. Research is necessary to determine the actual status of these animals, some of which may be endangered or even extinct.

The setting aside of land for refuges and parks was the first measure taken to protect vanishing forms of wildlife. Some 16 of our endangered species are currently surviving in refuges and parks. More could be saved if more lands were added to the system. Many species can be managed successfully in the wild, even to the extent of allowing some hunting, fishing, and collecting. Other species require more support if they are to survive. The whooping crane is being hatched in captivity, from eggs taken from wild nests, to improve the survival rate of the cranes.

In some cases the reason for a species' decline is not even known for sure. For example, the southern bald eagle is decreasing in numbers. The reason may be loss of its habitat or poaching or poisoning by pesticides or all three. Research is needed so that protective measures can be devised. Some animals can be transplanted from a diminishing habitat to a similar one in another part of the country. A number of game species including Canada geese, wild turkeys, and pronghorns have been successfully transplanted. The careful regulation of hunting to protect scarce animals is beneficial. Another helpful measure involves the limiting of chemical pest control programs around the habitat of endangered species.

The situation is thus by no means hopeless. Species can be brought back from the brink of extinction. One example is the American bison, reduced to a few hundred animals 50 years ago and now numerous enough to allow very limited use as a food source. Another example is the trumpeter swan. It too was nearly extinct but is today protected on the Red Rocks Lake refuge in Montana.

Table R.3 is a summary of animals and birds of the continental United States which are extinct, endangered, or rare. Additional data on measures taken to protect the animals, as well as detailed descriptions of the species, their distribution, and their habitat, can be found in the original references.

REFERENCES

1. *Rare and Endangered Fish and Wildlife of the United States*, U.S. Dept. of the Interior, Bureau of Sport Fisheries and Wildlife, 1968.
2. "The Right to Exist," Resource Publication 69, U.S. Dept. of the Interior, Bureau of Sport Fisheries and Wildlife, 1969.

Reservoirs

Reservoirs are constructed for many purposes; water supply, irrigation, flood and erosion control, power generation, recreational use, and low flow augmentation. Low flow augmentation is the release of stored waters during periods of low river flow. Its purpose is primarily pollution control. Waste flows into a river during the dry season could put a heavy pollution load on a stream. Supplemental stored water is added to dilute the wastes and make the sewage load less harmful to aquatic life. Low flow augmentation is known in the trade as "the dilution solution to pollution." We have defined augmentation because it is a less-well-known use of reservoirs and also because, under certain circumstances,

Comerford Station, a hydroelectric plant on the Connecticut River between New Hampshire and Vermont. The plant began operation in 1930 and has a power rating of 162 megawatts. The reservoir behind the dam is 8 miles long and covers 1,093 acres.

reservoir releases could potentially impair rather than enhance water quality.

Many reservoirs combine several of the functions listed above. Thus, reservoirs whose central function is to supply water may also be used for boating and fishing. Reservoirs built to supply power may be used in a similar way. Recreation is certainly a valuable by-product of building a reservoir. But the value of the by-product must be weighed against the loss of the land as a recreational resource and wildlife habitat.

When waters back up behind a reservoir, there are numerous effects beyond the disappearance of lands. The most obvious effect is the dropping of the load of suspended sediment which is carried by the stream. The faster water flows, the more suspended material it is capable of carrying. Particles eroded over the length of the stream may be carried great distances in suspension if the water remains sufficiently turbulent. The slowing of waters behind a reservoir decreases the turbulence which has entrained the particles, causing the coarse material to settle to the bottom of the reservoir. When a reservoir level is low, one may see the edges of smooth mounds of sediment which layer what was once a river valley. This is the effect most often cited by conservationists in their frequent quarrels with the Army Corps of Engineers. There are many less obvious effects of water storage by reservoirs.

The passage of fish swimming upstream to spawn may be blocked by the reservoir. Fish ladders, concrete "staircases," may be utilized to provide fish with short heights which they can scale one at a time in order to reach the stream above.

A phenomenon that plays a key role in the quality of the released water is the temperature stratification of the water in the reservoir. Because cool water is more dense than warm water, the water may "stratify"—that is, form layers of different densities and temperatures which do not mix. Of course, cooler water is on the bottom[8] because it is more dense. There is a cycle of stratification which we may, for convenience, say begins in the fall. The water in late fall (in the northern latitudes) is well mixed by wind-induced circulation. Through the winter and into the early spring, a well-mixed character is sustained. With the coming of warmer weather the upper layers of the reservoir become warmed by two mechanisms. First, the warm water inflows stay in the surface layers, and second, radiation and convection transfer heat to the surface water. The result is stratification, which often creates layers so stable that circulation is minute. The situation may remain this way until the fall when the surface waters cool. When, with cooling, the density of the upper layer reaches a critical level, the wind induces an "overturn," and the waters mix thoroughly. The cycle then begins again. This phenomenon of stratification gives rise to several water quality effects.

First, since releases are typically from the bottom of the reservoir, the summer releases may be much colder than the normal seasonal temperature of the stream's flow. Cold flows could conceivably put a stress on aquatic life. The effect may be thought of as the inverse of **Thermal Pollution,** in which warm water discharges may have detrimental consequences.

The combination of thermal stratification and the dropping of sediment load gives rise to a second important effect. In the flowing stream the penetration of sunlight was blocked by particles in suspension. As particles settle to the bottom of the reservoir, the surface layers of water clear. Now algae, which require sunlight, may grow more extensively. The growing algae are confined to the upper layers of the reservoir —say, the first 30 to 40 feet, for sunlight penetration below such depths is negligible. When the algae die, they sink, and in the lower levels of the reservoir they are decomposed. The decomposition of the algae by bacteria depletes oxygen from the lower layers of the reservoir.

One consequence of thermal stratification is that bottom layers of water exist which do not mix with the upper layers. For this reason upper levels may be well oxygenated and lower levels deoxygenated. Oxygen is a key component influencing stream "health" (see **Organic Water Pollution**). Thus, the release of water from the reservoir may not be as beneficial as one would hope in pollution control since the waters released are most likely to be from the bottom of the reservoir and these waters may be severely deoxygenated. Nor will the waters be particularly beneficial in this regard to aquatic life even when pollution control is not an explicit objective. In defense of the practice, we may note that cooler water is capable of dissolving more oxygen. This capability causes the rate of replenishment of oxygen from the atmosphere to increase.

Whether relatively deoxygenated waters are released depends on the season and the temperature history within the season. In summer when the reservoir is stratified and inflows are warm and join the surface layers, releases from the bottom may well bring water which has been in the reservoir for many weeks and has lost much of its dissolved oxygen. (Mechanical arrangements are possible to prevent this situation.) In fall when the surface is still relatively warm but inflows are cool, the inflow may sink to the

[8] Winter may bring on an inverted stratification because of the lower density of water near the freezing temperature.

bottom of the reservoir and most of it may be withdrawn in but a short time (say, several weeks). In this event little water quality change is induced because storage is minimal. Once the fall overturn has occurred, the mixing of inflows with the reservoir contents is much more likely, and discharges differ less from usual upstream conditions.

One may argue that the release of cold, oxygen-poor water is largely restricted to several warm months. Although this is the case, those warm months are the critical ones from the standpoint of water pollution control. During that time, stream flows may be at their lowest levels, causing the normal pollution loads to exert a much larger effect than at other times of the year. Thus, since the reservoir releases may be oxygen poor during those months, real enhancement of water quality by flow augmentation may not necessarily be achieved.

One barrier to achieving enhancement of stream quality is the presence of algae in the reservoir. In the absence of algae, the oxygen level of the lower layers of the reservoir would be much higher, for cold water is capable of dissolving more oxygen than warmer water. Algal growth is stimulated by the presence of nutrients such as phosphates and nitrates. The effects and control of these substances are discussed in **Eutrophication.**

REFERENCE

1. J. Symons, R. Weibel, and G. Robeck, *Influence of Impoundments on Water Quality*, Public Health Service Publication 999-WP-18, U.S. Dept. of Health, Education and Welfare, 1966.

S

Smog

See Air Pollution, contaminants—photochemical pollution.

Solid Wastes

Two or three times each week in most urban and suburban areas the trash is collected from the curbside. This is a prosaic, routine operation. It should give one pause, however, to consider that in 1967, urban wastes from residences reached an annual quantity of 128 million tons. The figure is calculated by multiplying a population of 200 million by a daily waste production of 3.5 pounds per person. Several sources feel the per capita waste production is even larger than that. The composition of residential wastes is indicated in Table S.1.

In general, the per capita domestic waste production has been rising in quantity, with an accompanying change in the composition of waste material. The moisture and ash content have fallen (more prepared foods are purchased, and coal-burning heating systems have virtually disappeared). As plastic and paper refuse has risen, the heat derived from burning a pound of refuse has also risen.

There are five methods of solid waste disposal: landfill, incineration, ocean dumping, composting, and garbage grinding with disposal via waste water treatment. A community may often use more than one of these methods in

TABLE S.1
AVERAGE COMPOSITION OF RESIDENTIAL SOLID WASTES BY PERCENT OF TOTAL WEIGHT IN 21 CITIES, 1966–1969

Source	Percentage
Food	18.2
Garden	7.9
Paper products	43.8
Metals	9.1
Glass	9.0
Plastics, rubber, leather	3.0
Textiles	2.7
Wood	2.5
Rock, dirt, ash	3.7

From J. DeMarco, D. Keller, J. Leekman, and J. Newton, *Incinerator Guidelines, 1969*, Public Health Service Publication No. 2012, U.S. Dept. of Health, Education and Welfare.

Open dumping. The application of an earth cover would forestall the development of fires and the swarming of flies.

disposing of its solid wastes. Incineration followed by landfill is an especially common sequence. Reclamation and recycling of useful material may enter the train of methods at various stages.

SANITARY LANDFILLING

Of all the various processes of landfilling, the technique referred to as sanitary landfilling is to be recommended most highly. The practice of sanitary landfilling dates only to 1916 when a process known as "controlled tipping" was begun in Great Britain. Although the process was introduced in New York City and Fresno, California, in the 1930's, widespread adoption by United States communities awaited refinement by the United States Army. After World War II, the use of sanitary landfilling spread from a scant 100 communities in 1945 to over 1,400 communities in 1960.

It is important at the outset to note what distinguishes a sanitary landfill operation from its less desirable relations—important because inferior processes may be referred to as sanitary landfilling to give the impression of advanced practices. Four operations may be compared: "open dumping," "controlled burning dumping," "refuse filling," and "sanitary landfilling." They represent a succession of advancement in practices. The open dump is merely a locality set aside for receipt of refuse. While burning is not planned, open dumps are frequently afire. An earth cover over the refuse is not part of the disposal scheme. Open dumps may be infested with insects and rodents which carry disease. Controlled burning dumping utilizes fire to reduce the volume of the refuse. Refuse filling implies compaction and cover of the refuse, although a daily cover of earth is not utilized.

In sanitary landfilling the earth cover is added more frequently—in fact, at least on a daily basis. The daily application of an earth cover, the crucial procedure in sanitary landfilling, has been found to enhance a number of aspects of the disposal operation. Besides checking the spread of flies, fire and consequent air pollution are prevented. The earth layer also compacts the refuse into a smaller volume and prevents the wind from spreading it. Sanitary landfill operations are classified into three types: the area method, the trench method, and the ramp method, the last process bearing resemblances to the first two. These are illustrated in Figure S.1

The area method is most suitable to sites which are natural depressions in the land surface, e.g., shallow valleys. The wastes are spread on the land, compacted, and covered by earth; finally, the soil itself is compacted. The trench method involves deposit and compaction of wastes in long trenches. The earth cover is furnished during the digging of a parallel trench for the next day's wastes. Obtaining the earth cover for the area method may require extensive hauling, but the trench method provides a ready source of the earth cover. In the ramp method, wastes are deposited on a slope and are compacted; earth from the base of the slope is bulldozed up and over the wastes.

It is recommended that wastes be put down in 5-foot layers and then compacted. Total depth of the compacted wastes probably should not exceed 8 feet if surface fissures are to be avoided. The daily cover of compacted sandy loam soil (preferred covering material) should be about 6 inches thick and the final cover, 2 feet thick; however, 3 feet of cover soil seems to be needed for growing trees. The field should be inclined a degree or so and trenched to achieve proper drainage of surface water. Careful initial selection of the landfill is needed to prevent groundwater contamination.

The land which has been filled may become suitable for use as a recreational area, although settling and gas production from the decomposing wastes may delay such use.

FIGURE S.1
SANITARY LANDFILL METHODS

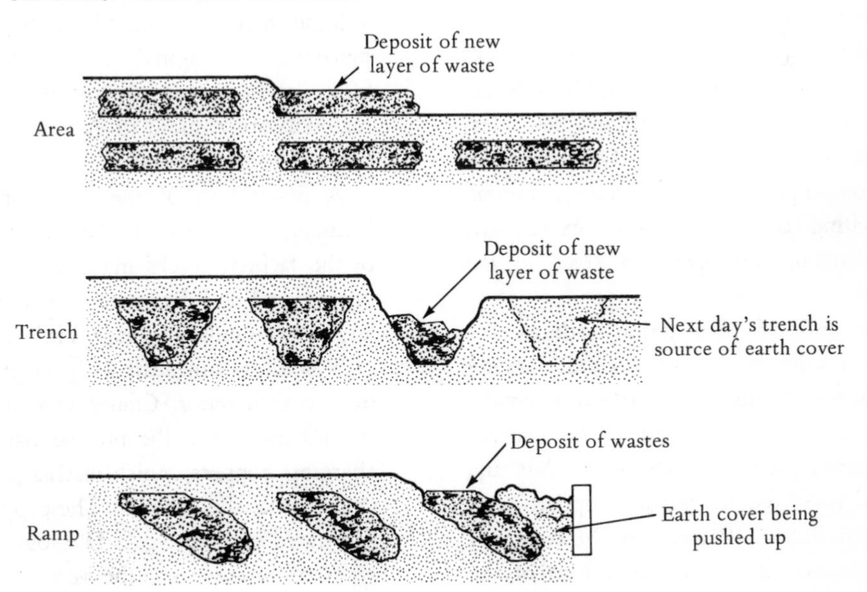

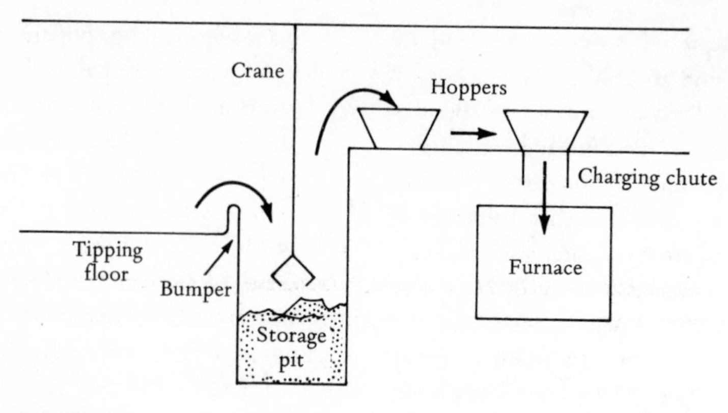

FIGURE S.2
OPERATIONS AT A LARGE INCINERATOR

INCINERATION

In situations in which the cost of transporting wastes to the nearest landfill site is excessive or where land for filling is in short supply, a community may choose to burn its wastes at a centrally located incinerator. The choice between incineration and landfilling may hinge on social and political considerations as well as economics, because the economically desirable sites for an incinerator may not be politically acceptable to the community. Incineration is a 24-hour operation which produces noise (trucks and operation) and smoke. Landfill sites are typically more remotely situated and hence less vulnerable to public clamor. The choice of incineration implies that the hauling, capital, and operating costs for landfill exceed the hauling, capital, and operating costs of the incinerator.

The process of incineration reduces the wastes to a solid residue and to gases—water vapor, **Carbon Monoxide, Carbon Dioxide, Sulfur Oxides,** etc. The solid residue consists of combustion products along with glass and metal, and these require further disposal, typically by landfilling. **Particulate Matter** may also be removed from the gas stream of the incinerator to prevent its entering the atmosphere as a pollutant; this material joins the solid residue. Glass bottles and metal cans may actually be desirable components of wastes which are to be incinerated; they provide greater porosity to the bed of burning wastes and enhance combustion. As much as 80 percent of the weight and 90 percent of the volume of the wastes are eliminated in incineration. With compaction of the residue, the final volume may be reduced to as little as 1.5 percent of the original volume. Of course, some bulky items may not be incinerated, and these materials are reduced in volume by only about half at the landfill site.

A description of the operation at a large incinerator is instructive. On the "tipping floor" of the facility, trucks maneuver to a foot-high bumper rail and discharge their contents into a large, rectangular "storage pit." The pit typically holds $1\frac{1}{2}$ times the daily weight consumed by the incinerator. Cranes mix and distribute the wastes within the pit and haul them into charging hoppers, which discharge them down chutes into the furnace. These operations are illustrated in Figure S.2. The four fundamental types of furnaces are the vertical circular furnace, the multicell rectangular furnace, the rectangular furnace, and the rotary kiln; none seems to be singled out as the best design. The devices are described and illustrated in *Incinerator Guidelines, 1969* [1]. The residue from the furnace consists of ash, clinkers, tin cans, rocks, and unburned organics. Operation may be by continuous feed or in batches.

Incinerators, however useful in solid wastes management, have been found to contribute significantly to urban **Air Pollution.** It has been estimated that the addition of air pollution devices to incineration systems could add 8 percent to the cost of the incinerator. On the positive side, the waste heat vented by the incinerator may prove to be usable for the generation of electric power.

OCEAN DUMPING

Ocean dumping of municipal refuse now constitutes but a minute fraction of ocean disposal activities. However, as population in the United States coastal zone grows, pressures to utilize the ocean as a waste sink will increase. Although municipal refuse is not widely disposed of via this route, it is important to evaluate current practices regarding other wastes in order to project the effects of future activities. In addition, we will examine regulatory facets of the problem.

In the late 1960's, about 60 million tons of wastes were being discharged into the ocean at over 200 disposal sites. *Dredge spoils,* obtained by the U.S Corps of Engineers in their maintenance of navigation channels, constituted over 85 percent of that total. Industrial wastes and sewage solids (see **Water Treatment and Water Pollution Control**) made up the bulk of the remainder at about 4.5 million tons each.

Although the term *dredge spoils* sounds innocuous, in fact, many harbor sediments are polluted in that they possess a high oxygen demand (see **Organic Water Pollution**). In general, the dredge spoils are dumped within three or four miles of the coastline; spoils from New York City harbors, however, may be transported as much as 30 miles. The turbidity caused by the spoils reduces photosynthetic activity of the marine organisms in the vicinity of the disposal site. By layering the bottom, the sediments may also destroy spawning grounds in localized areas.

Sewage solids are barged to the ocean from New York City and Philadelphia; the city of Baltimore has also applied for a permit. New York's fleet of five barges average a 30-mile round trip from the various sewage treatment plants to the disposal site. Philadelphia's barge travels 227 miles to a point 10 miles off Cape May, New Jersey. Besides contributing turbidity and oxygen demand, sewage solids may carry pathogenic bacteria. Fecal coliforms indicate the potential presence of the pathogens (see **Bacterial and Viral Pollution of Water**), and these indicator bacteria have been found in surf clams in the waters off New York City, apparently as a consequence of the dumping of the sewage solids. A study by the Corps of Engineers (which formerly had general responsibility for ocean disposal) suggested that clams not be taken in a 6-mile-radius circle around the disposal site, and the FDA has prohibited harvesting in the affected area.

Although sewage solids are not barged to the ocean in California, Los Angeles pumps its sludge out to the ocean through submarine outfalls. Sludge from the White Point outfall (65 to 195-foot depths) has destroyed the usual marine life along about 6 miles of coast. On the other hand, wastes from the Hyperion outfall, which discharges in about 320 feet of water, have seemingly had little effect on marine life abundance.

More than a quarter-century ago, the disposal of *municipal refuse* at sea was terminated because

The barging of New York City's garbage for disposal at sea off the coast of New Jersey. This practice was halted 25 years ago; today, sewage solids from New York are dumped into the ocean.

of public reaction to the floating litter which resulted, and no significant quantities have been dumped since that time. Nevertheless, interest in sinking refuse at sea continues. Of course, it would be unthinkable to discharge refuse which was not baled in order to make it sink. Nevertheless, if the baling should eventually disintegrate, undegraded, floatable plastics would rise to the surface to plague us. The disposal of incinerator wastes at sea has recently been investigated; incineration would take place on board the ships which dump the waste. A study by an engineering firm conducted for Westchester County, New York, found ocean disposal economically attractive over landfilling and incineration but did not recommend the practice because of its unknown environmental effects.

Discarded rubber tires and auto hulks are among the bulkier items in municipal refuse, and these have been used in the construction of artificial reefs. Fish flock to such topographic features on the otherwise relatively flat continental shelf. The congregation of fish, in turn, attracts the ever watchful fishermen, the pleasure of the latter being the removal of the former. At a placement cost of $70 to 100 per hulk, however, the practice may not be as economical as putting the vehicle back in the scrap cycle. Moreover, hulks may rust out in three to five years, making the cost of maintaining the reef quite high.

Regulatory aspects of ocean disposal are worthy of mention. The conveyors of solid wastes to the ocean were, until 1972, operating under the authorization of the U.S. Corps of Engineers. A letter to the Corps requesting permission to dump wastes was evaluated by that organization and circulated to other parties for comment. A letter of "no objection" gave effective permission to the conveyors.

The Corps's responsibility for ocean disposal stemmed historically from its responsibility to maintain the navigability of the nation's waterways. Navigation had been its sole concern in evaluating requests, and the authority to evalute on any other basis was in doubt until 1970. In that year, a court decision upheld the Corps's authority to refuse a permit on the basis of environmental considerations.

Nevertheless, with the passage of the Marine Protection, Research and Sanctuaries Act of 1972 [15], the role of the Army Corps of Engineers in issuing permits was drastically altered. While the Secretary of the Army is still empowered to issue permits for ocean dumping of dredged materials, the authority to regulate the dumping of all other wastes was transferred to the Administrator of the Environmental Protection Agency (EPA). In addition, permits issued by the Secretary of the Army are open to review by the EPA.

The law prohibits the ocean disposal of radiological, chemical, or biological warfare agents as well as high-level radioactive wastes. Except for dredged material, the EPA will issue permits for ocean disposal of materials after its determination that "such dumping will not unreasonably degrade or endanger human health, welfare, or amenities, or the marine environment, ecological systems, or economic potentialities." The EPA may also designate recommended sites for disposal.

Weeks after the United States legislation was signed into law, 91 nations agreed on an international convention to regulate ocean dumping [16]. Although enforcement and sanction were left to the agreeing nations, the convention forbids dumping of such materials as high-level radioactive wastes, biological and chemical warfare agents, crude and refined oils, mercury and cadmium compounds, certain slowly degradable pesticides, and durable plastics. Special permits are required for other potentially harmful substances such as arsenic or lead. All other materials are to be controlled by issuance of general permits.

COMPOSTING

Composting has been promoted and criticized with such intensity that it is difficult to set the process in perspective. It is neither a panacea nor a flop, but an alternative scheme for solid waste management. As with other alternatives, consideration should involve a careful weighing of economic and social factors.

The compost heap has been utilized for centuries by farmers to reclaim organic matter. From this practice, a set of mechanized biological processes known as composting has evolved to reduce organic municipal wastes to stable substances. From its introduction in the 1920's and 1930's to the present it has been the object of derision and hope.

There are two types of composting, outdoor and indoor. In the former, piles of refuse are laid in long lines called windrows; the piles may be 10 feet wide and 5 to 6 feet deep. Occasional turning of the piles promotes microbial activity and gas exchange. Two to three weeks of processing produces a stable humic material. The availability of nearby suitable land for the windrow method is not likely in the United States. In the indoor or factory method, the refuse may be spread in drums, which rotate at slow speeds (1 rpm or less) and agitate the wastes. As microbial action proceeds, the temperature of the refuse quickly rises to 140°–150°F where further biodegradation takes place. In the 7- to 10-day detention required to "cure" the wastes, most disease vectors are eliminated by the elevated temperature. No special inoculum appears to be needed to start the process. It should be noted that not all domestic wastes are capable of being composted. Metal and glass, in particular, may be removed prior to processing.

The product of composting is a brown to black humic material which finds use, especially in Europe, as a soil conditioner. As a soil conditioner, compost fosters the absorptive properties of soil for water. The substance has less than 1 percent nitrogen and less than 0.2 percent phosphate. Thus it is not a particularly good competitor with commercial fertilizer, and it finds little, if any, use in basic agriculture. There are those who argue that its use could be beneficial in that compost might restore trace elements to the soil. The argument falls on deaf ears, for compost is costly to apply and costly to obtain in comparison to fertilizers. The farmer in basic agriculture shows no preference for it.

Compost has found use in Europe, however, in luxury agriculture. Holland disposes of a sixth of its municipal wastes by composting. Much of the product goes into intensive gardening, such as bulb growing, and into city parks. Germany and Switzerland also engage in composting although to nowhere near the same extent as Holland. A use in Germany is in preventing erosion of steep hillsides in the German wine country. Still, less than 1 percent of the municipal wastes of Germany are composted. There is some luxury gardening in the United States; nursery gardening and park gardening are examples, but given the mass of wastes produced annually (an average of 3.5 pounds per person per day) it is unlikely that a sufficient market exists for the profitable sale of municipal compost. Profitable sale is made more unlikely by the costs of transporting the compost. Thus, from an economic standpoint, it would not be wise for a city to plan to compost its wastes without first carefully investigating market opportunities. A half-dozen United States cities began composting in the 1960's, only to cease their operations within a short time, principally for lack of a market.

This is not to say that composting is not an alternative worthy of consideration. Simply put, it is unwarranted to *assume* a market for compost. Instead, the choice of composting would need to be founded on the economics of the process within the total waste management system. It may be possible for a centrally located mechanized composting process to be competitive with a centrally located incinerator, if air pollution controls make incineration exceedingly costly. Furthermore, the landfill site to which compost has been applied has more potential future uses than does the land filled with untreated wastes. Nevertheless, the likelihood that composting will achieve the status of a major method of waste disposal seems remote.

GARBAGE GRINDING

The practice of grinding wastes in the home and discharging them to the sanitary sewer has grown in the last decade. Of course, this activity does not eliminate the need for collection and disposal of other household wastes, but it can change the composition of such wastes. Additionally, the notion of hauling wastes to a central grinding station has been raised. In this operation, as well, the wastes would enter the sanitary sewer. Both methods place a burden of added solids disposal on the sewage treatment plant (see **Water Treatment and Water Pollution Control**).

The cost of disposing of sewage solids is $1 to $3 per ton of sludge, but since sludge is only 3 to 5 percent solids, this works out to about $40 to $60 per ton of dry solids. Calculating the cost of disposing of a ton of refuse in this way is complicated by the percentage of moisture of the substance which is ground up and by the fraction of material solubilized by the grinding operation.

REFERENCES

1. J. DeMarco, D. Keller, J. Leekman, and J. Newton, *Incinerator Guidelines, 1969*, Public Health Service Publication No. 2012, U.S. Dept. of Health, Education and Welfare.
2. R. Black, "A Review of Sanitary Landfilling Practices in the United States," reprint from *Proceedings*, Third International Congress, International Research Group on Refuse Disposal, Trento, Italy, 1965.
3. T. Sorg and H. Hickman, *Sanitary Landfill Facts*, Public Health Service Publication No. 1792 CSW-4ts, U.S. Dept. of Health, Education and Welfare, 1965.
4. D. Brunner and D. Keller, *Sanitary Landfill Design and Operation*, SW-65ts, U.S. Environmental Protection Agency, 1971.
5. *Solid Waste Handling in Metropolitan Areas*, U.S. Public Health Service, 1964; reprinted in 1966 and 1968.
6. S. Messman, *An Analysis of Institutional Solid Wastes*, SW-2tg, U.S. Environmental Protection Agency, 1971.
7. "Solid Waste Management," President's Office of Science and Technology, 1969.
8. P. McGauhey, *American Composting Concepts*, SW-2r, U.S. Environmental Protection Agency, 1971.
9. J. Wiley and O. Kochtitzky, "Composting Developments in the U.S.," *Compost Science*, Summer 1965.
10. L. Rich, *Unit Processes of Sanitary Engineering*, Wiley, New York, 1963.
11. S. Hart, *Solid Waste Management/Composting: European Activity and American Potential*, Public Health Service Publication No. 1826 (SW-tc), U.S. Dept. of Health, Education and Welfare, 1968.
12. M. Jensen, *Observations of Continental European Solid Waste Management Practices*, Public Health Service Publication No. 1880, U.S. Dept. of Health, Education and Welfare, 1969.
13. "Ocean Dumping—A National Policy," President's Council on Environmental Quality, 1970.
14. D. Smith and R. Brown, *Ocean Disposal of Barge Delivered Liquid and Solid Wastes from U.S. Coastal Cities*, No. SW-19c, U.S. Environmental Protection Agency, 1971.
15. Public Law 92-532, 92nd Congress, H.R. 9727, October 23, 1972.
16. *New York Times*, November 4, 1972, p. 1.

Strip Mining of Coal

See also Fuel Resources and Energy Conservation, coal.

The preservation and enhancement of the environment is often an issue of scientific substance. It is also, however, in many instances an issue of conscience and of something akin to morality—morality between man and the land that nurtures him. No other practice, not even reservoir and power plant construction, rivals strip mining as an offense against the land. There are scientific aspects to the issue, but they are dwarfed by the monumental destruction of the land.

The discussion of **Electric Power** noted the projected demands for electrical energy in the year 2000 and indicated that the coal requirements for steam power plants would more than *treble* by the end of the century. The growth of strip mining has been fostered by the combination of this growing demand, by new equipment, and by laws to improve coal mine safety. It is not hard to see why an insistence on coal mine safety favors strip versus deep mining. The risk of cave-ins and explosions is negligible, and movement to strip mining means cost savings in the safety area to the operator.

A surface mining operation in Pennsylvania menacing rural homes.

Equipment includes a giant auger drill which bores into hillsides; the drill reaches 7 feet in diameter. A power shovel hauls 200 tons in a scoop and has a cab which is more than four stories in the air. The small operator, however, can get by with a bulldozer and truck. Because of the relative ease of stripping, the productivity of the surface miner may be double that of his fellow worker in a deep mine.

Stripping has increased to the point that in 1970 it accounted for 44 percent of the annual United States bituminous coal production of about 600 million tons, according to the Bureau of Mines. A quantity of land approaching 1.8 million acres has already been stripped, the equivalent of an area at least 50 miles by 50 miles. The Department of the Interior estimated in 1965 that, of the 1.3 million acres stripped so far, only about 18 percent of the land had been reclaimed by man's endeavors; another 16 percent of the land had been reclaimed by nature. Thus, at the time, nearly two-thirds of the lands which had been strip-mined were in need of reclamation. The land remaining to be strip-mined might total as much as 71,000 square miles. It will not be stripped next year, but it is an indication of the potential magnitude of the problem.

Strip mining is carried on in 26 of the states. From the record of reclamation, it is clear that many of the laws governing stripping in those states were toothless, or else no laws even existed. Nevertheless, in the early 1970's West Virginia, where coal has been king for a century, banned stripping in the 22 of her 55 counties which still remain unscarred by the power shovel and the auger.

Strip mining is characterized by a removal of vegetation and the overburden of soil and rock in order to "mine" the underlying coal. The overburden is laid up in "spoil banks," and the removal of the coal leaves vast depressions in

The orderly devastation produced by strip mining for coal. This land, near Nucla, Colorado, may lie barren for years. Ferrous sulfide in the rubble from coal operations makes soil and surface water in a strip-mined area highly acidic. Erosion from such land is extreme. In the East, strip-mined land has been successfully reclaimed. There is concern, however, that stripped land in arid regions such as the Southwest may resist efforts to re-establish vegetation.

the earth. Alternatively, auger drilling may gouge horizontal tunnels into the earth from which the coal comes out like shavings. Reclamation ideally would consist in replacement of topsoil and the planting of trees and other vegetation. Such efforts are not commonly made, primarily because of their cost. The price of stripped coal, then, does not reflect the cost of restoring the land.

The "moonscape" left behind by the strippers may lack vegetation of any sort. In addition, the growth of new vegetation on land without topsoil may be severely hindered. Present in the wasted coal remaining near the surface of the land is iron pyrite, one of the minerals known as "fool's gold." Iron pyrite is ferrous sulfide, a compound of iron and sulfur, and its oxidation in the presence of water produces sulfuric acid. Thus, in an area which has been strip-mined and left unreclaimed the water is likely to be highly acidic and unsuitable to the needs of plants. The barrenness produced by the stripping will probably persist for many years. It is difficult to say how long.

The discussion here touches on a related water pollution problem associated with both surface and deep mining. The water which comes in contact with the unreclaimed land and becomes acidic may drain away from the area, entering streams and rivers and the underground water. The same is true of water which collects in abandoned or working coal mines and drains away. The acid water may be yellow-brown in color because of the precipitation of ferric hydroxide from solution. The polluted waters are known as **Acid Mine Drainage** and may be undrinkable for man and uninhabitable for aquatic life. This is not a chemist's concept; it is an observed reality.

Because vegetation is missing from unreclaimed land, the loose rubble may be eroded away in storms and carried as sediment into the streams. One estimate places the erosion from strip-mined land at up to 1,000 times greater than the erosion from comparable forested land. If the stream bottoms fill with sediment, flood danger may be increased because of the stream's diminished ability to carry storm runoff. Further, the mounds of earth, if laid on hillsides, may be unstable without a tree covering. Landslides are possible under such conditions, placing land, property, and people in physical danger.

There are additional physical effects of surface mining. When a portion of a hillside is mined, a "highwall" may be left which physically isolates the hilltop from surrounding land and limits the access of both humans and animals. Strip mining has produced about 20,000 miles of highwalls in the Appalachian region.

In 1972, legislation was introduced in Congress to deal with strip mining, but no law was enacted. One bill asked states to set regulations and gave them two years to do so. The interval would leave strip miners free to continue their operations without reclamation. No funds for restoration would be provided by the bill. Another bill would have *banned* all strip mining within six months of enactment of the legislation and offered 90 percent matching grants for reclamation. Laws which simply require reclamation by the coal companies were also introduced. In 1973, pressure for legislation continued.

Powerful interests are involved. Not only coal companies but also, in many cases, large oil companies are affected. This is not such an anomaly as it seems, for in recent years many oil companies have purchased coal companies as another aspect of their energy-supplier function. Naturally the coal companies would prefer to go unregulated. They raise the economic issue of jobs for the 25,000 coal "miners" currently employed in stripping operations. If a law merely required reclamation, it is doubtful that massive unemployment would result. On the other hand, even if strip mining were banned, its loss would force the demand for coal to be filled with the output from deep mines. Since deep mining is only about half as productive per man-day as surface mining, employment might actually increase. Demand could diminish on account of the higher price of deep-mined coal, but serious long-term unemployment seems unlikely.

The issue is clear. Strip mining is an abuse, and state laws have failed to bring about sufficient reclamation. Unfortunately, the nation has become quite dependent on coal from strip mining. A rapid cessation of the practice could create problems in the generation of **Electric Power** since coal-fired power plants account for about 50 percent of our power generation.

However, lack of intelligent planning in the past should not be used as an excuse for taking no action in the present. The first step should be stringent *federal* laws which require complete reclamation of strip-mined lands and which prohibit stripping on steep grades where reclamation efforts may be ineffective. The requirement of reclamation ought to have two effects. The first is simply the restoration itself, but the second may be a slowing and even perhaps a contraction of the strip mining industry as energy companies exploit other energy sources. At the present time, since the cost of stripped coal does not include the cost of reclamation,

ımation will
stripped coal,
may thus be
... stops short of reclamation requirements is inadequate. Probably, however, the best of reclamation cannot adequately replace the natural growth removed from the land.

Recognizing the immediate needs for strip-mined coal should cause us to shy away from an immediate ban. It is possible to have stripping diminish on a gradual basis, however, with a complete cessation in, say, a period of five years. During the phaseout, coal companies would shift their production to deep mines, so that at the end of the period demands could still be fully met. The phaseout would be accomplished by levying taxes on strip-mined coal, to be paid by the coal companies; the taxes would increase year by year. The succession of increasing taxes would make strip-mined coal less and less attractive in the marketplace as the companies passed the tax burden along in the price.

REFERENCES

1. *Surface Mining and Our Environment*, U.S. Dept. of the Interior, 1967.
2. Paul Averitt, "Stripping-Coal Resources of the United States—January 1, 1970," *U.S. Geological Survey Bulletin* 1322 (1970).
3. "Surface Mining," *Hearings Before the Subcommittee on Minerals, Materials, and Fuels of the Committee on Interior and Insular Affairs*, Parts 1 and 2 (November–December 1971), Part 3 (February 1972), Serial no. 92-13, U.S. Senate, 92nd Congress, 1st Session.

Sulfur Oxides

Civilization's need for energy is at the root of sulfur oxide air pollution. Sulfur oxides in the atmosphere are derived principally from the combustion of the fossil fuels, coal and petroleum, which are used to generate electricity, to heat buildings, and to power machines.

THE FAMILY OF COMPOUNDS

Sulfur dioxide, sulfur trioxide, sulfurous acid, sulfuric acid, sulfite salts, and sulfate salts comprise a family of compounds which exist as gaseous and particulate pollutants in the air. During the combustion of coal and the products of petroleum, sulfur compounds in the fuels are oxidized to sulfur dioxide and to sulfur trioxide. The portion of the sulfur oxides produced in the trioxide form is variable, ranging from 1 to 3 percent.

The sulfur dioxide produced in the combustion process is oxidized gradually in the air to the trioxide form; sulfur trioxide immediately combines with water vapor to form an extremely corrosive sulfuric acid mist. Although the mechanism of the oxidation of sulfur dioxide is still unclear, sunlight and high relative humidities may play a role. Because of its extreme solubility, sulfur trioxide may exist only in trace quantities in the air. Besides reacting with water to form sulfuric acid, sulfur trioxide may react with the oxides of calcium or other metals to produce sulfates. The sulfuric acid droplets and the sulfate compounds can make up between 5 and 20 percent of the **Particulate Matter** in urban air. Particulates contribute to soiling and to a reduction in visibility; they accelerate corrosion and are strongly implicated in deleterious health effects.

SOURCES

The burning of coal, oil, and gasoline for power and heat accounted for about 80 percent of the 28 million tons of sulfur dioxide emitted in the United States in 1966. Coal combustion for power generation alone contributed approximately half of this amount. The automobile appears relatively innocent with regard to sulfur oxides; automobiles generated only about 4 percent of the total quantity of emissions. Natural gas, also, does not contribute significantly to sulfur oxide emissions. The remaining 20 percent of the United States total is generated by industrial processes—smelting, petroleum refining, and sulfuric acid manufacture, to name a few.

The U.S. Public Health Service has monitored the atmospheric concentrations of sulfur dioxide in a number of major cities. Urban areas of the northeastern United States exhibited particularly high concentrations, probably due to the use of the high-sulfur-content coal which is available to that region. New York City and Chicago headed the list of cities with high sulfur dioxide concentrations.

Sulfur dioxide is present naturally in the atmosphere at very low concentrations. It comes from the air oxidation of hydrogen sulfide, a gas which arises from bacterial action on organic matter in the absence of oxygen. One estimate places the background concentration of sulfur dioxide at about 0.5 microgram per cubic meter.

Average annual concentrations for cities may range from about 27 micrograms per cubic meter (Denver, 1966) to about 470 micrograms per cubic meter (Chicago, 1964) (see **Air Quality Standards**). There are other ways than average annual concentrations to report the concentration of air pollutants. One way is to determine the maximum [illegible] over the 24 hours of a given [illegible] the number of days in which the maximu[illegible] the hourly averages exceeded a certain level. [illegible] Chicago, the maximum of the hourly averages exceeded 270 micrograms per cubic meter for nearly 45 percent of the days in the interval from 1962 to 1967. Philadelphia had concentrations in excess of that level on nearly 30 percent of the days in the same period.

EFFECTS

ON MATERIALS

It was noted earlier that the conversion of atmospheric sulfur dioxide to the trioxide and the subsequent formation of sulfuric acid mist were hastened by higher relative humidities. The corrosive atmosphere generated under such conditions affects a wide variety of materials. The surfaces of steel, zinc, copper, aluminum, and other metals may be corroded to the corresponding metallic sulfates. Building materials composed of carbonate compounds (limestone, dolomite, marble, and mortar) are especially vulnerable to attack by the acid mist. Fabrics such as cotton, rayon, nylon, and leather are also susceptible. It is, as one might imagine, difficult to assess the economic impact of sulfur dioxide damage to materials, and few firm data are available.

ON PLANTS

High concentrations of sulfur dioxide (perhaps 1 part per million[1] or more) over periods as short as an hour cause *acute injury* to plants. Thus, even though a region experiences a low *average* annual concentration, peak concentrations caused by meteorological conditions may

[1] 1 part per million by volume = 2,860 micrograms per cubic meter at O°C and 760 millimeters of mercury pressure.

still cause plants to be damaged. Acute injury appears to be caused by rapid absorption of sulfur dioxide. The injured plant tissue is characterized at first by a dry, bleached appearance, then later possibly by a reddish-brown discoloration. In the vicinity of smelters one may observe injured pine trees with reddish-brown needles and less than full foliage.

Chronic injury may occur where concentrations never exceed 0.1 part per million, with annual average concentrations as low as .03 part per million. In chronic injury the green color of leaves may fade gradually to yellow, and high concentrations of sulfate may be noted in the plant tissue. It is thought that the sulfate results from oxidation in the leaf of the absorbed sulfur dioxide to the trioxide form, followed by combination of the trioxide with water to give sulfuric acid. Apple, pear, ponderosa pine, larch, and mountain ash trees are especially susceptible to chronic injury; alfalfa, barley, and cotton are susceptible nonwoody plants.

A synergism between sulfur dioxide and ozone has been noted. Combinations of the two components have caused injury which neither pollutant in the same concentration alone was able to produce. A similar synergistic effect has been observed between sulfur dioxide and nitrogen dioxide. (See **Photochemical Air Pollution** for more information on ozone and nitrogen dioxide.)

ON MAN

Air pollution incidents occurred in New York City in 1953, 1962, 1963, and 1966. During the 1953 episode, respiratory and cardiac illnesses were increased (measured by visits to emergency clinics). The episode of 1966 also saw an increased number of clinic visits, this time for bronchitis and asthma. During and after these incidents, deaths in excess of the number expected for the time of year were noted. Levels of sulfur dioxide and concentrations of **Particulate Matter** were elevated during all the incidents. The sulfur dioxide level rose above 2,500 micrograms per cubic meter (average concentration for an entire day) during the episode in 1962.

A study of mortality in Japan saw an increase in deaths during an episode in Osaka when the particulate level reached 1,000 micrograms per cubic meter and the sulfur dioxide level exceeded 285 micrograms per cubic meter (average concentrations for an entire day). One investigator of air pollution episodes in Rotterdam felt that the relation between excess deaths and mean daily concentrations of sulfur dioxide began at a concentration of about 500 micrograms per cubic meter. On the other hand, an analysis of episodes in London suggested that the association can be discerned at sulfur dioxide levels above 715 micrograms per cubic meter and particulate levels greater than 750 micrograms per cubic meter.

A number of investigations have found relations between certain types of illnesses and the levels of sulfur dioxide and particles. A study in New York City associated sulfur dioxide with eye irritation; the same researchers also discerned an increase in illness not simply with sulfur dioxide levels but with the *increase* in those levels. A study of bronchitis deaths in rural boroughs in England found a meaningful association with sulfur dioxide levels. The number of bronchitis illnesses was shown to be related to sulfur dioxide levels in seven sections of Genoa, Italy. One area of the city with an average annual sulfur dioxide concentration of 80 micrograms per cubic meter was observed to have discernibly less respiratory illnesses per unit population than an area with a sulfur dioxide level of 105 micrograms per cubic meter. Even in summer such a relation could be noted. Several other studies in England have noted that lung cancer deaths in adjacent communities are related to both particle and sulfur dioxide concentrations. In various sections in the city of Buffalo, New York, researchers were able to link deaths from respiratory disease to particulate sulfate levels, but socioeconomic factors made the result debatable. A Japanese study noted a diminishing frequency of respiratory illness from more industrialized to less industrialized areas.

In summary, numerous studies of illnesses and absences have firmly linked respiratory illness (such as bronchitis) with sulfur dioxide and particulate concentrations; some have attempted to separate effects of the two pollutants; others have not. Some have simply related illness to industrialization. Especially strong conclusions on the link between sulfur dioxide and ill health have been drawn from investigations of illnesses during and following air pollution episodes.

CONTROL OF EMISSIONS

There is no single approach to control of sulfur oxide emissions. In combustion processes, three basic alternatives exist. The first is the substitution of fuels of lower sulfur content for those currently being used. The second is cleansing the stack gases of sulfur oxides by chemical processes. The third is specific for power generation and involves switching to other energy sources such as hydroelectric power plants and **Nuclear Power Plants**. All three of these alternatives are being utilized by the electric power industry, although the movement to nuclear power appears related more to the cost of generation than to the need for emission controls. It is true also that simply increasing combustion efficiencies would decrease sulfur oxide quantities since less fuel would be utilized.

The first method, fuel substitution, depends on the availabilities and prices of low-sulfur coal and low-sulfur oil. There are claims that low-sulfur coal is concentrated in surface coal deposits and thus that **Strip Mining** is necessary. There are claims to the contrary as well. The authors have as yet seen no definitive evidence. It is also possible to cleanse the coal of some of its sulfur compounds by crushing followed by a flotation process; the operation is called coal desulfurization. In addition, several pilot studies are being conducted to determine the feasibility of manufacturing from coal a gas of the quality of natural gas; the process is called coal gasification and could potentially extend the supply of clean fuel. See **Fuel Resources and Energy Conservation, coal.**

Of the crude oils produced in the United States in 1966, about 80 percent had sulfur contents of 1 percent or less. In the process of refining crude oil, the highest sulfur concentrations are shifted into the denser grades, known as residual oils. With sulfur contents between 0.5 and 5 percent, these oils find principal use in industrial and commercial activities. It is residual fuel oil which is used in oil-fired power plants. The lighter distillate fuel oils have concentrations between 0.04 and 0.35 percent sulfur and are used primarily in home and apartment heating. Refineries may also use chemical processes which reduce the sulfur content of fuel oils. The product of those processes may be blended with the heavy oil to lower the sulfur content of the mixture. Natural gas yields very little sulfur dioxide on combustion, but reported shortages of the fuel are expected to result in higher prices.

Although chemical removal of sulfur oxides from stack gases was cited as the second alternative, many of the removal processes are still in the research stages. Two, the alkalized alumina process and the catalytic oxidation process, have advanced to the development stage, and one, the limestone injection process, is already being installed in several plants. These processes for sulfur dioxide removal are thought of as being applicable to controlling the extensive emissions of coal-fired power plants.

LIMESTONE INJECTION PROCESS

There are two versions of the limestone injection process, a dry process and a wet process. In both, limestone or dolomite is injected directly into the furnace, where calcium and magnesium oxides are produced by the intense heat. In the dry process, these oxides react with sulfur dioxide and oxygen to yield a dust of calcium and magnesium sulfates. The electrostatic precipitator removes the sulfate dusts, unreacted material, and the other **Particulate Matter.**

The wet process begins in similar fashion with conversion of limestone or dolomite to oxides in the furnace heat, but then, to obtain additional sulfur dioxide removal, the calcium and magnesium oxides are slurried in water. The calcium and magnesium hydroxides which are formed in the water react further with sulfur dioxides to give sulfite and sulfate salts. These, along with other particulates, are removed by allowing them to settle from the water. The flue gas, though diminished in sulfur dioxide, now requires preheating prior to entering the stack in order to provide sufficient buoyancy to the plume of smoke. A lack of buoyancy would lead to poor dispersion in the atmosphere and settling of the plume near ground level.

ALKALIZED ALUMINA PROCESS

The alkalized alumina process adds a new dimension to the removal of sulfur dioxide from flue gases. Whereas the limestone injection process creates a new waste product (calcium sulfate) to be disposed of, the alkalized alumina process recovers elemental sulfur as a product which may be sold.

Sulfur dioxide is removed from the hot flue gas by reacting with activated sodium aluminate pellets. The product of the reaction is sodium sulfate, which adheres to the surface of the pellet. Dust removal precedes discharge of the gas to the stack, and the gas remains hot enough to achieve a buoyant smoke plume without preheating. Producer gas (carbon monoxide and hydrogen) is used to remove sulfur from the pellets as hydrogen sulfide as well as to regenerate the aluminate. The hydrogen sulfide is converted to elemental sulfur by the Claus process. The alkalized alumina process may be quite costly because of losses of the aluminate which must be made up.

CATALYTIC OXIDATION PROCESS

The catalytic oxidation process also produces a potentially saleable by-product—in this case sulfuric acid at about 75 percent strength. The process begins with the removal of particulate matter from the flue gas; the gas is then passed through a heated bed of vanadium pentoxide catalyst. Particulate removal aids in preventing poisoning or deactivating of the catalyst. Sulfur dioxide is oxidized to sulfur trioxide in the presence of the catalyst, and the trioxide immediately combines with water vapor in the gas to yield sulfuric acid, which is condensed and collected. As may be imagined, corrosion-resistant equipment is required wherever the acid is in contact. As in the alkalized alumina process, the flue gas remains sufficiently hot to provide a buoyant smoke plume.

These processes are relatively untested in full-scale operations, and it is not clear whether any one of them will be widely adopted. The limestone injection process appears at this stage to be the leading candidate primarily because of its low equipment cost and its suitability to both new and existing power plants. There are, however, over 50 processes in lesser states of development than the ones mentioned here.

REFERENCES

The discussion here is summarized from two recent and authoritative books from the national Air Pollution Control Administration, now the Office of Air Programs of the U.S. Environmental Protection Agency:

1. *Air Quality Criteria for Sulfur Oxides*, AP-50, 1969.
2. *Control Techniques for Sulfur Oxide Air Pollutants*, AP-52, 1969.

A more recent discussion of processes for control of sulfur oxides emission from Power Plants is:

3. A. V. Slack, "Removing SO_2 from Stack Gases," *Environmental Science and Technology* 7, no. 2 (February 1973): p. 110.

The EPA has compiled a summary of the installations of sulfur oxide control processes, which appears in:

4. *Federal Register* 37, no. 55 (March 21, 1972): p. 5768.

T

Thermal Pollution

See also Electric Power; Fossil Fuel Power Plants; Nuclear Power Plants.

Although there are a number of sources of waste heat, the dominant portion is contributed by the **Electric Power** industry. Over 80 percent of the water used for cooling by industry stems from electric power plants. For this reason the discussion is focused on steam electric power plants.

Three principal technologies are at present being utilized for generating the electric power our civilization requires: hydropower plants (see **Reservoirs**), **Fossil Fuel Power Plants,** and **Nuclear Power Plants.** The lack of suitable and acceptable sites for hydroelectric power generation pushed utilities into building steam power plants. Whereas the hydroelectric plant uses water to turn its turbine in order to generate power, the steam plant uses steam under pressure and at an elevated temperature to turn its turbine. Here begins the problem of waste heat, for not all of the heat required to raise the temperature of the steam is convertible to electrical energy. In fact, at present the maximum conversion efficiency of steam power plants is about 40 percent, this being the "thermal efficiency" of a fossil fuel power plant with superheating of the steam. The remainder of the heat is wasted either to the atmosphere or to the water or to both.

STEAM POWER PLANTS

Steam is generated in a boiler and sent to turn the turbine (see Figure T.1). The steam exiting from the turbine is condensed to water in a heat exchanger. The heat exchanger, or condenser, consists of parallel pipes inside a large closed cylinder. Cool water from a lake, river, or ocean flows through the pipes; the steam from the turbine occupies the space around the pipes. The cooling water leaves the condenser at a temperature higher than it was when it entered; the steam condenses to liquid water and is pumped back to the boiler to be vaporized again. The flow is continuous. The same water is continually cycled from the boiler to the turbine and condenser and back because water of high purity and exceedingly low hardness is needed to prevent the boiler tubes from choking with deposits. So that great quantities of water need not be continually purified, the boiler water is recycled. The water in the boiler-turbine loop of the nuclear plant, however, is also maintained within the plant because quantities of radioactive material may be dissolved and suspended in it, and they must be prevented from being discharged. The concept of the primary loop with steam that turns the turbine and the secondary loop which condenses the steam is shown in Figure T.1.

The main point is that cool water is warmed during the condensation of the steam. The quantity of heat discharged in the cooling water depends on the electric power rating of the plant (the number of megawatts it can produce) and the plant's thermal efficiency. *Thermal efficiency* refers to the fraction of heat energy which is converted to electric energy (per unit time). Heat energy is released in the combustion of fuel to boil water and in the fission process, which also boils water. A plant that is 30 percent efficient wastes to the environment 70 percent of the heat energy produced by the fuel and converts 30 percent of that heat energy to electrical energy. By the phrase *waste to the environment* is meant the discharge of heat either as water at a warmer temperature or as stack gases at high temperatures or both. The fossil fuel power plant wastes heat in both directions; that is, it produces both heated stack gases and heated water. The nuclear power plant wastes heat essentially only to the cooling water.

Nuclear plants now in operation have efficiencies in the range of 30 to 32 percent. The most efficient fossil fuel plants are those which superheat steam, and they may have an efficiency of about 40 percent. About 15 to 25 percent of the fossil fuel plant's waste heat goes up the stack as hot gases. Comparing the nuclear plant at 32 percent efficiency and the

FIGURE T.1
OPERATIONS AT A STEAM POWER PLANT

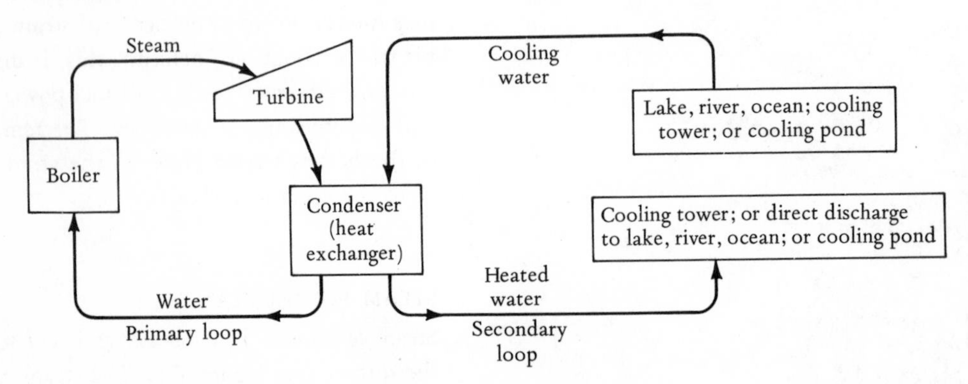

fossil fuel plant at 40 percent efficiency, one can calculate that the nuclear plant may discharge up to about 90 percent more heat to the water per kilowatt-hour of electricity than does the fossil fuel plant.

Consider a nuclear plant and fossil plant which both produce heat for conversion to electricity at the rate of 1,000 megawatts. The rate of electrical energy production of the fossil plant is 400 megawatts or 40 percent of 1,000. Its rate of waste heat production is 600 megawatts. The rate of electrical energy production of the nuclear plant is 320 megawatts or 32 percent of 1,000. Its waste heat production is 680 megawatts. The rate of waste heat production per unit of electrical energy of the fossil plant is 600/400 or 1.5 units of waste heat per unit of electrical energy. The rate of waste heat production per unit of electrical energy for the nuclear plant is 680/320 or 2.125 units of waste heat per unit of electrical energy. All this waste heat is in the heated water.

If we assume that 75 percent of the waste heat of the fossil fuel plant is directed to the water, then its rate of waste heat discharge to the water is 0.75 (600) or 450 megawatts. The ratio of the rate of heat wasted to the water to electric power production is 450 to 400 or 1.125. The comparable ratio for the nuclear plant is 2.125. Thus, the nuclear plant directs approximately 90 percent more waste heat to the water per unit of electrical energy than does the fossil fuel plant.

If both plants are designed for the same rate of electrical energy production, one would note a temperature rise 90 percent greater for the nuclear plant if the same volume of cooling water were used by both. Actually, the fossil fuel plant may use only about two-thirds to five-sixths of the volume the nuclear plant does. At the two-thirds rate, the temperature rise across the condenser in the nuclear plant is only about 25 percent larger than the rise in the fossil fuel plant. The average temperature rise across the condenser in the nuclear plant is about 18°F; in the fossil fuel plant it is about 15°F. The rise may vary from 10° to 30°F.

CONTROL

When heated discharges from the condenser mix with surface water, the temperature of the mixture is elevated. Depending on how well mixed the discharge is, there may occur "hot spots," localized areas of high temperature. Since the temperature conditions of hot spots may be extremely disadvantageous to aquatic life, several engineering concepts have been developed to control the mixing. Clearly, if the volume of cooling water passing through the condenser could be made greater, the temperature rise of the cooling water would be less. This is one avenue open to the designer, but it is doubtful that the volume could be made large enough to *alone* solve the heat dissipation problem. Another approach is to mix unheated water with the warmed water from the condensor *prior to discharge* into the lake or stream. A third approach is to discharge the condenser flow beneath the water surface through multiple spray diffusing devices which spread the warm water over a wide area and promote turbulent mixing. In addition, it is possible to withdraw cooling water from deeper (and thus colder) layers of the body of water. Because the water was colder to begin with, after it has been used to condense the steam, it will not have quite so high a temperature as water which had been drawn from higher (warmer) levels. The discharge may then be made with less effect on surface waters. Of course, combinations of these approaches may be utilized. All the concepts are aimed at alleviating hot spots or concentrated areas of high-temperature water. None of the four methods diminishes the actual quantity of heat being discharged to the water.

A fifth technique for discharging condenser water has both advantages and disadvantages. In this method the warmed condenser water is "floated" on the surface of the receiving body; a discharge canal may be used to obtain this effect. Because the warm water is in direct contact with the atmosphere, radiation, convection, and evaporation provide more rapid cooling. The procedure utilizes the principle that the rate of heat transfer increases with a greater difference in temperature between the surfaces. Unfortunately, an abrupt temperature change is unavoidably produced at the boundary between the surface and deeper layers of the receiving body. Thus, although heat is dissipated rapidly, the heated condenser water could prove detrimental to aquatic organisms. The effects of this method are discussed more fully below, under the section on biological effects of thermal pollution.

The technologies just described are typically utilized in conjunction with "once-through" cooling. In once-through cooling, the water used to condense steam is discharged without the benefit of any heat removal prior to discharge. There are, however, two types of devices which are capable of removing heat from the condenser water and dissipating it to the atmosphere, rather than to the body of water. These are the cooling tower and the cooling pond or artificial lake. Their use on new steam power plants is likely to increase considerably in the coming decades because the volume of water required to accomplish cooling is increasing in proportion to the dramatic growth in the demand for **Electric Power.** The Department of the Interior projects cooling water needs in the year 2000 at 850 billion gallons per day to yield a 12°F temperature rise or 640 billion gallons per day to yield a 16°F temperature rise. The average daily runoff in the United States is about 1,200 billion gallons per day. Comparison of cooling water requirements and the level of average daily runoff leads to the conclusion that extensive use of cooling devices will occur and that ocean-side plants will also increase markedly.

There are numerous types of cooling towers, but a basic division by principle of operation is into two classes: "wet" (or evaporative) and "dry." The wet cooling tower allows a portion of the condenser water *to evaporate* into the air in order *to cool* the bulk of the condenser flow. The dry tower transfers heat from the water to the air *by conduction* and *convection* and does not allow the condenser water to contact the air directly.

In the operation of the wet cooling tower, condenser water is sprayed into the device in fine droplets, and a portion of the water (about 1 to 1.5 percent) vaporizes. A relatively large quantity of heat is removed from the remaining liquid water and transferred to the water vapor. The process of vaporization of 1 pound of water requires about the same amount of heat as is required to raise the temperature of 100 pounds of water by 10°F. Thus, in the tower if 1 pound out of 100 pounds were to vaporize (1 percent), the remaining 99 pounds would be cooled by about 10°F. A great deal of heat is therefore removed by the vaporization.

After appropriate treatment the cooled water may be recycled to the condenser or discharged into the receiving body. Water is lost from the condenser stream through the vaporization processes, and a small amount of mist (water droplets) may drift from the tower; the rate of water loss due to drift is probably less than 0.2 percent of the circulating water rate. When the water is recycled, the buildup of solids within the circulating water must be prevented. To accomplish this end, there is a periodic or continuous withdrawal and replacement (referred to as "blowdown") of a small portion of the circulation. The total of the losses from vaporization, drift, and blowdown may amount

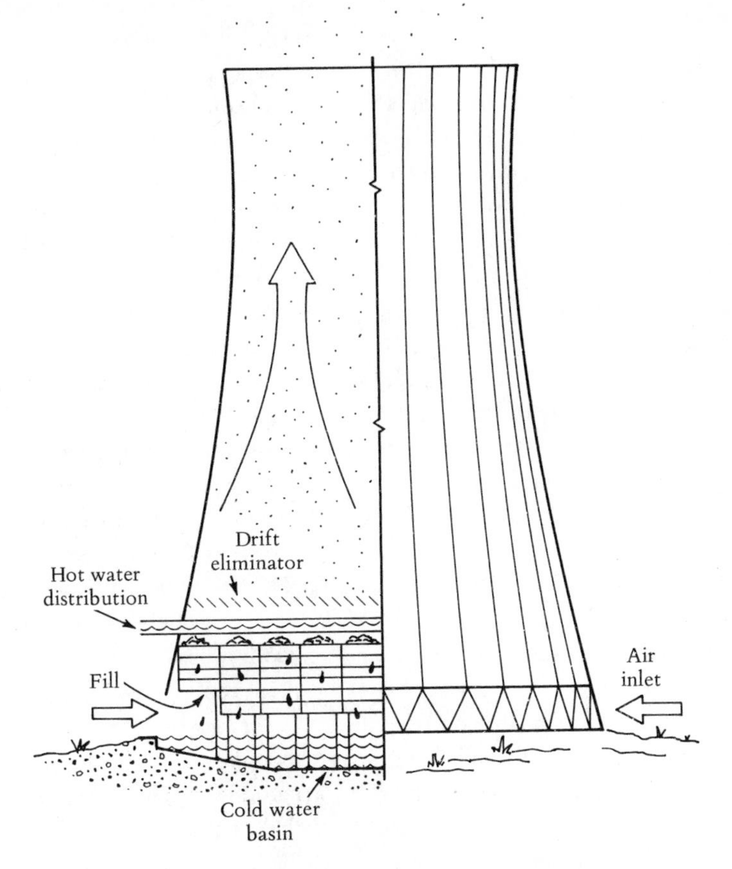

FIGURE T.2

NATURAL-DRAFT, EVAPORATIVE COOLING TOWER

From *Industrial Waste Guide on Thermal Pollution,* Federal Water Pollution Control Administration, U.S. Dept. of the Interior, 1968.

to 4 percent of the circulating stream, and this quantity must be continuously replaced.

There are several different types of wet cooling towers. Natural-draft cooling towers may reach 300 feet in diameter (300 feet is the length of a football field) and stand over 450 feet high (at 10 feet per story this is the height of a 45-story building) (see Figure T.2). They are often referred to as "hyperbolic" cooling towers because of the shape of their concrete shells. Air circulates in the tower by a natural convection which is induced by the evaporation. Thus, the tower has no requirements for electricity. The packing over which the water is distributed is restricted to the bottom 10 to 20 feet of the enormous tower. The falling warm water heats the air in the packing region of the tower. Since the air decreases in density because of its temperature increase, it becomes buoyant and rises up through the tower. Cooler air is drawn in at the base of the tower resulting in a continuous upward current which carries off the moisture-laden air.

Two evaporative natural-draft cooling towers on the Monongahela River at Fort Martin, West Virginia, will serve to illustrate these devices. In 1967, when the first of the towers was completed, it was the world's largest. Cooling the waters from two 540-megawatt coal-fired power plants, the tower and its twin stand 370 feet high. They are 378 feet across the base, 180 feet across at the top, and the throat diameter is 160 feet. The shell of the tower is 5-inch-thick reinforced concrete. The tower cools 250,000 gallons per minute of water from 114°F to 90°F. In the lower portion of the tower known as the fill section are tiered layers of Douglas fir slats onto which droplets are sprayed from large pipes.

To prevent mist from settling in the 300-foot-deep valley in which the plant is located, the towers extend above the surrounding hills. The stack of the power plant reaches some 200 feet above the hills to prevent smoking out the valley and to prevent mixing of the mist and stack gases. The need for the towers at this location becomes evident when one notes that the dry weather flow of the Monongahela River would have been barely sufficient to supply all the cooling water of *one* of the plants. One plant then, using once-through cooling, would have consumed all of the dry weather flow.

Another commonly used wet tower is the mechanical-draft tower in which air is forced or pulled in by fans. Such a tower is substantially smaller, perhaps only 50 feet high by 280 feet long and 70 feet wide, although a number of units may be needed. The electrical power requirements for this type of tower are considerable, and fan noise may be severe. Although the capital cost of the device is less than that of the natural-draft tower, the fans are subject to breakdown, and maintenance costs may be high. In addition, redwood is a favored material of construction.

One type of tower not now in common use is the dry cooling tower. It does not rely on evaporation for cooling but utilizes conduction and convection instead. The principal advantage of the dry cooling tower is that no moist air is produced. As a direct consequence, water losses are much smaller, although they may still amount to 1.5 percent because of equipment leaks. A 120-megawatt electric plant in Rugeley, England, has such a device which cost several million dollars; it has been in successful use since 1962. Several dry towers are being installed in Europe and South Africa; they will cool water from plants in the range of 150 to 200 megawatts. These pilot installations may eventually lead to technological developments which will allow the use of the dry tower on the new 1,000-megawatt nuclear plants.

The Rugeley plant utilizes the Heller dry cooling system in conjunction with a natural-draft cooling tower. At the time the Rugeley tower was built, it was the world's largest—350 feet tall. Unfortunately, the capital cost of a natural-draft dry tower may be as much as triple that of the natural-draft wet tower for comparable plants on account of the larger air flows needed to accomplish cooling. In fact, the dry system may cost as much as the turbine-generator system itself.

Cooling towers are utilized in two ways, described as closed-cycle and supplemental cooling. The closed-cycle system returns water, cooled by a tower or pond, to the condenser for reuse; it is a recycling with some small losses due to evaporation, etc. The heat absorbed by the condenser water is discharged to the atmosphere rather than a body of water. The supplemental cooling system in contrast, *discharges* a stream cooled by a tower or pond into the body of water. In such a system it may be possible for condenser water to bypass the tower or pond

A bank of mechanical draft cooling towers used to cool condenser water at the Peach Bottom nuclear power plant in Pennsylvania. In operation since 1967, the plant is a "high temperature gas-cooled reactor" (HTGR), one of two in the United States.

and be discharged directly to the body of water. This procedure might be utilized in winter, while in summer the condenser water would be directed through the tower.

There are a number of disadvantages to cooling towers. Not the least of these is the possible massive size of the towers and their dominance of the landscape. One percent evaporation could amount to the evaporation of between 7,000 and 10,000 gallons of water every minute for a 1,000-megawatt nuclear plant. The mist could be a considerable aesthetic blight, and ice might form on roads and power lines in cold weather. The use of the cooling tower in conjunction with a coal-fired plant may lead to the formation of a corrosive mist unless care is taken to separate the stack gases and water vapor. This condition would occur if **Sulfur Oxides** from the stack gases dissolved in the water mist forming sulfurous and sulfuric acid. The use of sea water in a wet cooling tower could result in the drift of a salty mist onto the land.

Cooling ponds may also be utilized to remove heat from condenser water. The amount of pond surface area required depends on the average air temperature in the locale. A 700-megawatt (electrical rating) plant owned by the Carolina Power and Light Company will utilize a 2,250-acre artificial lake, and two plants generating a total 1,300 megawatts in Michigan are to discharge their water into a cooling pond of about 900 acres. The availability of land at reasonable prices thus becomes a crucial factor in the decision to use cooling ponds. Cooling ponds may be utilized in closed systems or for supplemental cooling.

The cost to cool the condenser water by

ponds or towers does not appear to be prohibitive, except perhaps for the dry cooling tower, which consumes a great deal of electric power. Compared to the once-through alternative, the use of natural-draft, evaporative cooling towers may add in the neighborhood of 5 percent to the capital cost of the plant. This increase would result in only about a 1 percent increment in the residential electric rate. The natural-draft tower may be half as costly as the mechanical-draft tower in this regard.

BIOLOGICAL EFFECTS OF THERMAL DISCHARGES

Temperature is one of the most important factors governing the well-being of an organism. While man has been very ingenious about regulating the temperature of his own environment, aquatic forms of life do not have his options. In addition, unlike man, most aquatic creatures are poikilothermic; that is, they cannot maintain a constant body temperature by internal means. Rather, their body temperature is regulated by the environment in which they live. Such organisms maintain an appropriate body temperature by shifting their locale to find their necessary environmental temperature; this mechanism is referred to as "behavioral regulation." Species are found, therefore, only in areas in which the natural temperatures do not rise too high or fall too low for their survival. The competition for living space has led to the evolution of organisms able to live even at extremes of the temperature scale. Thus, algae in the hot springs at Yellowstone grow in waters registering 185°F, while blackfish found in Alaskan waters are reported to withstand freezing.

A particular organism, however, has only a relatively limited temperature range in which it can survive and reproduce. For example, the small organisms which build coral reefs are killed by temperature rises of only two or three degrees. For this reason man must be careful about altering the temperature of the environment where desirable creatures live. The addition of waste heat to natural waters is a form of pollution if it decreases the area in which such creatures live or if it is actually lethal to organisms that are unable to move to a cooler region.

Concern is usually centered on fish and shellfish, which form a portion of man's direct food supply. However, all the residents of a body of water are related to one another in a food chain or food web. The smaller organisms such as the plankton, worms, snails, insects, and water plants are eaten by each other and by fish. Thus, the effect of heat on all members of the aquatic community must be considered.

THE WELL-BEING OF FISH

Lethal Temperatures

The most dramatic of possible thermal effects is probably a fish kill. Actually, although it is a possibility to be taken into account, thermal fish kills have been relatively rare.

The Federal Water Pollution Control Administration has promoted a system of voluntary reporting of fish kills due to heated waste water discharges since 1962. The reporting is done by state and local water pollution control and conservation agencies. The number of incidents reported may be low, both because it is a voluntary activity and because the agencies may not be aware of all the fish kills that occur. Between 1962 and 1968, 18 incidents were reported, ranging in severity from 150 to over 300,000 fish killed. Three of the 18 fish kills occurred on the Sandusky River in Ohio. On January 1, 1967, over 300,000 fish were killed there, almost 250,000 more were destroyed one year later on January 2, and another 3,000 were killed in December, 1968, in the same area.

A great deal of laboratory study has been directed toward determining temperatures which will kill fish. An often used measure is the TL_M, or the temperature which will kill 50 percent of the exposed fish in some specified time period, usually 24, 48, or 96 hours. This is not the most useful measure since it does not provide information on the effects of sustained exposure to particular temperatures. Nor can a single TL_M be used to predict the effects of a shorter period of exposure to high temperatures such as can occur when organisms are entrained in condenser cooling waters. In this case an exposure might last only 10 to 20 minutes.

A preferable experimental method depends on exposing fish to different temperatures for much longer periods of time. Several general responses become evident with this approach. First, there are temperatures which the fish under examination can tolerate for an indefinite length of time; this is called the zone of thermal tolerance. Second, in an increasing or decreasing series of temperatures, one temperature will be reached at which 50 percent of the fish will die when brought to it *rapidly;* this may be either the upper or the lower incipient lethal temperature. However, the incipient lethal temperature is not a fixed number but varies with the temperature at which the fish has been acclimated before the experiment. A fish kept at 60°F will acclimate to that temperature, changing its metabolic functions. Acclimation processes will also establish the current upper and lower incipient lethal temperatures for that fish. When the fish is moved to a different temperature—for instance, 70°F—acclimation again occurs and new incipient lethal temperatures result. To continue the hypothetical example, the fish kept at 70°F might withstand a rapid change to 90°F while the same fish kept at 60°F would be killed by the change to 90°F. The goldfish can be subjected to what is possibly the most extreme range of upper incipient lethal temperatures, which may vary from approximately 80° to over 100°F.

Acclimation to a new temperature may require a rather long time (days or weeks) especially in going from high temperatures to low ones. Acclimation in an upward direction is faster. As an example of how fish survival might be affected in a natural setting, we note that in winter when fish are attracted to warm discharge waters, plant shutdowns could result in fish kills if the fish cannot acclimate rapidly enough to the suddenly cold water.

Although the incipient lethal temperature is not a fixed number but will vary with the temperature the fish have been kept at, as the water is made warmer and warmer, a temperature is reached which the fish cannot tolerate with any amount of previous conditioning. This is called the ultimate incipient lethal temperature.

The usefulness of all laboratory measures of lethal temperatures is limited by the fact that experiments are carried out in aquariums where fish are quiescent and not fed. Ample supplies of oxygen are usually present. Waste products may contaminate the water, or all pollutants may be absent. Conditions are often not similar to those found in nature.

Lethal temperatures vary widely with the species of fish; the goldfish, as already mentioned, can have an incipient lethal temperature of over 100°F, while that for the pink salmon is near 75°F. In addition, body size, season, sex, diet, hormones, and day length all have an effect. Nevertheless, graphs have been prepared for the different fish species using variables of temperature, time, and acclimation temperature. They are useful in predicting the survival of organisms entrained in cooling waters and discharge canals as well as the effects of temperature rises in whole bodies of water. Illustrations of such predictions are given by C. C. Coutant [3].

Temperature Preferences

Of equal importance to lethal temperatures are the temperatures at which fish can be found in nature. Most fish show a decided preference for a particular narrow range of temperature. Field studies tend to show a somewhat lower preference than do laboratory studies (see Table T.1). The lowest preference determined is about 54°F, which is still higher than the temperature of many natural waters in the winter. This correlates with the fact that fish are often attracted to thermal discharges in the winter. Although fish will tend to move to new locations when their normal areas are warmed above their preference, it must be kept in mind that competition or geographic features sometimes keep other locations from being available.

A special situation occurs in rivers which are cooler than would be predicted considering their geographic location—for instance, being fed by melting ice and snows. Such rivers warm up as they flow away from their source until they reach the temperature of other natural waters in the area. Adding heat to them may not have an effect at the point of discharge but rather downstream. The river temperature will rise above the level suitable for cold-water fish sooner than if no heat were added. Many miles of suitable habitat may be lost in this manner. The Columbia River is an example of this type of stream. Dams on the river which hold back the water also allow it to warm up and thus make the river less suitable for such fish as salmon.

TABLE T.1
TEMPERATURE PREFERENCES OF FISH (IN DEGREES FAHRENHEIT)

Species	Laboratory Determination	Natural Habitat
Salmon	58.1	51.0
Brook trout	57.2–60.8	53.6–68.5
Largemouth bass	86.0–89.6	80.0–81.9
Yellow perch	75.6	54.0–69.8
Smallmouth bass	82.4	68.5–70.5

From "The Extent to Which Environmental Factors Are Considered in Selecting Power Plant Sites, with Particular Emphasis on the Ecological Effects of the Discharge of Waste Heat into Rivers, Lakes, Estuaries and Coastal Waters," *Hearings Before the Subcommittee on Air and Water Pollution of the Committee on Public Works*, U.S. Senate, 90th Congress, 2nd Session, February 1968.

Survival and Reproduction

It is not necessary to kill fish outright to eliminate species. Factors which prevent successful reproduction can just as easily destroy a population. Fish reproduction is very dependent on temperature, both in the formation of eggs and sperm and in the actual spawning and hatching of the eggs.

Yellow perch must have low temperatures during the fall and winter months for maturation and the formation of eggs and sperm. Growth during late spring and summer is accelerated by higher temperatures. Trout, on the other hand, spawn in the fall; thus, late summer and fall temperatures are critical for maturation of eggs and sperm and for spawning. The fathead minnow spawns in the summer and needs to be protected only then. Fish exposed to warmer than normal waters may spawn too early, with loss of the young fish. Some fish require a cold winter period alternating with warm summer temperatures for reproduction while others need be protected only from very high summer temperatures. Clearly, then, in order to decide whether a body of water can accept additional heat, the species of fish which are to be protected and their seasonal temperature requirements must be known.

This is the basis for one objection to the heat discharged by the Connecticut Yankee Power Plant on the Connecticut River. Four states, Connecticut, Massachusetts, Vermont, and New Hampshire, the U.S. Bureau of Sport Fisheries and Wildlife, and the Bureau of Commercial Fisheries have been trying to reestablish the once flourishing Atlantic salmon and American shad fisheries in the river. It is feared that since the summer river temperatures are already rather high for these species, additional heat will cause the project to fail.

Temperature has effects on the well-being of fish other than influencing their ability to reproduce. For instance, the TL_M for trout is 82° to 83°F. Although they are not killed at 73°F, they are unable to catch small fish for food. A better survival temperature would be 68°F. Further, below 60°F, they will grow more rapidly.

Another indirectly lethal effect of high temperatures is immobilization; this may prolong the period of exposure of the fish to unfavorable temperatures. Behavioral responses induced by high temperatures may make the fish more likely to be eaten by predators. Even though such behavioral changes may not be obvious to the investigator, predators will recognize and choose thermally shocked fish over nonshocked ones. Predators, including gulls and large fish, are often attracted to thermal discharges, presumably for this reason. Fish may also become more susceptible to disease at higher temperatures. The warming of the Columbia River is believed to have caused more severe outbreaks of the usually rare columnaris disease of salmon.

Migrational Blocks

Warm water may act as a block preventing fish from migrating up or down a river during spawning seasons. If fish are repelled by the temperature of the discharge and if the discharge occupies the major portion of the river or normal migrational channels, serious effects on reproduction and fishing can be envisioned. In the 1960's a thermal block of 70°F prevented the migration of salmon from the Columbia River into the Okanogan River in Washington. The salmon prefer a temperature of 45° to 60°F.

Other aquatic organisms migrate vertically in the water. For instance, the opossum shrimp stays on the bottom during the day but moves to the top at night. It may well be affected by thermal discharges "layered" on top of the receiving water. Aquatic insects which must come to the surface to complete their life cycle or lay eggs may be affected too.

ORGANISMS OTHER THAN FISH

Population Shifts

The types of algae which will grow in a body of water differ, depending on its temperature, as illustrated in Figure T.3. Blue-green algae are a less desirable type of algae, since they are not food for many other organisms and because they may cause unpleasant odors and tastes in water. In one instance, an experimental flume of water at 115°F in the Delaware River produced large growths of the blue-green algae *Oscillatoria.* Oxygen bubbles trapped in the dense mat of algae floated it to the surface, where it died and decayed. A condition equivalent to pollution by a sewage outflow with a large oxygen demand[1] was thus created by the addition of heat alone. Besides the blue-green algae, another group of odor-causing organisms, the actinomycetes, grow best at 86°F. Temperature increases generally accelerate bacterial growth, especially in organically polluted streams.

The species of insects found may vary with temperature. Certain insects are stimulated to hatch by particular temperatures. If the hatching temperature is reached earlier in the year than normally, because of thermal additions to the water, the insects may hatch prematurely. This situation could prove fatal to the insect population. If the insect is a food organism for other aquatic creatures, effects could be far-reaching. The species of fish found in an area will also change with temperature, as pointed out previously. Preferred species like salmon

[1] See **Organic Water Pollution** for the effects of low dissolved oxygen.

and trout may be replaced by less desirable fish as temperature is increased.

Not only do the types of organisms found change with increasing temperature, but the number of species found declines. In one study a 54 percent decrease in number of species was observed as the temperature went from 80° to 87°F. A further 24 percent of this number was lost as the temperature rose from 87°F to 93°F. Such simplification of the ecosystem in a body of water can have adverse effects. Simple ecosystems are often less stable than complex ones. Hence, large oscillations in the populations of organisms present may occur.[2]

If no supplemental cooling is provided for condenser water, it may be desirable from a *technological* standpoint to withdraw cooling waters from the cooler, lower layer of lake water, called the hypolimnion.[3] The water, when heated and returned to the lake, then becomes part of the upper, warmer water layer called the epilimnion and may cause problems. The growth of algae is generally limited to the epilimnion, where sufficient sunlight penetrates for photosynthesis. Use of the hypolimnetic waters for cooling in such a system will both increase the volume of the epilimnion and possibly prolong the growing season for algae. In addition, nutrients which have settled out into the hypolimnion will be recirculated into the epilimnion, becoming available for use. These factors may well combine to increase the rate of **Eutrophication** in a lake. Such reasoning is part of the stated opposition to the Bell Station Nuclear Power Plant proposed for construction on Cayuga Lake, one of the New York State Finger Lakes.

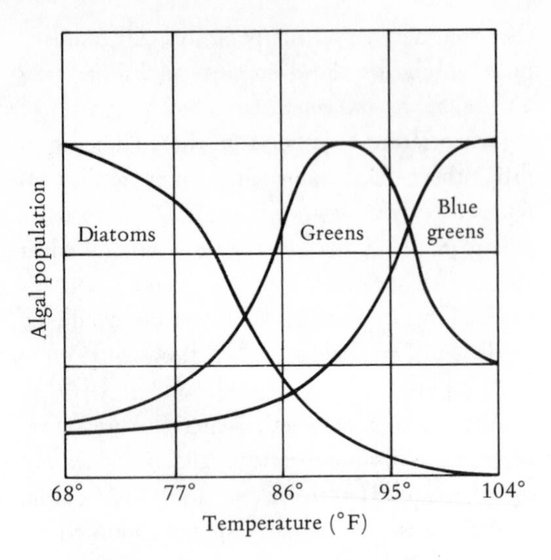

FIGURE T.3
EFFECT OF TEMPERATURE ON TYPES OF PHYTOPLANKTON

From J. Cairns, Jr., "Effects of Increased Temperature on Aquatic Organisms," *Industrial Wastes* 1, no. 4 (1956): 150. Used by permission of Scranton Publishing Co.

Enhancement of Toxic Effects

Higher temperatures seem to enhance the toxic effects of trace pollutants such as heavy metals and **Pesticides.** Fish are adversely affected by nitrogen and carbon dioxide at lower concentrations as temperatures rise.

Oxygen-Temperature Interrelationships

The addition of heat to surface waters may reduce the quality of the water by changing the parameters of two key processes in natural waters. The elevation of temperature has several effects on the degradation of organic wastes and the depletion of dissolved oxygen in the water. In the first place, the rate at which organics from sewage and industrial wastes are oxidized is increased at higher temperatures. Thus, the

[2] Ecosystem dynamics is a fairly complex subject and beyond the scope and intent of this discussion. See E. P. Odum, *Fundamentals of Ecology*, 3rd ed., Saunders, Philadelphia, 1971.

[3] This would be the case if the heated water discharge were limited to a specified temperature. The lower the temperature of the water entering the condenser, the lower the effluent temperature.

same quantity of waste would deplete oxygen from the water more rapidly. Secondly, the saturation concentration of dissolved oxygen *decreases* at higher temperatures. By saturation concentration is meant the greatest quantity of dissolved oxygen that the water can hold. This effect by itself is not very important since streams are "healthy" at concentrations lower than the saturation concentration. However, the rate at which oxygen is replenished to the water—the rate of reaeration from the atmosphere—depends strongly on the quantity of oxygen the water is capable of dissolving before reaching the saturation level. The greater the quantity of oxygen that can be dissolved, the greater the *rate* of oxygen replenishment from the air. A lower saturation concentration means a lower rate of replenishment of oxygen. This subject is further discussed in **Organic Water Pollution.**

The first factor was an increase in the rate of oxygen removal. The second was a decrease in the rate of replenishment of oxygen once it is removed. The net effect is that the same pollution level causes worse conditions in terms of lower dissolved oxygen concentrations, diminishing the quality of the water.

Some concern has been voiced that warmer discharge waters themselves will be depleted in oxygen since oxygen gas is less soluble in warm than in cold water. Experimental findings, however, seem to indicate that turbulence in the condensers may actually increase the oxygen content of waters used for cooling. The heating of such waters does not drive the oxygen out but apparently causes the water to become supersaturated with oxygen.

The oxygen requirements of aquatic organisms increase as they become more active and as levels of carbon dioxide rise. When fish are first placed in warmer water, their oxygen needs increase. If the fish can acclimate to the new temperature, this requirement will decrease again. A good review of the biochemical and physiological processes known to be involved in acclimation is presented by Jensen et al [5]. Not all fish acclimate, however. Thus, the effect of temperature on rising oxygen requirements will vary depending on the species of fish involved.

Chemicals in Discharged Cooling Water

Biocides such as chlorine and sodium hydroxide are sometimes added to cooling waters to prevent fouling of the condenser pipes. They are often lethal to entrained organisms and can seriously affect the receiving waters. Sodium hydroxide used at the Cousins Island power plant resulted in very alkaline receiving waters and low oxygen saturation of the waters. A study by Milhursky and Kennedy showed that up to 95 percent of the organisms entrained in the cooling waters of the Chalk Point power plant on the Patuxent River were killed [7]. This event was attributed both to thermal effects and to the use of chlorine as a biocide. Mechanical methods to clean condenser pipes are available which might eliminate this problem.

Condenser pipes sometimes corrode, releasing heavy metals into cooling waters. This is believed to be the cause of "greening" of oysters noted in the Patuxent River after the power plant began operating. The oysters are apparently accumulating copper released into the water from the condenser piping.

ENTRAINMENT AND DISCHARGE CANALS

Organisms which are small enough can be sucked into the condensers along with the cooling water and are thus exposed to short periods of very high temperature. This is known as entrainment. Plankton, including algae, small crustaceans, eggs and larvae of many fish and shellfish, and small fish may all be entrained. Losses in the summer are greater since the natural water temperature is already at its highest point of the year. Many organisms will be under stress, and the additional heat may well be lethal.

The effect of entrainment on the population as a whole is related to the amount of river or stream flow used as cooling water. The larger the portion of the stream used, the more organisms are likely to be destroyed and the more serious is the probable effect on the population. A detailed discussion of the effects of entrainment on various organisms is presented by Coutant [3].

Thermal discharges are sometimes passed down a canal before being returned to the original body of water. This procedure cools the discharge somewhat but prolongs the exposure of entrained organisms to high temperatures.

Discharge canals may also develop their own populations. The high temperatures, constant addition of nutrients and removal of wastes, and high input of sunlight combine to create a situation in which certain organisms such as blue-green algae grow rapidly, without normal checks. When the algae die and are washed out into the receiving waters, a net addition of organic water pollution is the result.

MARINE ENVIRONMENTS AND ESTUARIES

Estuaries are areas where freshwater rivers flow into bodies of salt water. If the cooling water intake is closer to the source of salt water than is the outflow, as occurs at the Chalk Point plant, heated discharge water, being more saline and thus more dense than the receiving water, will sink. This situation has the effect of disturbing bottom-dwelling organisms, which are accustomed to cool temperatures. It also means that discharge waters will take longer to cool than if they remained on the surface.

Important commercial species and key food chain organisms in estuaries cannot survive temperatures far above 90°F. Some are under thermal stress in the range 86° to 88°F. Estuaries in the Chesapeake Bay area may reach such temperatures naturally during the summer. A power plant situated on an estuary in South Wales has raised the water temperature in its vicinity to 73°F in winter and 99°F in summer. This may be compared to the normal sea temperature of 45°F in winter and 63°F in summer in the area.

A number of estuarine species such as oysters are fixed to the bottom and are unable to leave the area if their environment becomes overheated. Like fish, shellfish have more restrictive temperature requirements for spawning than for survival. They may thus survive but not breed above certain temperatures. At high temperatures certain oysters react by decreasing the ciliary movement which allows them to filter water. The oyster cannot feed without filtering water and so loses weight.

Estuaries are known as the "nurseries" of the oceans. The early stages in the life cycle of many marine organisms require the special water qualities found there. The siting of power plants in estuaries must therefore be approached very cautiously. In salt waters additional heat may not only increase oxygen demands but also increase the salinity of the water, leading to a higher mortality of eggs and larvae.

BENEFICIAL EFFECTS AND USES OF THERMAL ADDITIONS

Areas which usually ice over in winter may be kept open by thermal discharges—an advantage to waterfowl as well as allowing replenishment of the oxygen content of such waters in the winter. Limited additions of heat have been shown to increase the growth of fish food organisms. This is a somewhat tricky proposition, however, since above a certain point deleterious effects will cancel the benefit.

Plans to purposely use heated water have

been advanced. Aquaculture or mariculture, the farming of aquatic organisms for food, is one such plan. Warm-water species may be grown at optimum rates in waters warmed by thermal discharges. Carp are grown in this way in central Russia although they are found naturally only in southern Russia. Experimenters in Great Britain are examining the possibility of growing carp, flounder, trout, oysters, sole, and *Tilapia* in heated outflows. Experimental oyster farms in Long Island Sound and Puget Sound are using power plant discharge waters to maintain proper culture temperatures. It is hoped that warm-water outflows will improve lobster production in Maine waters to offset a natural cooling trend which seems to be decreasing normal production.

One especially appealing project in Dorset, England, at the Poole Electric Generating Station involves the construction of a complete ecosystem for the use of waste products. Flue gas which is cleaned of impurities is used as a carbon dioxide source for algae. Other essential nutrients are provided by domestic sewage. The algae act as a food source for the American hard clam (quahog), which is grown in the heated water discharged from the plant. The clams are otherwise at the northern limit of their range. Better growth and earlier and more successful spawning are noted in the warmer water.

Several problems should be noted with respect to such schemes. Aquatic organisms, especially shellfish, concentrate many toxic chemicals. Biological magnification of radioactive elements may militate against the use of cooling water from the older **Nuclear Power Plants.** Oysters are very good at concentrating pesticides or heavy metals such as copper. Discharges must, therefore, be free from such materials if they are to be used in mariculture. The aqua-farms, if operated on a large scale, can generate large amounts of waste. Methods for disposal of this sewage must be developed or a serious water pollution problem will result. Further, a strictly controlled temperature will have to be maintained in order for warm-water species to survive in water normally too cold for them. Power plants which might be subject to shutdowns for repairs or due to worker strikes may not be a completely dependable source for hot water. The oyster farming project in Long Island Sound which uses warmed water from Long Island Lighting Company plants has been threatened by a jurisdictional dispute over control of and revisions of standards for thermal discharges.

Some other suggestions include the use of waste heat to increase the rate of biological treatment processes for sewage in treatment plants, to aid in purification of drinking waters, to heat homes, and to prevent frost damage to crops. The use of waste steam for heating and industrial purposes, however, is hampered because the steam, after it exits from the turbines, is not hot enough for these purposes. It is possible that systems could be designed *initially* to both generate electricity and provide heat for buildings or industry. Such a system in Tapiola, Finland, provides electricity, heat, and hot water for 20,000 people.

EFFECTS OF POWER PLANTS IN OPERATION

A number of studies have been conducted to determine the ecological effects of power plants currently operating. It has also become the policy of regulatory agencies to require studies both before and after operation begins in order for power companies to obtain operating licenses.

Adverse biological effects have been noted at some plants. The almost complete destruction of marine life entrained in cooling waters at the Chalk Point plant was mentioned previously. In 1966 some 40,000 blue crabs were found dead in and around the discharge canal at the same plant.

Ecological changes are apparent around the Cousins Island plant (fossil fuel is utilized here). Two seaweeds have disappeared, along with several aquatic species that live in or consume them. Changes in the shell structure of native mussels are seen, and a species of oyster which has never grown this far north has become established.

A large fish kill took place at the site of the Turkey Point plant located on Biscayne Bay in Florida in 1969. High temperatures in the bay are blamed for lethal effects on bottom dwellers. Dead crabs were found from the mouth of the canal 1,000 yards into the bay. Dead shrimp, sponges, coral, and algae, among other species, were found.

Examples of operations which studies have shown were not deleterious to aquatic organisms include the fossil fuel plant at Waukegan on Lake Michigan and most of the plants in the TVA system. One of the latter group, the Paradise plant on the Green River in Kentucky, was adversely affecting fish food organisms. Cooling towers were installed, and the problem was solved.

Preliminary studies at the Connecticut Yankee power plant on the Connecticut River have revealed no serious harm to natural species except for catfish which live in the discharge canal. They are subjected to crowding and exhibit increased metabolic rates because of the heated water.

Since alternative cooling methods such as towers and ponds are available for use in areas where the waters cannot accept additional heat, it does not seem necessary to tolerate adverse thermal effects on aquatic life. It is essential, however, to site plants carefully and to study the environment before a plant begins operation. Such an investigation will provide a baseline for later comparative purposes as the operating plants are monitored.

LAWS RELATING TO THERMAL DISCHARGES

The Water Quality Control Act of 1965 requires each state to submit water quality standards for its waters to the Department of the Interior for approval. A National Technical Advisory Committee, acting as an advisory committee to the former Federal Water Pollution Control Administration and composed of many recognized experts in the field, proposed a set of standards in 1968 on which most of the states' laws have been based. The committee's recommendations include the following:

1. Discharges into *streams* should cause an increase of no more than 5°F over natural water temperatures (based on the average, for that month, of maximum daily water temperatures). To ensure that this 5° rise will not raise summer water temperatures to lethal levels, the committee recommends that heat added should in any case not raise the water temperature over 83° to 86°F in normally cool areas, or over 86° to 90°F in normally moderate regions, or over 90° to 96°F in warm sections of the country. Limits are also set for the duration of such peak temperatures.

2. Discharges into *lakes and reservoirs* should not raise the water temperatures more than 3° above normal (based, for each month, on an average of the maximum daily temperatures for that month). Pumping water from the hypolimnion is not recommended.

3. Waters which are inhabited by trout and salmon require special protection. Thus, no heat should be added in the vicinity of spawning areas, and winter temperatures should not be raised above 55°F.

4. Discharges into *estuaries* should not raise water temperatures more than 4° above normal for fall, spring, or winter months or more than 1.5° above normal for summer months.

5. Discharges into marine waters should not raise temperatures more than 1° per hour with a maximum of 7° to 9° in any 24-hour period.

6. Mixing zones should be as small as possible. When they occur in rivers or streams they should not occupy more than 50 percent of the passageway, and preferably 25 percent or less.

7. Maximum temperatures for exposure of certain fish are also recommended.

The standards have been criticized on the basis that they are too general. Each body of water with its particular aquatic population may require separate study and individual standards. In some cases—for instance, Lake Michigan—this step has been taken. The Department of the Interior has recommended that discharges into the lake be no more than 1° over the ambient temperature.

Finally, it must be noted that a need exists for further study to determine temperature requirements of many organisms. Although quantities of data of predictive value are available, by no means have all the desirable species been identified and studied.

REFERENCES

1. "The Environmental Effects of Producing Electric Power," *Hearings Before the Joint Committee on Atomic Energy*, 91st Congress, 1st Session, October–February 1969–1970.
2. "The Extent to Which Environmental Factors Are Considered in Selecting Power Plant Sites, with Particular Emphasis on the Ecological Effects of the Discharge of Waste Heat into Rivers, Lakes, Estuaries and Coastal Waters," *Hearings Before the Subcommittee on Air and Water Pollution of the Committee on Public Works*, U.S. Senate, 90th Congress, 2nd Session, February 1968.
3. C. C. Coutant, "Biological Aspects of Thermal Pollution: 1. Entrainment and Discharge Canal Effects," *Critical Reviews in Environmental Control* 1, no. 3, Chemical Rubber Publishing Co. (1970).
4. D. Merriman, "The Calefaction of a River," *Scientific American*, May 1970, p. 42.
5. L. D. Jensen et al., *The Effects of Elevated Temperature upon Aquatic Invertebrates*, Edison Electric Institute, New York, 1969.
6. R. H. Strand and P. A. Douglas, "Thermal Pollution of Water," *Sport Fishery Institute Bulletin*, no. 191, 1968.
7. J. A. Milhursky and V. S. Kennedy, "Water Temperature Criteria to Protect Aquatic Life," *Symposium on Criteria to Protect Aquatic Life*, special publication no. 4, American Fisheries Society, 1967.
8. J. Cairns, Jr., "Effects of Increased Temperature on Aquatic Organisms," *Industrial Wastes* 1, no. 4 (1956): 150.
9. J. Cairns, Jr., "Ecological Management Problems Caused by Heated Waste Water Discharge into the Aquatic Environment," *Water Resources Bulletin* 6, no. 6 (1970): 868.
10. J. Cairns, Jr., "Thermal Pollution: A Cause for Concern," *Journal of the Water Pollution Control Federation* 43 (1971): 55.
11. C. L. Tarzwell et al., "Water Quality Criteria for Fish, Other Aquatic Life and Wildlife," Interim Report, National Technical Advisory Committee to the FWPCA on Water Quality Criteria, U.S. Dept. of the Interior, 1967.
12. "Industrial Waste Guide on Thermal Pollution," U.S. Dept. of the Interior, Federal Water Pollution Control Administration, 1968.
13. A. W. Eipper et al., "Thermal Pollution of Cayuga Lake by a Proposed Power Plant," Ithaca, N.Y., 1968.
14. R. R. Garton and P. E. Christianson, *Beneficial Uses of Waste Heat—An Evaluation*, U.S. Environmental Protection Agency, National Thermal Pollution Research Program, Water Quality Office, Pacific Northwest Water Laboratory, Corvallis, Ore., 1970.

15. D. A. Andelman, *New York Times*, October 29, 1971.
16. "Engineering for Resolution of the Energy-Environment Dilemma," Committee on Power Plant Siting of the National Academy of Engineering, Washington, D.C., 1972.
17. "Industrial Waste Guide on Thermal Pollution," Federal Water Pollution Control Administration, U.S. Dept. of the Interior, 1968.
18. P. Leung and R. Moore, "Thermal Cycle Arrangements for Power Plants Employing Dry Cooling Towers," *Journal of Engineering for Power*, April 1971.
19. R. Woodson, "Cooling Towers," *Scientific American*, May 1971, p. 70.
20. J. Clark, "Thermal Pollution and Aquatic Life," *Scientific American*, March 1969, p. 19.
21. C. E. Warren, *Biology and Water Pollution Control*, Saunders, Philadelphia, 1971.
22. M. Eisenbud and G. Gleason, eds., *Electric Power and Thermal Discharges*, Gordon and Breach, New York, 1969.
23. T. E. Langford and T. E. Aston, "The Ecology of Some British Rivers in Relation to Warm Water Discharges from Power Stations," *Proceedings* of the Royal Society of London, Series B, 180 (1972): 407.
24. P. R. O. Barnett, "Effects of Warm Water Effluents from Power Stations on Marine Life," *Proceedings* of the Royal Society of London, Series B, 180 (1972): 497.

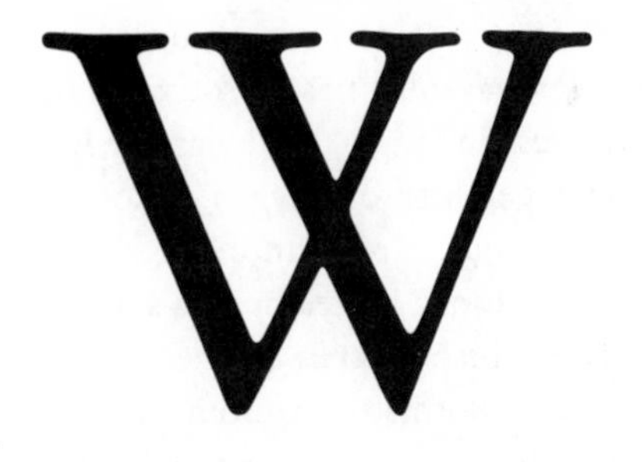

Water Pollution

There are commonly considered to be seven important kinds of water pollution: (1) bacterial and viral, (2) organic, (3) radioactive, (4) toxic and trace chemical, (5) eutrophication (enrichment), (6) thermal, and (7) oil. Each of these kinds will be discussed here, and legislation on water quality control will also be described.

BACTERIAL AND VIRAL POLLUTION

It was not many years ago that many communities in this country had untreated water supplies. Their water was frequently taken from streams and rivers into which untreated sewage was discharged by other communities. The result was periodic epidemics of typhoid and other water-borne diseases. Since the wastes of individuals infected with typhoid or other intestinal diseases may contain the bacteria which produce those illnesses, the result could have been expected. The most well-remembered epidemics of typhoid were in Plymouth, Pennsylvania (1885), the Schenectady-Albany area of New York (1890), and the Lowell-Lawrence area of Massachusetts (1890). All were established as having been spread by water supplies drawn from rivers.

With the advent of filtration and chlorination around the turn of the century, a marked and steady decrease in the number of waterborne epidemics began. Filtration through beds of sand *removes* most disease-causing bacteria; chlorination of the water *destroys* most disease-causing bacteria. Descriptions of these processes are found in **Water Treatment and Water**

Pollution Control. By the period 1946–1960, the diligence of the public utilities in providing safe supplies had reduced the annual incidence of water-borne disease to 1.1 illness per 100,000 population per year (including diarrhea). The story of epidemic control by water treatment is largely forgotten today, but it was one of the first instances of engineering's being directed to controlling the environment.

Drinking Water Standards are designed to prevent the presence of fecal material in drinking and recreational waters in order to ensure the absence of the disease-causing microorganisms. Currently used tests rely on detecting certain bacteria, in the coliform group, which inhabit the intestine of warm-blooded animals. The presence of these bacteria is regarded as evidence of fecal contamination, although the coliform organisms themselves are not harmful. For a more detailed discussion, see **Bacterial and Viral Pollution of Water, bacterial standards.**

In the last several decades, we have become aware of the potentiality for the spread of virus diseases through water. As with water-borne bacterial diseases, the wastes of infected individuals contain the offending virus. However, only infectious hepatitis and poliomyelitis have been traced as water-borne epidemics, although nearly 100 human enteric viruses have been identified.

ORGANIC POLLUTANTS

High concentrations of soluble organic waste eventually cause a low dissolved oxygen concentration in the stream (see **Organic Water Pollution**). The low dissolved oxygen causes severe changes in the animal population of the stream. Upstream from the pollution discharge where dissolved oxygen may be as high as 8 or 9 milligrams per liter a large array of species exist; sports fish, minnows, mayflies, snails, and the like are characteristic of the clean stream, prior to pollution. The entrance of organic wastes and the lowering of the dissolved oxygen create an environment suitable for what are termed "pollution-tolerant" organisms. One such is the sludgeworm, which consumes sludge (sewage solids) as its source of nutrients and thrives in water with as little as 0.5 milligram per liter of oxygen. Another is the rat-tailed maggot, a resident in the sludge at the bottom of streams; it breathes via a long tubular appendage which reaches to the water surface. Many miles downstream from the pollution discharge the dissolved oxygen will be restored by natural reaeration, and the characteristic populace of the clean stream will return. A rough guide for stream "health," then, is the concentration of dissolved oxygen. A concentration of about 5 milligrams per liter is generally regarded as necessary for maintaining fish in the stream, although the needs of individual species vary.

Organic solids also contribute to the deterioration of stream quality. Besides exerting a demand for oxygen, as the soluble organics do, the solids eventually layer the stream bottom downstream from the point of discharge. In this layer the solids decompose in the absence of oxygen, producing gases such as hydrogen sulfide, whose odor is noteworthy for its lack of appeal.

The sources of organic solids and the soluble organic wastes are domestic and industrial effluents. Organic solids are removed efficiently by the process known as settling or sedimentation. Soluble organics are removed by either the trickling filter or the activated sludge process. Sedimentation is referred to as "primary treatment"; the latter processes as "secondary treatment." Usually secondary treatment accomplishes no better than 90 percent removal of the wastes. As an added precaution, the treated waste water leaving the treatment plant is commonly chlorinated to destroy any disease-causing bacteria. These processes are described in **Water Treatment and Water Pollution Control.**

RADIOACTIVE SUBSTANCES

Radioactive substances enter the water in small quantities from the wastes of uranium mills, laboratories, and **Nuclear Power Plants.** Because of an early recognition of the hazards involved, standards for handling and disposal have been set up, and monitoring has been widely utilized. The safety record has been excellent.

Trace amounts, however, do reach the water in the cooling water from nuclear power plants. In addition, the stack gases from such plants contain radioactive substances. Plants and animals may ingest the water with trace quantities of radioactive substances and retain and accumulate those substances; this is the process known as biological magnification. There is no evidence that marine life is in danger from the current concentrations of radioactive substances from power plants. Furthermore, the permissible concentrations in water are set in such a way that the dosage which man is to receive from all sources, including marine sources of food and from air, is below a level initially recommended by the Federal Radiation Council. However, all possible routes of biological magnification which lead to man may not yet be known.

In 1970, the standard setting function was transferred to the U.S. Environmental Protection Agency (EPA), and the standards are presumably undergoing review. Even though the EPA has been given responsibility for setting radiation standards, the U.S. Atomic Energy Commission has recently *proposed* new design objectives for nuclear power plants. If the design objectives are finalized and implemented, the expected radiation exposure of sizable population groups from power plant effluents will be less than 1 percent of natural background. For further details see **Radiation Standards** and **Nuclear Power Plants.**

TOXIC SUBSTANCES AND TRACE CHEMICALS

A number of toxic substances enter waters in very small or even undetectable amounts. Included in this group are **Pesticides** such as DDT and heavy metals such as **Mercury** and **Arsenic.** Although water concentrations may be well below levels directly hazardous to humans, organisms living in the water have the ability to concentrate the toxic substances to levels which are dangerous; this again is the process of biological magnification. Thus, fish in several areas of the TVA reservoir system were recently found to contain more DDT than the allowable maximum of 5 parts per million, although water levels were nowhere near this amount. Shellfish have been shown to concentrate mercury up to 100,000 times.

Of course it is possible for industrial and shipping spills to introduce levels of toxic chemicals which are directly lethal into waters. A number of fish kills occurring here and in other countries are attributable to these sources. The biological magnification of trace amounts of such poisonous compounds appears at the moment to present a widespread, difficult to control, and insidious hazard.

EUTROPHICATION

When the nutrient-rich materials in industrial or municipal wastes and in agricultural or urban runoff reach lakes and rivers, conditions may be created which are suitable for large nuisance growth of algae. The blooms, as they are called, can blanket lakes, ponds, and streams in late summer, making the water unfit for recreational use and imparting off colors, odors, and tastes to public water supplies. Dissolved oxygen in the body of water, important in maintaining a desirable community of sport and commercial food fish, may fluctuate widely as algae go through alternate cycles of growth and decay; low levels of dissolved oxygen may interfere with the survival of the fish populations.

This rapid nutrient enrichment is referred to as Eutrophication. Phosphates, nitrates, and organic matter have all been cited as controlling factors in eutrophication. The phosphates in detergents have been particularly indicted in this regard. Extensive discussions of the role of phosphates and other substances are in **Eutrophication.** Tertiary treatment of sewage to remove phosphates and nitrates (see **Water Treatment and Water Pollution Control**) and good farming practices to reduce agricultural runoff are believed to provide a possible hope for prevention of eutrophication.

HEAT

Both **Nuclear Power Plants** and **Fossil Fuel Power Plants** use surface waters to condense steam. In nuclear power plants, the temperature of the water stream returned after its use in cooling may be increased by about 18° to 30°F. This warm water enters the river or lake and raises the temperature of the receiving water.

Several adverse environmental effects are associated with the use of surface waters to condense steam. Small fish, the eggs and larvae of many creatures, and plankton may be drawn directly through the plant along with the cooling waters. The organisms may be killed directly if cooling water temperatures are raised high enough or may be incapacitated in some way.

Organisms which do not pass through the plant directly may be affected by the cooling water which is returned to the main body of water at a new, higher temperature. Effects may be either beneficial or detrimental, depending on the species. Reproduction may be improved or impaired; organisms may grow faster and larger, or their metabolism may speed up, causing them to require more food than is available. Oxygen levels in the body of water may be altered. The concentration of dissolved oxygen depends in part on the temperature of the water (less is dissolved at higher temperatures) and in part on the turbulence of the water (which increases the level of oxygen). More rapid corrosion is also possible at higher temperatures. These effects must be considered in the design and operation of power plants which discharge condenser water into natural bodies of water. Their sum total must be carefully weighed when plants are designed and sited.

Two approaches to the control of thermal pollution are possible. The first involves changes in the cooling and discharge design of the plant so that lower-temperature waters are discharged to the receiving body even though their total quantity of heat is unchanged. The second utilizes cooling ponds or cooling towers. The towers are huge concrete and wooden structures into which water is evaporated in order to remove some of its heat content. However, without an increase in plant efficiency, the environment is still the ultimate recipient of the waste heat. For a more detailed discussion, see **Thermal Pollution**.

OIL

The daily consumption of oil in the United States is expected to grow from 13.1 million barrels in 1968 to 18.6 million barrels by 1980. Refined oil is used as a lubricant for machinery and as a source of energy. A growing use is in electric generation from **Fossil Fuel Power Plants** in order to alleviate **Air Pollution** by such contaminants as **Sulfur Oxides** and **Particulate Matter.** The oil used there is residual oil, and a large portion of the quantity is imported by tanker.

A significant amount of oil finds its way into the oceans. The oil stems from accidents during tanker transport and from offshore drilling operations; it is contained in the bilge water pumped out of tankers and other ships; it escapes in refinery effluents. The principal source of the oil which pollutes the oceans, however (almost 70 percent), is believed to be used motor and industrial oil. About 5 million metric tons of oil were being added annually to the oceans from all sources as of the late 1960's.

Oil spills from tankers and from offshore drilling accidents may destroy recreational areas if they coat beaches and foul harbors. In addition, certain types of oils, notably the fuel oils, which contain a high proportion of low-boiling components, are extremely toxic to marine life. But the combination of oil with the detergents used to disperse oil slicks may exhibit even greater toxicity. Birds are adversely affected by oil if it coats their feathers; the coating may prevent the birds from floating on the water surface or flying. Birds can also be blinded by oil; it may destroy their feeding grounds, and it can be directly toxic to them. Long-term effects of oil pollution are a cause for concern as suspected carcinogenic chemicals are found in oil. Furthermore, the ocean surface is accumulating lumps of tar which are believed to be formed from the more persistent high-boiling fractions in oil. A more thorough treatment is in **Oil Pollution**.

LAWS

The Water Pollution Control Act Amendments of 1972 were passed over presidential veto on October 18, 1972. They amended the Federal Water Pollution Control Act of 1965, which was also amended in 1966 and 1970 (the latter change is known as the Water Quality Improvement Act). The new legislation includes a timetable for the elimination of the discharge into U.S. waters of all polluting wastes, by 1985. A permit system to govern pollutant discharges into U.S. waters between now and 1985 is authorized. Such a system has been operating under the authority of the Refuse Act of 1899. The new legislation confirms the legality of such a system and transfers the power of granting permits from the Corps of Engineers to the Environmental Protection Agency. A grant of $18 billion to the states for the building of water treatment plants is authorized. Whether the funds will be released by the President for spending was not clear in April 1973.

REFERENCES

1. J. Walsh, "Environmental Legislation: Last Word from Congress," *Science*, November 10, 1972, (p. 593).
2. "Water Pollution Law—1972 Style," *Environmental Science and Technology* 6, no. 13 (1972): 1068.
3. C. Warren, *Biology and Water Pollution Control*, Saunders, Philadelphia, 1971.

Water Treatment and Water Pollution Control

The term *water treatment* describes the processes which are used to purify water for public consumption. The term *water pollution control* refers commonly to the typical sequence of processes which removes pollutants from the waste streams of factories or cities. Thus, the former deals with treatment of water before its use; the latter deals with treatment after the water has been used.

WATER TREATMENT

Water being supplied to a community from a river or reservoir may first undergo spray aeration. Jets of water are sprayed into the air in a large basin filled with fountains. Aeration removes carbon dioxide and hydrogen sulfide gases from the water. Hydrogen sulfide removal diminishes odor in the water. The aerated water is then mixed with activated carbon prior to "coagulation." The carbon is removed subsequently in a sedimentation step, and with it comes some of the colloidal material responsible for the color in the water.

Coagulation proceeds in several stages; the first involves mixture of the aerated water with alum (hydrated aluminum sulfate) or copperas (hydrated ferrous sulfate) or certain other ferric compounds. Untreated water contains colloids such as clay particles and proteins which are suspended in water because of the negative charges surrounding them; these charges repel the particles from one another. The negatively charged colloids are destabilized by the presence of the positively charged aluminum and iron ions. Additionally, the insoluble oxides of the iron and aluminum form colloidal material which is positively charged. Large heavy particles form from the oxides and from the original colloidal material; these are called "floc," and the process of particle formation is called "flocculation." The flocculated material settles out of the water in the sedimentation basin. A preliminary chlorination step may precede sedimentation to kill algae and bacteria, which then will also settle out in the sedimentation process (see **Bacterial and Viral Pollution of Water**).

The water leaving the sedimentation basin enters a "rapid sand filter" where colloidal materials, bacteria, and viruses may be further removed. While the mechanism of removal is debated by scientists, we note that the filter is an effective device whether it strains out these materials or adsorbs them. The filter commonly consists of concrete basins containing several layers of porous substances. A typical combination of layers in the filter is sand, graded from smallest at the top to largest at the bottom, followed by gravel. Water is passed downward through these layers, and except for a final stage of chemical treatment it is then ready for distribution. Periodic backwashing of the filters with clean, filtered water regenerates the sand filter to effective operation. The backwashing removes material which clogged the filter, and the water used for washing is wasted to the drain.

The water is now treated prior to distribution by a final chlorination in order to destroy most remaining microorganisms. The solution of chlorine gas in water produces hypochlorous acid, which is primarily responsible for the disinfection. The hypochlorous acid, however, combines with organics and amines to effectively "tie up" the chemical disinfectant. Thus, the presence of such compounds (which is common) requires larger quantities of chlorine to be used in order to produce sufficient *free* hypochlorous acid to accomplish disinfection.

Many variations in the processes are possible. Water softening (or removal of hardness) is included in some water treatment plants.

WATER POLLUTION CONTROL

Water pollution control embraces a wide collection of devices or actions. For instance, mechanical in-stream aeration provides dissolved oxygen to a stream and may be considered pollution control. The use of biodegradable detergents is a form of water pollution control. From the standpoint of the engineer or technologist, though, water pollution control means "sewage treatment" or "waste treatment." These are the older and more earthy descriptions of the chemical, physical, and biological processes used to upgrade the quality of waste water.

The waste flow from a city, town, or factory may contain high concentrations of both dissolved organics and organic solids (see **Organic Water Pollution**). It first passes a rack or screen for removal of coarse floating matter, which may be burned or buried. A grit chamber may then be used to remove coarse settleable grit, which can be disposed of as landfill material. An alternative to the rack and screen and grit chamber is the comminutor, which shreds and beats coarse solids into small particles.

PRIMARY TREATMENT

The "primary treatment" or "settling" of the waste flow removes settleable solids. In this process, which may take place in a rectangular or circular basin, waste water is detained so that the turbulence which suspends solid particles is reduced. The sludge of particles collected at the bottom of the basin may be mechanically pushed into a sludge drain. The fate of this

A plant for primary sewage treatment in Yakima, Washington. The settling of solids (primary treatment) occurs in the four circular clarifiers. The resulting sludge is pumped into three anaerobic digesters for additional treatment. Only about 35 percent of the oxygen-demanding organics are removed by primary treatment. Waste water from the clarifiers still has a high biochemical oxygen demand. Secondary treatment is required to achieve a higher rate of removal of organic waste—up to 90 percent.

material is dealt with in the section on disposal of suspended solids, below. The clarified water enters the next phase of treatment.

SECONDARY TREATMENT

Whereas primary treatment is a physical process, secondary treatment is mainly biological. There are two alternatives in secondary treatment. Both processes require the presence of air for their oxygen and hence are termed "aerobic." The first is the activated sludge process, in which the clarified stream from the settling basin is contacted with a suspension of microorganisms, protozoa, and metazoa in an aerated tank. The dissolved organics provide food for the growth and division of the activated sludge population. Thus, the dissolved organics are largely removed from the water and incorporated into the biological material of the sludge itself. A detention time of four to eight hours for the waste water is common for the activated sludge process. The resulting suspension, richer in solids and poorer in dissolved organics, passes to a thickening tank, where solids are removed in a thick suspension at the bottom and the clarified liquid is taken from the top. The bottom suspension may be separated into two flows, the major one being returned to the activated sludge tank to mix with the entering stream from the settling basin. This recycling of biological solids ensures a sufficient and adapted biological population to accomplish the desired removal. The portion not returned joins the sludge from the primary settling basin in a further phase of treatment. A number of variations of the activated sludge process are possible.

A new development in the activated sludge plant is aeration not with air but with high-purity oxygen. Demonstrated at Batavia, New York, by the Linde Corporation, the process appears to accomplish slightly greater removal of organics with a shorter retention time in the basin. The shorter retention time means that an activated sludge basin may be built considerably smaller in size and, hence, at lower cost. A reduction in solids produced was also noted at the demonstration plant. The process holds promise not only for new construction but also for upgrading the performance of some existing plants.

The second major process and alternative to activated sludge is the "trickling filter." The filter, commonly a circular concrete basin, is filled with crushed stone with large void spaces between the stones to assure adequate contact with air. The flow from the settler is distributed over the stone bed and trickles down through the tank. On its passage through the filter the

A trickling filter. Waste water that has undergone primary treatment is distributed over a bed of stones by rotating arms. The water trickles down through the basin and contacts microorganisms growing on the stones. Dissolved organics are removed from the water by the organisms.

stream contacts a biological slime which grows attached to the stones. The slime is composed of organisms similar to those found in activated sludge, and the dissolved organics in the flow are used as the food for growth of the species in the slime. The resulting flow has lower dissolved organics but has more solids because of the presence of slime masses which slough off the stone. A second settling unit separates a solid suspension from the effluent of the filter. The suspension of solids joins the suspended-solids stream from the primary settling basin.

The clarified flows from the activated sludge process or from the trickling filter may have been reduced in organic material by up to 80 to 90 percent. Additionally, a large portion of the original bacterial population present in the waste flow has been removed or has died away by the time secondary treatment is complete. However, to prevent potentially hazardous microorganisms from entering the stream or river, the effluent from secondary treatment may undergo chlorination. The remaining dissolved organics may be partially removed by a third phase of treatment, referred to as tertiary treatment, a new sequence of processes which removes not only resistant organics but also phosphates and nitrogen compounds. Although tertiary treatment is not now widely utilized, it is expected to become quite important as emphasis on control of **Eutrophication** increases. Before discussing tertiary treatment, however, we still need to consider disposal of the several streams of suspended solids from primary and secondary treatment (see Figure W.1).

FIGURE W.1
OPERATIONS AT A WATER POLLUTION CONTROL PLANT

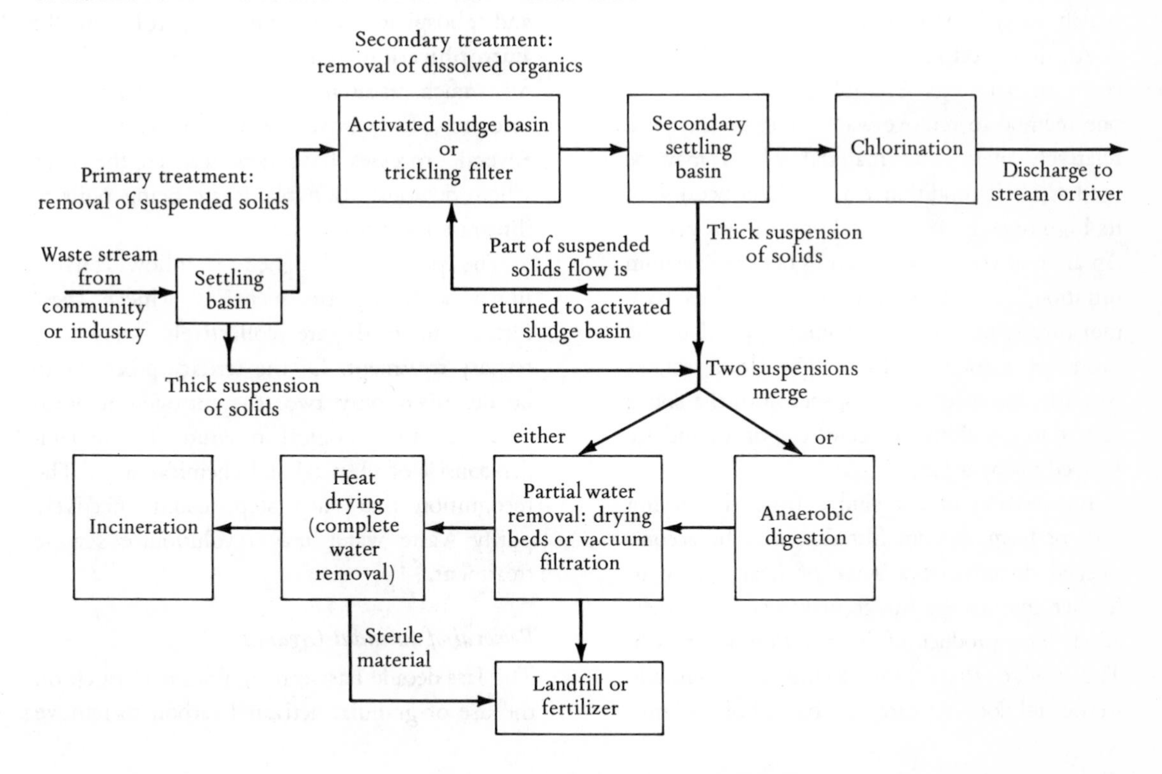

DISPOSAL OF SUSPENDED SOLIDS

The suspension of solid particles drawn from the primary settling basin and from the secondary settling basin which follows the activated sludge process and the trickling filter are combined into a single waste flow. This combined flow may undergo a variety of operations. A possible first stage in the treatment of the sludge solids is anaerobic digestion, a biological process which takes place in the absence of air. A different set of microorganisms convert the organic material to stable end products and "marsh gases." Digestion is accomplished in a large closed cylindrical tank which is usually heated by the burning of the marsh gases given off by the process. The elevated temperature (85° to 95°F) and the absence of air are responsible for the specialized microbial population that digests the sludge. A retention time of the order of 15 days is common. Digestion may, however, be bypassed entirely; in that case the treatment of solids begins with a phase in which water is partially removed.

Air drying on sand beds, possibly under glass cover in buildings resembling greenhouses, is one method to remove water. The product is a relatively inert solid material which may be used as a soil conditioner or fertilizer because of its high organic content and physical properties. An alternative to sludge drying beds is "vacuum filtration," a process in which the solids have moisture removed by a vacuum applied to the inside of a rotating drum. The drum rotates partially immersed in the suspension; the solids adhere to the cloth surface of the drum and are scraped off by a knife blade.

Incineration of the sludge from the drying beds or from vacuum filtration may be accomplished directly, or a stage of heat drying to further dewater the sludge may precede incineration. The product of incineration is an ash. The sludge from heat drying is essentially sterile and does not carry the hazard of contamination when used as a fertilizer, as partially dewatered sludge may.

TERTIARY TREATMENT

Resistant organics, nitrogen compounds, and phosphates are the three main classes of substances which tertiary processes are designed to remove. It was indicated earlier how the combination of primary treatment (settling) and secondary treatment (biological oxidation) could remove up to 90 percent of the organic material entering the plant. Unfortunately, this appears to be an upper limit for that combination of processes as they exist today. In addition, most conventional plants do not effect the removal of nitrogen compounds and phosphates.

Further removal of organics may be warranted when the water is to be reused or when the volume of the waste flow is substantial in comparison with the receiving body. Because nitrogen compounds—nitrates in particular—and phosphates play important roles in the **Eutrophication** (enrichment) of natural waters, much research and development has gone into how to remove them from waste water. Several processes have now reached the stage where demonstration plants are being built to illustrate their feasibility.

The processes designed to follow conventional secondary treatment to achieve these further removals are collectively known as tertiary treatment. Of the tertiary processes to be described only two, the nitrogen removal processes, are biological in nature. The remainder consist of physical and chemical steps. The recognition that such steps could effectively purify waste water may revolutionize sewage treatment.

Removal of Resistant Organics

The last decade has seen significant research on the use of granular activated carbon to remove organics which are resistant to removal by conventional treatment. Water from secondary treatment may be passed through 25-foot-tall towers containing these carbon particles. The surface properties of the particles make it possible for them to adsorb (attach to their surface) the molecules of many of the resistant organics.

Eventually, the adsorptive capacity of the particles in a tower will be exhausted. The flow will then be diverted to a second tower with particles still capable of removing organics. While the second tower is in use, the particles from the first tower will be regenerated by being heated to near 1,700°F. The regenerated particles may then be reused for organic removal after the capacity of the second tower has been exhausted. Some particles are destroyed in the regeneration process, and makeup carbon of the order of 5 to 8 percent may be necessary after each regeneration.

Costs of resistant organic removal by carbon treatment range from 8.5 to 12.5 cents per 1,000 gallons. To set this expense in perspective: the cost of water treatment (preparation of water for drinking) is between 30 and 40 cents per 1,000 gallons. The most notable installation of carbon treatment is at Lake Tahoe, where 7.5 million gallons per day of secondary effluent are treated. We will be mentioning this plant again for it combines most of the advanced concepts to be discussed here. The purpose of the plant was to save Lake Tahoe's sparkling clarity from the effects of eutrophication. The size of the plant is such that, when a process works there, it is essentially a full-scale demonstration for the technology.

A new process, called a physical-chemical process because of its steps, utilizes the carbon towers and does away with secondary biological treatment. The first stage of treatment consists of settling aided by polymers or settling with phosphate precipitation (due to addition of iron or aluminum salts). The second stage involves passing the waste water through towers of activated carbon. Periodic regeneration of the waste carbon by heating is necessary to keep its adsorptive capacity up. Organic removals of 93 percent and phosphate removals of approximately 90 percent have been demonstrated. A number of potential advantages are cited for the physical-chemical process. First, because less land and smaller pieces of equipment are involved, the capital costs may be 20 to 30 percent less than for the conventional plant. Second, the effects on system operation of occasional large inflows or occasional toxic materials or very low temperatures should be much less than on the conventional plant. Toxic substances or low temperatures could destroy the microorganisms of the activated sludge process or the trickling filter. Large inflows (due to storms) could overwhelm these biological processes. Portions of such flows might then be allowed to short-circuit the conventional biological processes. Last, the volume of sludge in the physical-chemical process may be reduced by half, allowing for considerable savings in disposal.

Removal of Nitrogen Compounds

Organic nitrogen compounds and ammonia are the two nitrogen-containing substances found in raw (untreated) waste water. The liquid effluent from secondary treatment contains nitrogen, mostly in the compound ammonia, and the solids from secondary treatment contain most of the organic nitrogen. Our concern here is with the removal of the ammonia nitrogen from the process effluent.

It was mentioned previously that nitrogen compounds may foster eutrophication, a condition of natural waters which is characterized by excessive algal growth. Ammonia, which is indicted in this regard, has additional undesirable characteristics. It is toxic to fish and corrosive to copper. Furthermore, when the effluent from secondary treatment is chlorinated in order to destroy bacteria, the ammonia will combine with the chlorine to give chloramine compounds. These compounds are less effective in destroying bacteria than is free chlorine, and the water therefore requires a higher dose of chlorine for effective disinfection.

Thus, even if eutrophication were not considered, there would be some justification for removing ammonia. Under such a circumstance —i.e., eutrophication not a problem—the conversion of ammonia to the nitrate ion may be sufficient action. The conversion is referred to as nitrification and can be accomplished by a separate aeration stage after activated sludge. In this stage, a bacterial population suitable for converting ammonia nitrogen to nitrate nitrogen will grow up.

Yet nitrate in the water may still not be recommended if the water is to be reused eventually for drinking by some downstream community (see **Nitrates and Nitrites**).

If nitrates may not be present either, there are two basic alternatives. The first is a direct removal of ammonia without prior conversion to nitrate. Ammonia may be "stripped" from the water by having air and the ammonia-rich water pass in opposite directions through a packed tower. This direct removal process was built at the Lake Tahoe plant and was observed to suffer from ice formation, which caused shutdown of the tower during freezing weather. Further, the water was made basic for the process by lime addition, and the resultant scale in the tower required periodic removal.

The second alternative is to remove the nitrate by the action of bacteria. Methyl alcohol is added to the water, and the bacteria convert the nitrate and methyl alcohol to carbon dioxide and nitrogen gases. The reactions are called denitrification. Costs in the order of several cents per 1,000 gallons are reported for these two processes.

Removal of Phosphates

A phosphate removal process may be easily added to most conventional sewage treatment plants. The addition of iron salts or aluminum salts to the activated sludge tank or to the settling tank (when secondary treatment is by means of a trickling filter) will serve to precipitate metal phosphates. Removals up to 80 or 90 percent appear to be possible. Costs range from 2 to 5 cents per 1,000 gallons. Where waste pickling liquor from the steel industry is available to be used as the source of iron salts, the cost is only that of transporting the liquid to the plant. At Lake Tahoe, lime addition is used to precipitate phosphate as calcium phosphate in a tertiary treatment stage. In general, though, lime addition might be applied at the primary settling tank as well.

OTHER TECHNOLOGIES

While the systems described in the sections above are the most common processes for conventional treatment, other arrangements for treatment of municipal and industrial wastes are sometimes utilized. They are, in general, not as efficient as the conventional systems. Their appeal is due largely to their lower costs. Some do not properly qualify as processes. Oxidation ponds and septic tanks are alternative processes, but cesspools and the use of sewage for irrigation are merely alternatives. In another category altogether is "low-flow augmentation."

Oxidation ponds are wide shallow ponds of several acres or more used to detain waste water until its quality is sufficiently improved for discharge. They are most widely used in industrial settings if land is inexpensive. As the name implies, the process requires oxygen from the air. Bacteria and algae cooperate in the removal of the dissolved organics. Bacteria ingest the organics and oxidize them for energy, the products of that oxidation being carbon dioxide and water. Algae utilize carbon dioxide as an input to photosynthesis, a by-product of which is oxygen. The oxygen is used, in turn, for the oxidation of the organics by bacteria. The design of oxidation ponds is not entirely on a scientific basis and involves several rules of thumb. The efficiency of the ponds fluctuates extensively. Ponds may be referred to as waste stabilization lagoons in settings where anaerobic conditions occur.

The septic tank is a watertight tank of concrete or metal which ranges in volume between 500 and 2,000 gallons; it detains sewage for periods between 12 and 72 hours depending on the tank capacity and the flow entering it. Sewage solids settle to the bottom of the tank, where anaerobic bacteria degrade and liquefy the organic solids. An overflow of settled sewage exits from the tank to a distribution box and a buried leaching system. The distribution box slows the rate of flow and breaks the directionality of the flow from the septic tank in order to achieve even distribution among the lateral arms of the leaching system.

Septic tanks are widely used in rural areas and subdivisions to partially treat wastes prior to their release into the soil. They are warranted where population is sparse and sewer lines pose high per capita costs. But where population is reasonably dense, as in a suburb, and the water supply is obtained from wells, septic tanks should not be used as they constitute a potential health hazard.

Cesspools and leaching pits are constructed of materials such as fieldstone, cinder blocks, or concrete blocks. Both have open joints—that is, no mortar between the stones or blocks. Leaching pits receive the settled sewage from a septic tank; well-water supplies should be kept suitably distant. Cesspools receive raw, unsettled sewage. They should not be used in conjunction with well or spring water supply.

The use of untreated sewage for irrigation, while utilizing the high organic content to advantage, may be inadvisable because of the creation of odors, the diminished visual quality of the landscape, and the possible spread of disease.

Low-flow augmentation is the descriptive term applied to releasing water stored in reservoirs to mix with streams of impaired quality. If the released water is rich in dissolved oxygen (and reservoir releases may not be—see **Reservoirs**), the resulting mixture will have a higher dissolved oxygen. If the released water has low organic content (and it probably does), the resulting mixture has a lower concentration of organic wastes. The cost of low-flow augmentation is primarily the capital cost of the reservoir. In the trade, augmentation is known as the "dilution solution to pollution."

REFERENCES

1. G. Fair, J. Geyer, and D. Okun, *Water and Wastewater Engineering*, vols. 1 and 2, Wiley, New York, 1968.
2. L. Rich, *Unit Operations of Sanitary Engineering*, Wiley, New York, 1961.
3. L. Rich, *Unit Processes of Sanitary Engineering*, Wiley, New York, 1963.
4. "Current Status of Advanced Waste-Treatment Processes, July 1, 1970," U.S. Dept. of the Interior, Advanced Waste Treatment Research Laboratory, Federal Water Quality Administration, Cincinnati, Ohio, 1970. (The Federal Water Quality Administration has become the Water Quality Office of the U.S. Environmental Protection Agency.)
5. Metcalf and Eddy, Inc., *Wastewater Engineering: Collection, Treatment, Disposal,* McGraw-Hill, New York, 1972.
6. C. E. Warren, *Biology and Water Pollution Control*, Saunders, Philadelphia, 1971, ch. 21.

Photo Credits

8 U.S. Environmental Protection Agency/Documerica/Bill Gillette
11 A. Devaney, Inc., New York
23 Rotkin/Photography for Industry
42 U.S. Dept. of Health, Education and Welfare/National Institute for Occupational Safety and Health
58 Grant Heilman, Lititz, Pa.
69 D. Tenenbaum/Editorial Photocolor Archives
70 © National Geographic Society
89 Daniel S. Brody/Stock, Boston
102 Daniel S. Brody/Stock, Boston
130 Grant Heilman, Lititz, Pa.
137 Jeff Albertson/Stock, Boston
139 Daniel S. Brody/Editorial Photocolor Archives
171 Daniel S. Brody/Stock, Boston
176 A. Devaney, Inc., New York
184 Franklin Wing/Stock, Boston
201 Carson Baldwin, Jr./Freelance Photographers Guild
204 Clean Seas, Inc.
212 Richard F. Conrat
216 U.S. Environmental Protection Agency
230 U.S. Environmental Protection Agency/Documerica/Gene Daniels
246 U.S. Dept. of Agriculture/Soil Conservation Service
273 Black-footed ferret: Edward Bonn/National Audubon Society
273 Timber wolf: John Gerard/Monkmeyer
273 American alligator/Les Line/National Audubon Society
274 California condor: John Borneman/National Audubon Society
274 Humpback whale: Karl W. Kenyon/National Audubon Society
280 Fairchild Aerial Surveys photo supplied by New England Electric
284 A. Devaney, Inc., New York
288 A. Devaney, Inc., New York
292 G. & M. Heilman from A. Devaney, Inc., New York
293 U.S. Environmental Protection Agency/Documerica/Bill Gillette
307 Daniel S. Brody/Editorial Photocolor Archives
325 U.S. Environmental Protection Agency
326 Authenticated News International